Metal-Filled Polymers

PLASTICS ENGINEERING

Series Editor

Donald E. Hudgin

Princeton Polymer Laboratories

Plainsboro, New Jersey

1. Plastics Waste: Recovery of Economic Value, *Jacob Leidner*
2. Polyester Molding Compounds, *Robert Burns*
3. Carbon Black-Polymer Composites: The Physics of Electrically Conducting Composites, *Edited by Enid Keil Sichel*
4. The Strength and Stiffness of Polymers, *Edited by Anagnostis E. Zachariades and Roger S. Porter*
5. Selecting Thermoplastics for Engineering Applications, *Charles P. MacDermott*
6. Engineering with Rigid PVC: Processability and Applications, *Edited by I. Luis Gomez*
7. Computer-Aided Design of Polymers and Composites, *D. H. Kaelble*
8. Engineering Thermoplastics: Properties and Applications, *edited by James M. Margolis*
9. Structural Foam: A Purchasing and Design Guide, *Bruce C. Wendle*
10. Plastics in Architecture: A Guide to Acrylic and Polycarbonate, *Ralph Montella*
11. Metal-Filled Polymers: Properties and Applications, *Edited by Swapan K. Bhattacharya*

Other Volumes in Preparation

Metal-Filled Polymers

PROPERTIES AND APPLICATIONS

edited by

Swapan K. Bhattacharya

School of Chemical Engineering
Georgia Institute of Technology
Atlanta, Georgia

MARCEL DEKKER, INC. New York and Basel

Library of Congress Cataloging-in-Publication Data

Metal-filled polymers.

(Plastics engineering ; 11)
Includes bibliographies and index.
1. Metal-filled plastics. I. Bhattacharya, Swapan K.
II. Series: Plastics engineering (Marcel Dekker, Inc.) ;
11.
TA455.P55M48 1986 620.1'923 86-16438
ISBN 0-8247-7555-4

MARCEL DEKKER, INC.
270 Madison Avenue, New York, New York 10016

Current printing (last digit):
10 9 8 7 6 5 4 3 2 1

PRINTED IN THE UNITED STATES OF AMERICA

Preface

Polymers and metals, by virtue of their respective insulating and conducting properties, have a long history of use in diverse applications. However, the first attempt to combine these two classes of materials was made only in the later half of this century. Among the resulting composites, metal-filled plastics (sometimes referred to as metallo-plastics) have emerged as a new-generation material with tremendous potential as an alternative to metals in specific applications. This potential derives primarily from the substantial modification of the thermal and electrical characteristics of plastics when combined with metallic fillers. In addition, metal-filled plastics retain some of the advantages of plastics (e.g., cost, fabrication, maintenance, density, and aesthetics). These properties, collectively, have widened the range of applications of metal-polymer composites.

In recent years the demands on plastics have been growing in many directions. Research has been directed to improve the dielectric strength of plastics so that they can be better insulators than ever before. On the other hand, because of the rapidly growing use of electronic devices, plastics manufacturers are looking into ways to make housing and molded parts conductive so that they can be used extensively and yet meet the restrictions imposed by the regulatory agencies (e.g., the Federal Communications Commission in the United States) regarding the marketing of electronic equipment. Since regulations concerning radiation emission are becoming increasingly stringent all over the globe, it is expected that the need to develop materials for better electromagnetic interference (EMI) shielding will increase steadily. Various U.S. reports indicate that the average annual growth in the consumption of metal- and carbon-filled polymers will be around 10% in the next five years.

At present, there are three routes to make plastics conductive. The first approach is to apply a conductive coating on the molded part, a two-step process which is becoming obsolete because of the cost involved and the difficulties in obtaining a good coating. Making accessories from conductive polymer compositions could be an alternative. Although polyacetylene- and polyphenylene-based compounds have proved to be candidate materials for this conductive family, problems still remain with processability, mechanical properties, and stability. The third option is to incorporate conductive fillers in the resin matrix. Metals, carbon, or metal-coated fillers in the form of particles and fibers are widely used conductive fillers and are discussed in appropriate sections of this volume.

The basic purpose of this book is to review the present status of metal-filled polymers. Literature in this developing interdisciplinary field has been extremely disorganized and widely scattered. We have, therefore, attempted to present the development of metal-filled plastics in the style of a monograph. Keeping this in mind we have tried to design each chapter as an overview, describing the evolution of the respective subjects to their current state. The first chapter introduces this new member of the composite family with an overall view of the science and technology involved in combining two dissimilar materials, describing important properties and stating the potential use of this new-generation material. The later part of this chapter is devoted to specific applications. The following chapters, in sequence, provide a more detailed account of the processing aspects, electrical behavior, mechanical and thermal properties, mode of crack propagation, and importance of the interphases in multiphase systems. Theoretical models in developing boundary layer concepts and evaluation of various elastic constants are addressed briefly in Chapters 4 and 5. The last chapter provides a comparative description of how some of the properties discussed in the preceding sections may be obtained with fillers other than metals.

In summary, this volume on metal-filled polymers provides a comprehensive description of the latest technologies in this field, with a historical perspective. We believe that it is the most complete single-source reference in this important area of composites.

I would like to express my sincere appreciation to the authors contributing to this volume. I am indebted to my wife, Muna, for her participation in this project. Finally, I wish to thank Dr. Eileen Gardiner of Marcel Dekker, Inc., for initiating this challenging and gratifying project.

Swapan K. Bhattacharya

Contents

Contributors

Swapan K. Bhattacharya School of Chemical Engineering, Georgia Institute of Technology, Atlanta, Georgia

Donald M. Bigg Polymer Science Division, Battelle Columbus Laboratories, Columbus, Ohio

T. S. Chow Webster Research Center, Xerox Corporation, Webster, New York

John Delmonte Delsen Testing Laboratories, Inc., Glendale, California

Robert P. Kusy Dental Research Center and Department of Physics, University of North Carolina, Chapel Hill, North Carolina

Richard C. Penwell Advanced Materials & Manufacturing Processes, Computer Integrated Engineering Manufacturing Systems, Xerox Corporation, Webster, New York

Jean-Pierre Reboul Laboratoire d'Electricité, Université des Sciences et Techniques du Languedoc, Montpellier, France

P. S. Theocaris Department of Engineering Science, Athens National Technical University, Athens, Greece

Metal-Filled Polymers

1

Applications

ROBERT P. KUSY

Dental Research Center
and Department of Physics
University of North Carolina
Chapel Hill, North Carolina

INTRODUCTION TO COMPOSITES

If formulated and fabricated correctly, composites can offer a combination of properties and a diversity of applications unobtainable with metals, ceramics, or polymers alone. Throughout the countless life cycles of plants and animals, nature has capitalized on this fact, for example, through the use of wood, bone, or teeth (Figure 1) (Refs. 1–3). The structure of these composites is composed of two or more discrete materials which have evolved to provide the proper balance of physical and mechanical properties. Whether it is the cellulose and lignin which form the woody cell walls of plants and in effect act as the cementing material between cells, the combination of a ceramic (calcium phosphate crystallites) and organic material (collagen fibers) to form the dense, strong cortical bone or the spongy, shear-resistant cancellous bone, or the combination of a hard, wear-resistant enamel composed of 95% inorganic hydroxyapatite crystals supported by the more vital and tougher dentin of tooth structure, each of these composites possesses properties which the constituents alone could not offer. Ancient metalworkers recognized that the properties of relatively more noble metals had limited usefulness and, in the most general sense, they formed composites called alloys. One of the most important structures of ferrous alloys, pearlite, is composed of alternating laminations of ferrite (a soft, ductile phase) and martensite (a hard, brittle phase). Together they form a strong, tough structural steel (Ref. 4).

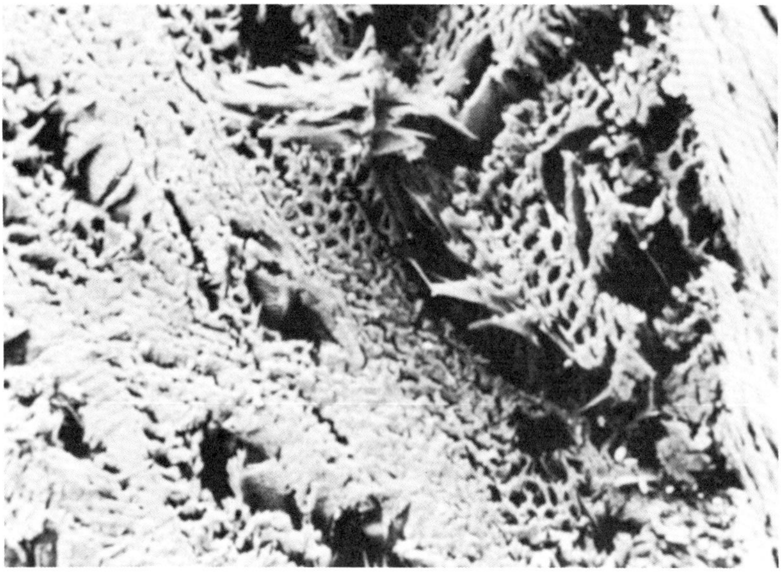

(a)

FIGURE 1 Composite structures which occur in nature (a) wood cells; (b) human bone; (c) tooth enamel. (Photomicrographs reproduced from Buchanan (Ref. 1), Urist (Ref. 2), and Jongebloed et al., Ref. 3, respectively.)

Brief History of Plastic Composites

Although there are isolated examples of composites being used throughout antiquity, the 20th century really marks the beginning of man's conscious effort to devise true composites. Because plastics are used so frequently in composites today, the development of plastics technology had to precede the development of suitable composites. The first plastics were phenolics (1920s), followed by urea and melamine (1930s) (Ref. 5). These thermosetting molding compounds consisted of a resin and a filler (either wood flour, mica, or glass) which was used to increase strength and stability. Just prior to World War II, cellulosic plastics such as cellulose acetate were introduced. These were the first thermoplastics which were relatively tough enough so that no filler additions were necessary. Shortly after the war, polystyrene and polyethylene entered the marketplace, displacing cellulose acetate. Polystyrene had good optical properties and polyethylene was tough and flexible. Neither required fillers. Then in 1951 the first modern reinforced composite was designed for

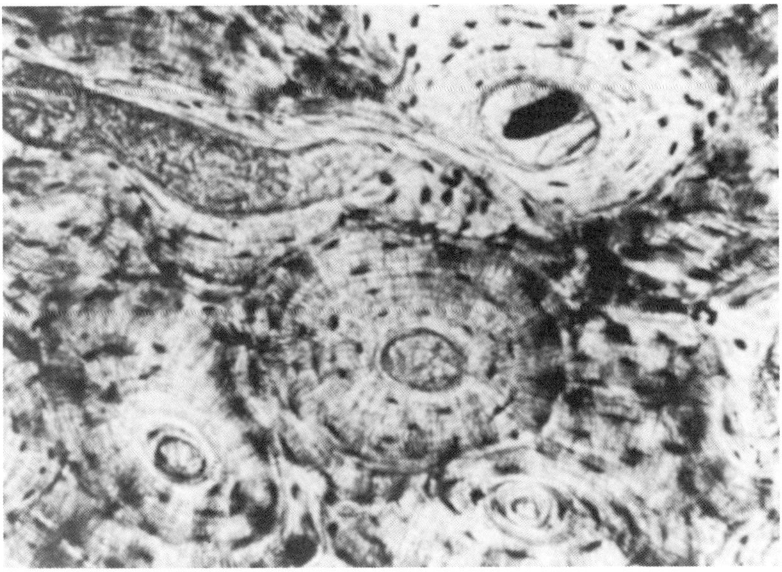

(b)

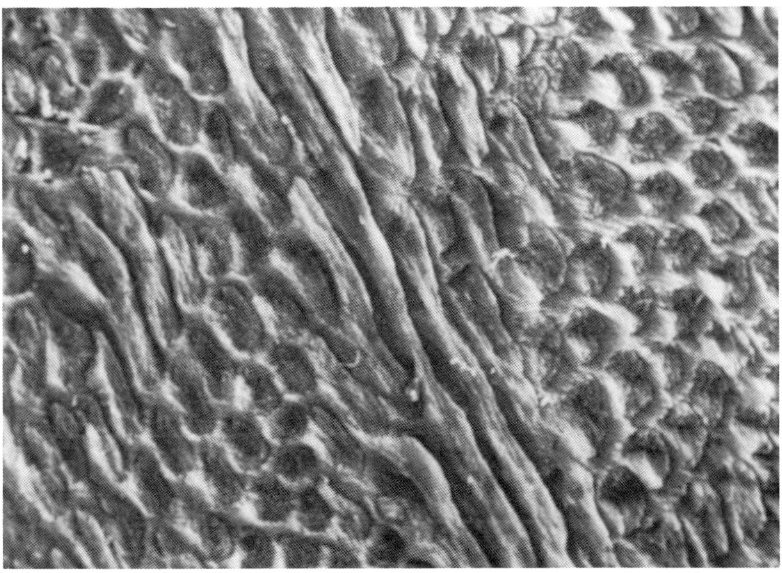

(c)

FIGURE 1 (Continued)

the Army by Bradt—a glass fiber-reinforced polystyrene mine case. Finally, with the first of the so-called engineering plastics, Nylon 6/6, a high-toughness, low-friction polymer was available which, when reinforced with glass, made a high wear-resistant composite. Over the next 30 years aramid, carbon, and hybrid fiber composites became prevalent in the automotive, aerospace, electronics, and construction industries.

Typical Properties

Today thousands of material combinations are marketed as composites. Often the base or *matrix* is a polymer and the filler is an inorganic in either a fiber or particulate form (Table 1). Properties of the more common mineral fillers and the more recent fibrous materials are shown in Tables 2 and 3, respectively (Refs. 5–7). How some of these fillers are likely to affect the properties of a composite are detailed in Table 4 (Ref. 8).

Filler Types

Fibers

Normally, fillers are used in either fiber or particulate form. Additions of fibers tend to enhance stiffness, strength, hardness, abrasion resistance, heat deflection temperature, and lubricity while reducing shrinkage and creep. The fiber aspect ratio (that is, the ratio of the length L to the diameter D of a fiber) and the orientation of fibers (that is, whether the fibers are globally oriented in one direction or laminated in alternating layers, plies, or mats) can have a profound influence on the properties of the composite. Generally speaking, the most effective reinforcing fillers are fibers which intrinsically have a high elastic modulus and tensile strength. Being the most important short-term properties in engineering applications, the superior composite will be that material which can transfer the load best to the fibers. Table 5, Figure 2, and Table 6 illustrate the typical properties of fiber-reinforced composites (Ref. 9) (for a detailed look at glass fiber-reinforced plastics, see data sheet, Ref. 10), the influence of carbon fibers on Young's modulus and the tensile strength of construction cement (Ref. 11) and the effect of ply orientation on Young's modulus, the torsional modulus, and the coefficient of thermal expansion in graphite/epoxy composites (GFRP) Ref. 12) respectively. Note that isotropic composites have the best balance of compliance under complex stress situations.

Particles

Unlike fibers, particulate fillers have low aspects ratios, often approximating those of spheres or plates. None of the particulate

TABLE 1 Some Matrix Polymers and Filler Materials Used in Composites Today

Matrix	Fillers
Acetal	Antimony trioxide
Epoxy	Asbestos (actinolite, amosite, anthophyllite, chrysotile, crocidolite, tremolite)
Polyacrylonitrile styrene butadiene (ABS)	Aluminum flakes
Polyamides (Nylon 6, 6/6, 6/10, 11, 12)	Bronze beads
Polybutylene terephthalate (PBT)	Calcium carbonate
Polycarbonate (PC)	Carbon black
Polyether sulfone (PES)	Cellulose
Polyethylene (PE)	Clay (kaolin)
Polyethylene terephthalate (PET)	Glass beads and fibers (E– or S–)
Polyimide	Graphite (Type 1—high modulus; Type 2—high strength)
Polypropylene (PP)	Mica
Polyphenylene oxide (PPO)	Silica
Polyphenylene sulfide (PPS)	Steel
Polystyrene (PS)	Talc
Polystyrene acrylonitrile (SAN)	Whiskers (alumina, potassium titanate, silicon carbide, silicon nitride)
Polysulfone (PSUL)	Wollastonite
Polytetrafluoroethylene (PTFE)	
Polyurethane (PU)	
Polyvinyl chloride (PVC)	

TABLE 2 Properties of Commonly Used Mineral Fillers

	Mica	Calcium carbonate	Wollastonite	Glass beads	Alumina trihydrate	Talc	Silica	Franklin fiber	Clay
Water content (%)	<5.0	<2.0	0.5	<0.1	34.6	4.8	<0.1	<1.0	<0.5
Specific-gravity	2.74–2.95	2.60–2.75	2.9	2.48	2.42	2.7–2.8	2.65	3.0	2.50
Hardness (mohs)	2.4–3.0	3.0	4.5	5.5	2.5–3.5	1.0	7.0	2.0	4.0–6.0
Melt point (°C)	1300[a]	900[a]	1540	1200	200–600[b]	Stable to 380	Stable to 573	—	1810
Index of refraction	1.54–1.69	1.49	1.63	1.51	1.58	1.59	1.55	1.56	1.56
Color	Gold-brown	White	White	Clear	White	Gray-white	White	White	White
Shape	Platelike	Prismatic	Fibrous	Spherical	Platelike	Platelike	Spherical	Fibrous	Platelike
Resistance to acids/ alkalis	G/G	P/F	P/F	G/P	G/G	G/G	E/P	G/G	G/G

[a]Decomposes.
[b]Loses 34.6% water.
Applicable ASTM procedures were used.
Source: Adopted from Theberge (Ref. 6).

TABLE 3 Properties of Some Fibrous Materials

Type	Tensile strength (GPa)	Young's modulus (GPa)	Specific gravity
Alumina			4.0
Single crystal whiskers (A-axis)	10–20	700–1500	Less compacted forms are of lower specific gravity.
Crystals from melt	2	460	
Bulk (microcrystalline)	0.3	350	
Sintered fibers	0.2–0.7	140–300	
Aromatic polyamide fibers	⩽3.6	60–131	1.44–1.45
Boron			2.3
Fibers deposited by thermal decomposition	2.75	400	Fibers have higher specific gravity depending on proportion of metal core.
Tungsten core	3.45	414	2.6
Boron nitride			Theoretical 2.27
Fibers	0.3–1.4	28–80	Practical values 1.8–2.0
Carbon			Theoretical 2.3
Whiskers	17–21	689–999	2.2
High-performance fibers (pyrolysis)	2–3	280–400	1.7–2.0
Low-performance fibers (carbonization)	0.3–0.7	28–69	1.0–1.5
Glass filaments			
E-type	2.8–3.7	72.4–75.8	2.55
S-type	4.5–4.8	86.2	2.48

TABLE 3 (Continued)

Type	Tensile strength (GPa)	Young's modulus (GPa)	Specific gravity
Silicon carbide			3.19
Whiskers	9.5	480–550	
Fibers deposited by thermal decomposition	1.7–2.5	400–450	Fibers have higher specific gravity depending on proportion of metal core.
Silicon nitride			3.2
Whiskers	5–7.7	350–380	
Stainless steel			
Wires	4.0	200	7.9

Source: In part, adopted from Parker (Ref. 7) and Titow and Lanham (Ref. 5).

TABLE 4 Properties Affected by Fillers

General purpose	Heat transfer	Compressive or flexural
Calcium carbonate	Aluminum	Silica
Calcium sulfate	Alumina	Alumina
Silica	Beryllia	Iron
Alumina	Silica	Aluminum
Talc	Gold	Silicon carbide
Mica	Silver	
Zinc oxide		Compressive yield
Barium sulfate	Heat sink	Silicon carbide
	Iron	Iron
Flame retardance	Aluminum	Alumina
Alumina trihydrate	Silica	
Antimony oxide		Flexural modulus
Arsenic trioxide	Electrical resistance	Calcium carbonate
Ammonium bicarbonate	Alumina	Calcium sulfate
	Silica	
EMI shielding	Talc	Impact strength
Metal fibers	Mica	Iron
Metal flake		Aluminum
	Electrical conductivity	Fibers
Radiation shielding (nuclear)	Gold	Elastomeric particles
Lead	Silver	
Boron carbide	Electrolytic copper	Abrasion resistance
	Aluminum	Silicon carbide
Machinability	Graphite	Aluminum oxide
Metals		Silica
Calcium carbonate	Tensile strength	Metals
Organic polymers	Fibers	
Calcium sulfate	Metals	Corrosion resistance
		Zinc

Source: In part, adopted from Thomas (Ref. 8).

TABLE 5 Typical Properties of Fiber-Reinforced Composites

Material	Weight of fiber (%)	Specific gravity	Tensile strength (MPa)	Tensile modulus (GPa)
Carbon fiber/epoxy	65, continuous	1.55	1380	124
S-glass/epoxy	65, continuous	2.08	1660	48
E-glass/epoxy	65, continuous	2.08	1100	39
Kevlar/epoxy	65, continuous	1.38	1380	76
Sheet molding compound: glass, carbon fiber/ isophthalic polyester	24 glass, 26 carbon (chopped: 1 to 5 cm)	1.65	120	24
Steel	—	7.75	1380	200
Aluminum	—	2.77	560	68

Source: Adopted from Quynn (Ref. 9).

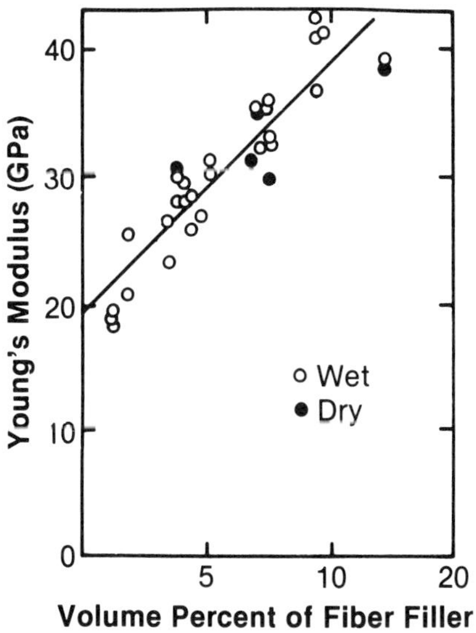

(a)

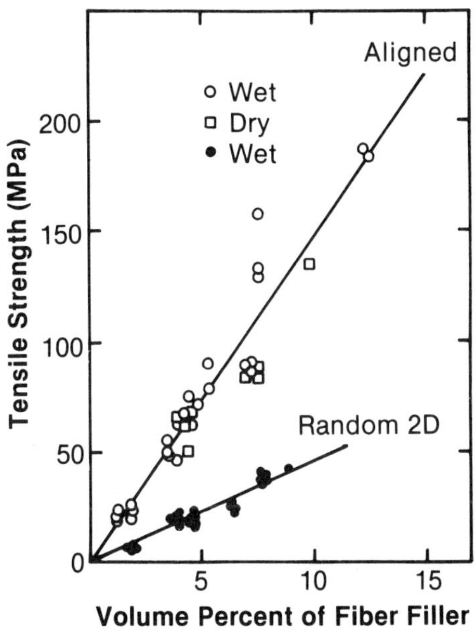

(b)

FIGURE 2 Influence of carbon fiber fillers (a) on Young's modulus and (b) on the tensile strength of construction cement. (Adapted from Aveston et al., Ref. 11, pp. 93–103, by permission of the National Physical Laboratory. British Crown copyright reserved.)

TABLE 6 Physical Properties of GFRP

Property	Lay-up		
	90° Cross-ply[a]	Isotropic	Unidirectional
Young's modulus (GPa)	117	73	138
Torsional modulus (GPa)	4.5	28	4.5
Coefficient of expansion (m/m-°C)			
(a) 24° to 150°C	8×10^{-7}	14×10^{-7}	-9×10^{-7}
(b) 24° to -143°C	1×10^{-7}	6×10^{-7}	-3×10^{-7}

[a]Two longitudinal plies and one transverse ply.

Source: Adopted from G.A. Smith, T. G. Berry, and L. Dibiasi, in *Designing with Fibre Reinforced Materials* (I. Mech. E. Conference Publications 1977–1979), Burlington Press, Foxton, G.B. (1977), p. 19.

fillers improves the ultimate tensile strength of thermoplastics. In fact, tensile, flexural, and impact strength are usually lowered, especially at higher filler contents. Although hardness, heat deflection temperature, and surface finish may be enhanced, the property which is mainly improved is stiffness. Thermal expansion, mold shrinkage, extensibility, and creep are reduced, too. Figures 3 and 4 illustrate the effects of talc and wollastonite filler loading in a thermoplastic (Ref. 13) and of glass beads in a thermosetting resin (Ref. 14) (see also Figure 1 in Ref. 15). Regardless of the bonding efficiency between particulate filler and matrix, the properties were consistent: the elastic modulus and the heat distortion temperature increased, while the structural strengths decreased.

Nonmetallic fillers are further discussed by Reboul in Chapter 6.

Some Applications

Applications of fiber-reinforced and particulate-filled composites are so extensive that volumes have been devoted to that subject alone (Refs. 5, 9, 12, 16–23). Some of the more exotic uses include graphite fiber-reinforced epoxy plastic hinges which provide secure deployment of and thermal stability for instrument packages in outer space (e.g., for the Space Telescope), filament-wound GFRP launch tubes for missiles some 21 m in length (one component of which is 14 m long by 2.5 m in diameter by 4 cm thick and weighs more than 5400 kg), the carbon-carbon composite nose cone and carbon-polyimide/epoxy composite cargo doors of the space shuttle Columbia, and the carbon fiber-reinforced epoxy fuselage of Lear Fan's business jet (everything except the engines and landing gears is a composite). Other uses include GFRP main and tail rotor blades (these must withstand 500 MPH constant tip velocity); composite sandwiches (honeycomb flanked by two GFRP sheets to resist shear and bending stresses, respectively) for building construction; GFRP hip prosthetic implants for high stiffness, strength, and biocompatibility; GFRP for leaf springs; glass/polyester and more recently aramid/epoxy boat hulls; steel fiber concrete for the fast reactor dome of the roof slab; aramid/polyurethane composite for the belt drive of motorcycles; versatile aramid fiber-reinforced composites for radial tires and bulletproof vests; and glass, aramid, and graphite-reinforced sports equipment (golf club shafts, skis, and tennis rackets). Within the present context of conductive composites, uses include high-density polyethylene/glass beads for conveyor rollers, nylon/glass beads for computer components, ABS/potassium titanate piping, talc-filled texile machinery components, chalk (calcium carbonate)-filled composites for electrical and radio parts, thick- and thin-walled glass spheres in epoxy resins as encapsulants in electrical applications, glass-reinforced Nylon 6 or 6/6 power tool bodies, glass-reinforced

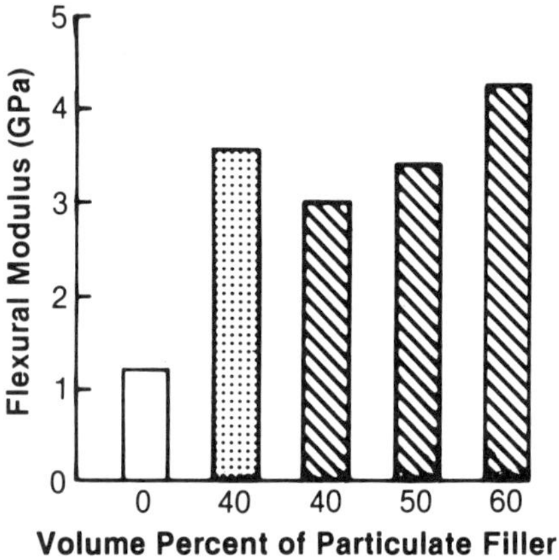

(a)

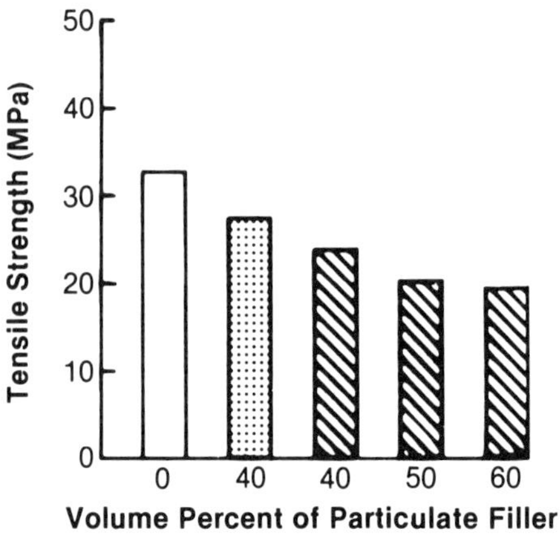

(b)

FIGURE 3 Influence of talc and wollastonite particulate fillers on (a) flexural modulus, (b) tensile strength, (c) impact strength, and (d) heat distortion temperature of a thermoplastic polymer, polypropylene. (Adapted from Fletcher and Tully (Ref. (13)).
□= Polymer alone, ⊡= Talc-filled, ▨= Wallastonite-filled.

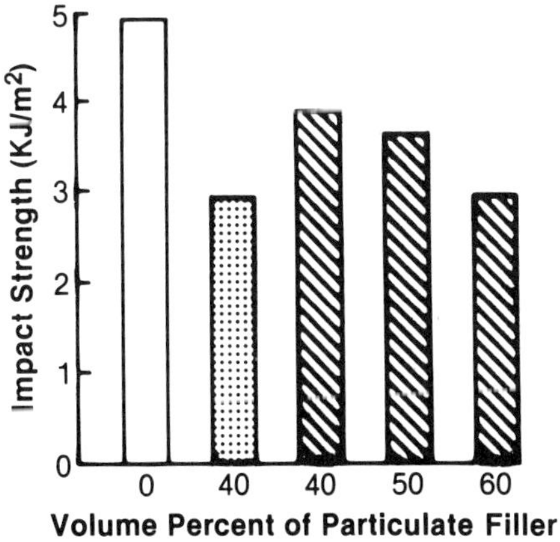

(c)

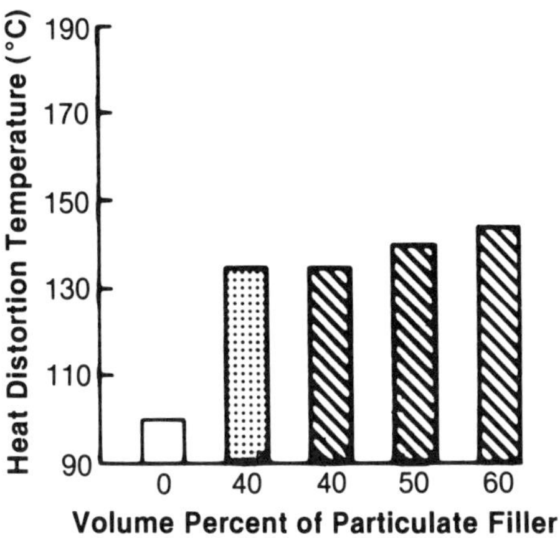

(d)

FIGURE 3 (Continued)

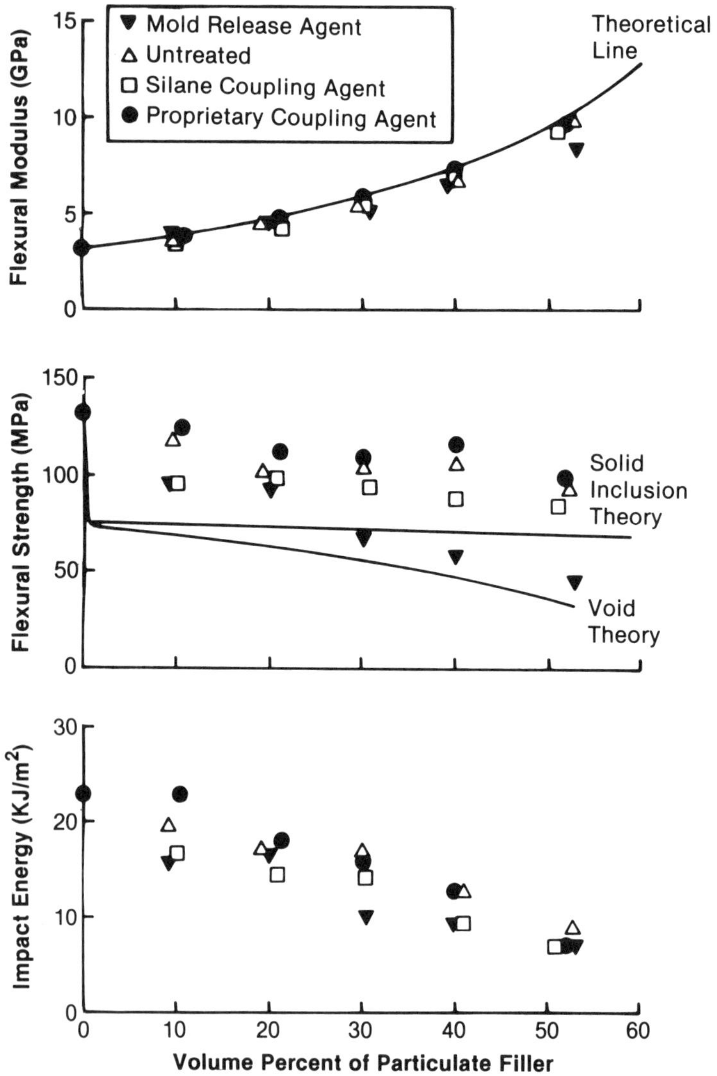

FIGURE 4 Influence of glass bead fillers on selected mechanical properties of a thermosetting epoxy matrix. When compared to the untreated filler particles, the addition of coupling agents had no effect on the properties investigated. The mold release agent did reduce the interphase adhesion, however. (Adopted from Sahu and Broutman, Ref. 14.)

polycarbonate and nylon equipment housings and tape spools, and glass fiber-reinforced polystyrene furniture panels.

TYPES OF METAL- OR METALLOID-FILLED COMPOSITES

When used as the matrix, polymers can enhance the adhesive properties, corrosion resistance, processing characteristics, and strength-to-weight ratios of composites while maintaining excellent electrical and thermal insulations. There are applications, however, in which polymers which possess high electrical and thermal conductivities would be desired. Unfortunately, conventional polymers inherently have an insufficient number of delocalized charge carriers. As a consequence, they must be modified so that the number of charge carriers and/or the charge carrier mobility is increased. To overcome this problem, metals or metalloids have been added to change the distribution of occupied states. Such inclusions can not only enhance the electrical and thermal conductivity but also absorb specific radiations, increase density, display permanent magnetism, and improve thermal stability. In decreasing filler content three approaches have been employed:

1. Dispersions of metal (or metalloid) powders or fibers in liquid thermosetting or in thermoplastic resins
2. Mixtures of metal (or metalloid) and polymer powders compacted by cold or hot pressing
3. Additions of metallic traces in the form of catalysts or organometallic complexes.

The first two are used for both carbon- and metal-filled composites, while the third is used to synthesize conductive polymers.

Carbon Black-Filled Polymers

Carbon black polymer composites have been used for over one-half century. Although they were first used to help reinforce gum rubbers (Table 7) (Ref. 24), their value as conductive composites to control electrostatic discharge (ESD), as resistive heating elements, and in electronic equipment (e.g., resistors and potentiometers) was acknowledged. Initially made by the channel process, today nearly all blacks are created from oil by the furnace process (Ref. 25). When residual refining oil is injected into a natural gas and air mixture, the polyaromatic molecules pyrolyze readily to form carbon black. When classified according to intrinsic electrical resistivities (Figure 5) (Ref. 26), conductivities may differ by 14–15 orders of magnitude, depending on the exact formulation of the composite vulcanizate, the type of polymer and black, its fractions of carbon

TABLE 8 Characterization of Various Carbon Black Particles

Carbon black[a]: Trade name	Surface area (m^2/g)	Mean particle size (nm)	Mean number of particles per aggregate	Mean diameter of equivalent sphere (nm)
A. Furnace blacks (rubber grade)				
Vulcan 9	138	20.9	135	--
Regal 660	109	26.6	74	136
Vulcan 6	112	24.7	193	207
Vulcan 6H	136	23.2	229	232
Regal 300	81	32.1	102	193
Vulcan 3	76	34.7	161	218
Vulcan 3H	87	30.5	239	262
Sterling SO-1	43	54.6	131	307
Sterling SO	42	60.9	160	480
Sterling SO-HS	48	52.5	193	469
Sterling SRF	27	101	73	--
B. Channel blacks				
Black Pearls 2	850	9.5	336	79
Black Pearls 71	380	8.5	165	--
Spheron 6	112	31.8	137	179
C. Conductive and special blacks				
Vulcan XC-72	238	21.9	303	245
Acetylene	51	56.6	493	628
Sterling FT	10	255	3.04	258

[a]All blacks are products of Cabot Corp., except the acetylene black, which was manufactured by the Shawinigan Corp. of Canada.

Source: In part, data obtained from unpublished work of F.A. Heckman and work of Medalia and Heckman (Refs. 28,29).

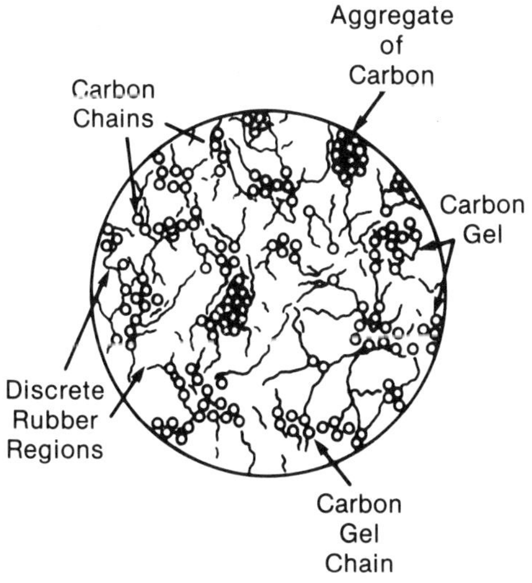

(a)

FIGURE 6 Structure and morphology of carbon black. (a) A schematic structure, and (b) an electron micrograph of a typical mix. (c) The fine structure of the carbon black will greatly influence the conductivity of the vulcanizate (all 50 parts carbon black in natural rubber). (Adopted from Blanchard, Ref. 30, Emma, Ref. 31, and Boonstra and Dannenberg, Ref. 26, respectively.)

fractions (Ref. 42). For more details concerning the influence of various polymers on the resistivity-loading curves refer to Chapter 3.

Four other characteristics of carbon black-filled conductive composites are detailed in Figure 9. A factor previously ignored, that of surface chemistry of the carbon black, is important if compounds are to be conveniently processed with the minimum disruption of the high structure, the optimum conductivity, and the most economical power requirements. Thus Figure 9a (Ref. 43) shows that a standard polyvinyl chloride compound containing 15% Ketjenblack surface treated with 1.5% stearoyl chloride reduces the apparent melt viscosity of the untreated Ketjenblack compound by as much as an order of magnitude. If, instead, the matrix is modified by various amounts of an active cross-linking agent (Figure 9b) (Ref. 44), not only the high and low plateaus of the sigmoid-shaped curve but also the critical volume fraction required for conductivity may shift. When the frequency dependence of a polyvinyl chloride-polypropylene copolymer/carbon composite is investigated (Figure 9c) (Ref. 45), a change is noted only when the carbon black loading is low. Apparently, higher frequencies facilitate the migration of charge by

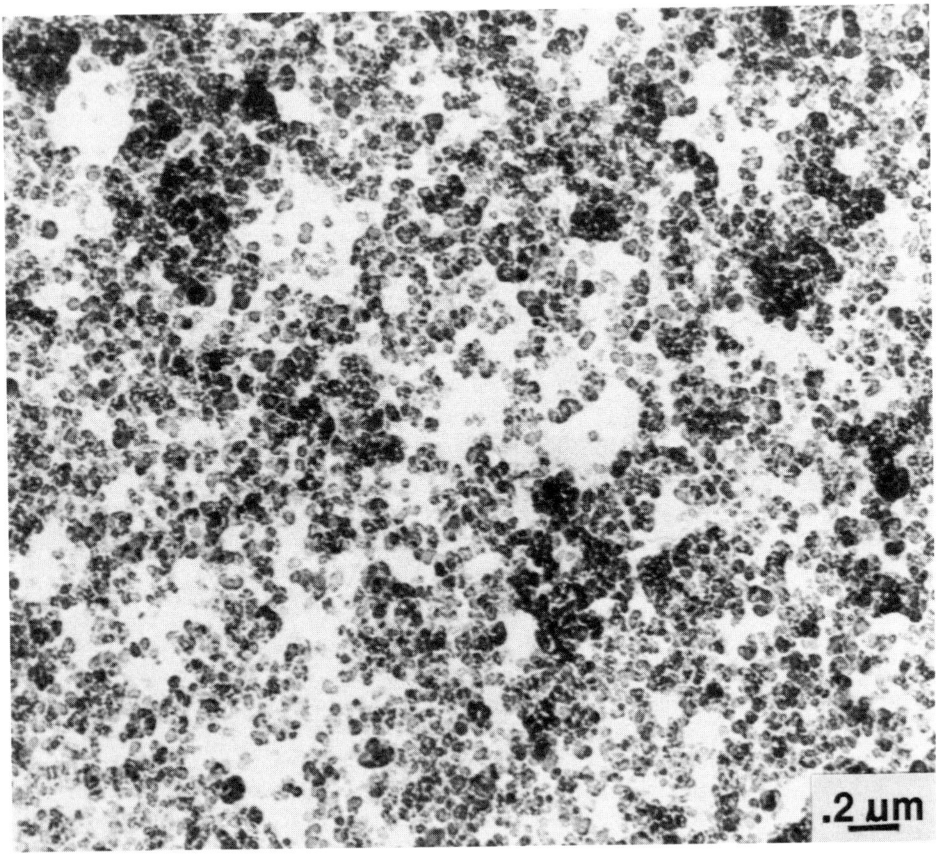

(b)

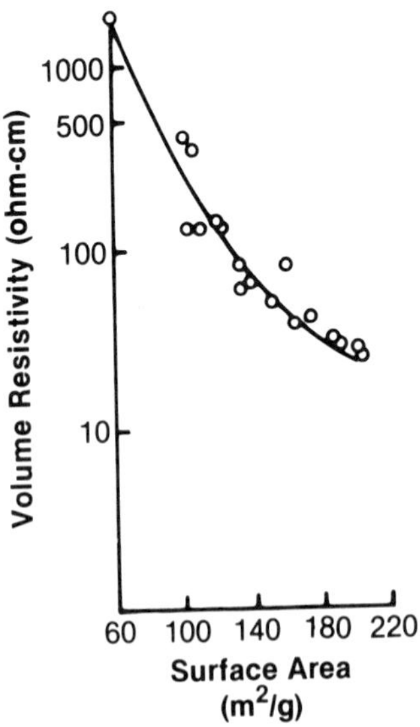

(c)

FIGURE 6 (Continued)

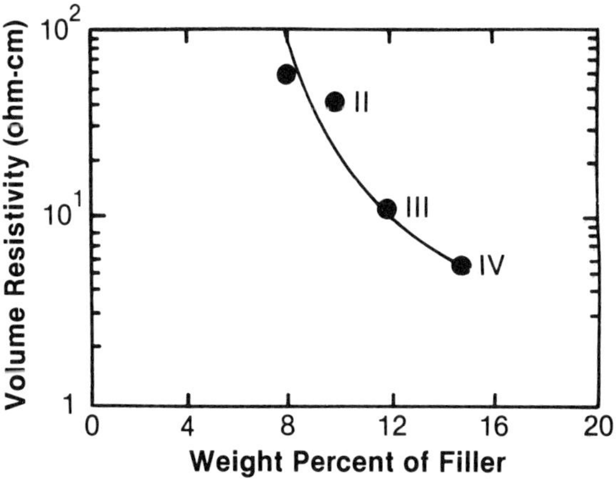

(a)

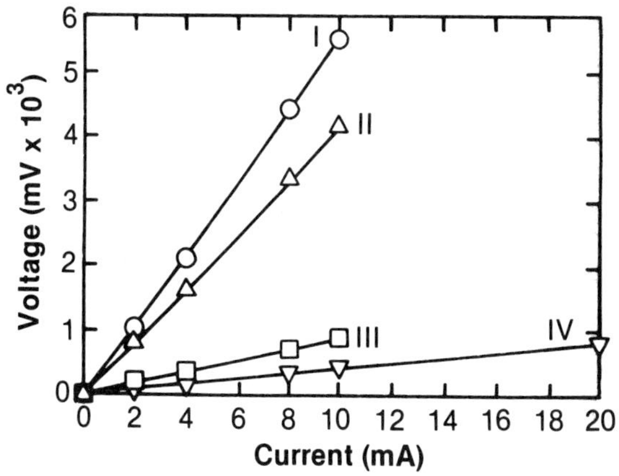

(b)

FIGURE 7 (a) The electrical resistivity and (b) the voltage-current characteristics of silicone compounds as a function of EC black loading. (Adopted from Kost et al., Ref. 38.)

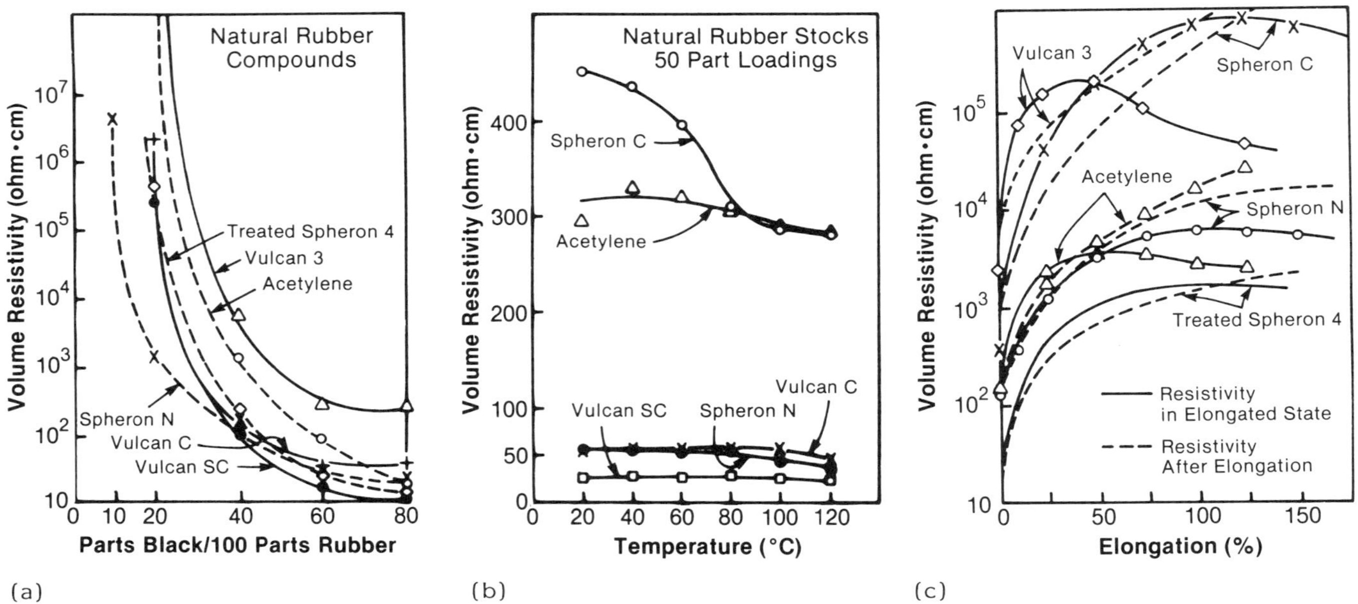

FIGURE 8 The dependence of the electrical resistivity of natural rubber compounds on (a) carbon black loading, (b) temperature, and (c) elongation. (Adopted from Boonstra and Dannenberg, Ref. 26.)

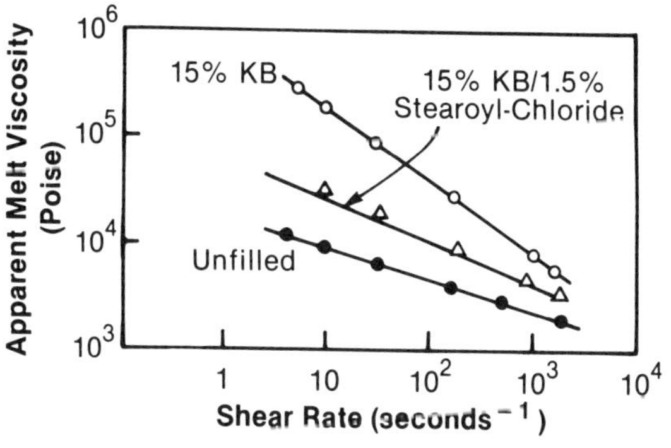

(a)

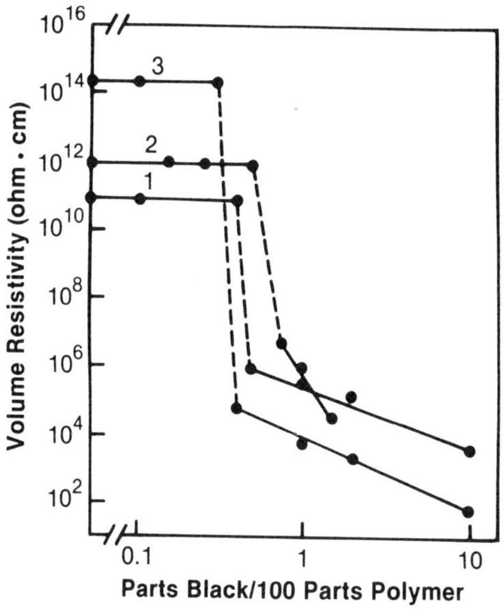

(b)

FIGURE 9 Other characteristics of carbon black-filled conductive composites: (a) influence of surfactant on the apparent melt viscosity at different shear rates; (b) influence of cross-linking agent [(1) with 45% methyl methacrylate, (2) with 5% styrene, and (3) with 55% styrene] on the volume resistivity at different carbon black concentrations; (c) influence of frequency on the volume resistivity at different carbon black concentrations; (d) influence of carbon black loading on the dielectric constant as a function of temperature. (Data of Datta and Friel, Ref. 43, Jachym et al., Ref. 44, Kawamoto, Ref. 45, and Narkis et al., Ref. 39, respectively.)

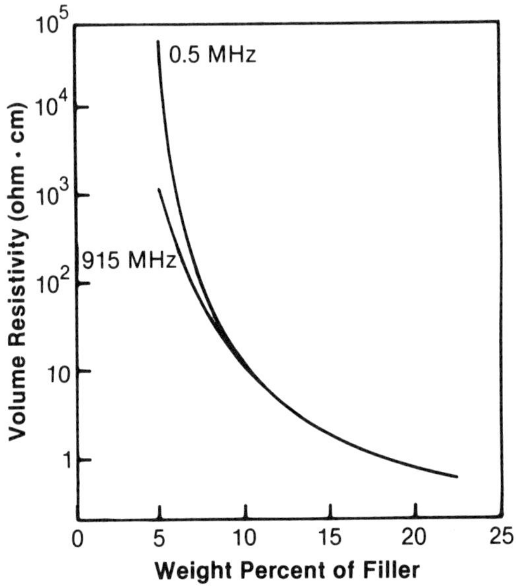

(c)

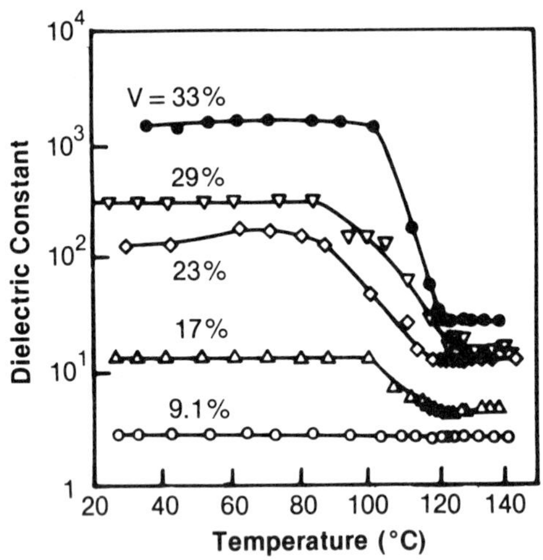

(d)

FIGURE 9 (Continued)

helping delocalized electrons hop across barriers. Finally, for a given loading at 1 kHz, the dielectric constant in polyethylene/MT black compounds is independent of temperature up to 90–100°C, although it does vary by a factor of 1000 as the amount of volume loading varies from 9.1 to 33% (Figure 9d) (Ref. 39). In the 5 wt % carbon black-loaded copolymer cited earlier, the dielectric constant decreased by a factor of 10 over the frequency range 1–1000 MHz at room temperature (Ref. 45).

Metal-Filled Polymers

Compared to carbon-filled polymer conductive composites, polymer/metal conductive composite technology is relatively young. Nevertheless, as long ago as the middle 1950s Hochberg (Ref. 46) obtained a patent which claimed that the frabrication of electrically conductive polymers could be achieved via the reduction of an aqueous solution of silver onto polystyrene beads. In principle, a structurally sound composite was to be fabricated by applying sufficient heat and pressure to crack the metallic coating, thereby permitting adjacent polymer beads to coalesce. About the same time, Coler (Ref. 47) obtained a patent for using fine laminar powders, e.g., flake- or disk-shaped silicon or copper, to render acrylics and vinyls conductive. He stated that, in order to form a reticulate structure, the metal particles should be <1/3 of the effective diameter of the plastic particles. This last point suggested an important geometric approach to reducing the metal filler necessary—that of varying the ratio of the radius of the polymer particle (R_p) to that of the metal particle (R_m).

In the 30-year period since that early work, two methods of packing metal particles in a polymer matrix have been employed, the so-called "random" and "segregated" distributions (Figure 10) (Ref. 48). In the random packing method (Figure 10a and b) continuity relies strictly on the chance contact between contiguous sites as governed by percolation theory (Refs. 49–51). Both particle species are similar in effective size and shape and may occupy any site—that is, either a metal or a polymer site—although the matrix may have been a liquid. In contrast, in the segregated distribution the likelihood of forming a network is increased by restricting the volume which any metal particle can occupy from the entire field of view (e.g., Figures 10a and b) to only the interproximal contacts or the interstices of the polymer particles (e.g., Figures 10c and d). The metallic particles are thereby segregated into a reticulated structure which is not unlike a highly structured carbon black-filled conductive composite (compare Figure 11 (Ref. 48) with Figure 6a). Random conductive composites are amenable to injection molding or extrusion techniques, whereas segregated conductive composites are normally processed by compression molding. Some typical

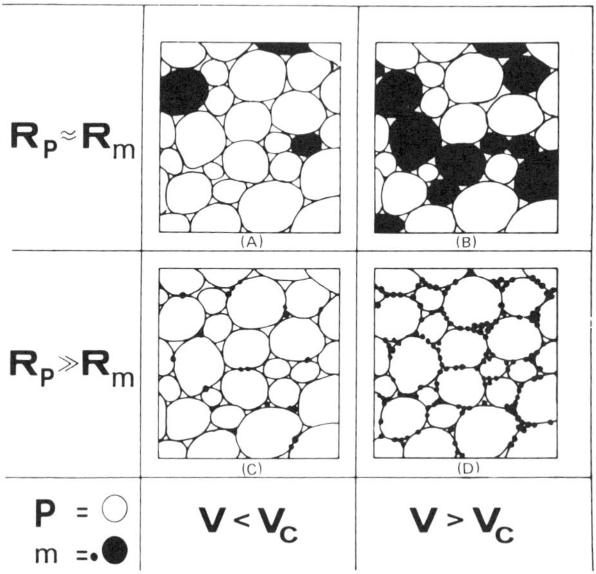

FIGURE 10 Schematic illustrations of "random" (A and B) versus "segregated" (C and D) particulate distributions both below (A and C) and above (B and D) the critical volume percent loading of metal (or metalloid) which is required for continuity (V_c). (Reproduced from Kusy, Ref. 48.)

conductive plastic compositions are shown in Table 9 (Ref. 52) along with some typical segregated composite morphologies in Figure 12 (Ref. 53).

As a direct result of the geometric arrangement, the loading required to achieve continuity is drastically reduced. Gurland (Ref. 50) showed that in a random composite the critical volume loading required for continuity (V_c) equaled 35–37% (effective $R_p/R_m \approx 2$), compared to 6% ($R_p/R_m = 30$) for the segregated composite of Kusy and Turner (Ref. 54) (cf. Figure 13). Figure 13 also illustrates how very different the geometry appears in these two composites which conduct electricity similarly (Refs. 50, 55). Figure 14 (Ref. 53) confirms that continuous runs of metal particles in the planar view are not a prerequisite for three-dimensional continuity, and do not occur even at $V \gg V_c$. The theoretical treatment of these networks (Refs. 48–50, 55–60) is presented in Chapter 3.

When preparing conductive polymer/metal composites at least seven factors must be considered:

1. Particle size ratios
2. Particle shape

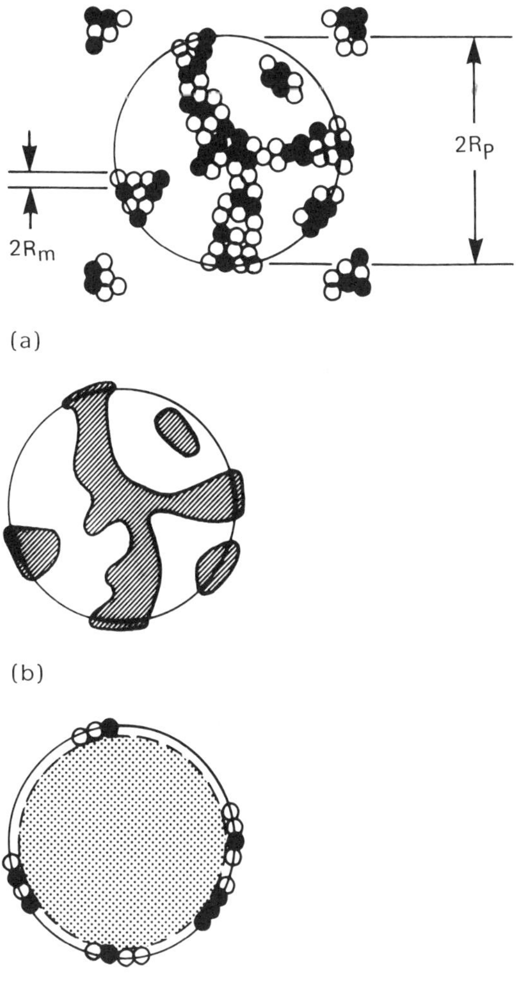

FIGURE 11 Schematic illustrations of the aggregate unit of a segregated particulate distribution at V_c. (a) A primary phase particle (of radius R_p) is covered by dispersed phase particles (of radius R_m) which are either associated with that unit (o) or associated with other adjacent units (•). (b) The effective surface fraction of primary phase particle covered by any contiguous dispersed phase particles is shaded. (c) A section through a primary phase particle indicates the volume percent of the circumference covered by the continuous dispersed phase. Note that not even a monolayer is required to achieve continuity. (Reproduced from Kusy, Ref. 48.)

TABLE 9 Conductive Plastic Compositions

Matrix	Filler	Molding pressure (MPa)	Molding temperature (°C)
		A. Curing	
Epoxy	Al, Fe	Atmospheric	100
Epoxy	Ni	Atmospheric	70
Epoxy	Steel	Atmospheric	30
Epoxy	C	Atmospheric	30-80
Epoxy	Pb	Atmospheric	20
Epoxy	Al, Cu	Atmospheric	100
Epoxy, polyurethane	Al_2O_3	Atmospheric	145
Epoxy, polyester	Al	Atmospheric	95
Furfural-acetone	C	Atmospheric	30-80
Polyamide-imide	Fe	Atmospheric	200
		B. Hot pressing	
Polyvinyl chloride	Ni	70-110	120-130

Polyvinyl chloride	Cu	70-110	120-130
Polyvinyl chloride	Cu	10-60	130-180
Polyvinyl chloride	Cu	45	130-140
Polyvinyl chloride	C	Not reported	Not reported
Polymethyl methacrylate	Ni	40	210
Polymethyl methacrylate	Cu	1	145
Polyethylene	Ni	70	140
Polycarbonate	Ni	40	150-190
Polyurethane	Ni	15-35	80-120
Polyvinyl acetate	Ni	40	120
Polytetrafluoroethylene	Mo, Cu, MoS_2, C	15-35	400
Polyphenyl formaldehyde	Ag	Not reported	Not reported
Nylon	Ni	70-110	260
Acrylics, vinyls, polystyrenes	Si, Cu Fe_2O_3, Cu_2O, Ag_2S	Not known	Not known

Source: Summarized from several sources by Bhattacharya and Chaklader (Ref. 52).

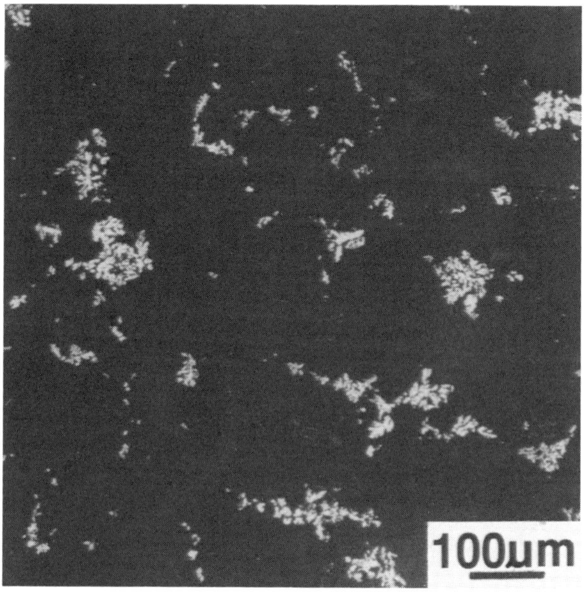

(a)

FIGURE 12 Typical morphologies of three conductive polymer-metal composites: (a) reflected light photomicrograph of a sectioned and polished amorphous polyvinyl chloride/6% copper dendrite mixture; (b) reflected light photomicrograph of a sectioned and polished semicrystalline nylon/3% spiky equiaxed nickel mixture (note the spherulites within individual polymer grains); (c) scanning electron micrograph of a liquid nitrogen-fractured polyurethane/18% allyl methacrylate (a tetrafunctional monomer)/8% spiky equiaxed nickel mixture.

3. Metal penetration
4. Prepolymer viscosity
5. Electrostatic attraction
6. Oxide layer thickness
7. Polymer bead shear

Each of these will be considered in turn followed by an overview of selected properties of experimental metal-filled polymers.

Particle Size Ratios

Earlier it was stated that the relative particle size ratio plays an important role in determining continuity. Figure 15 (Ref. 48) shows the inverse character of this relationship based on both the theoretical and practical minimum mean number of contacts per metal particle.

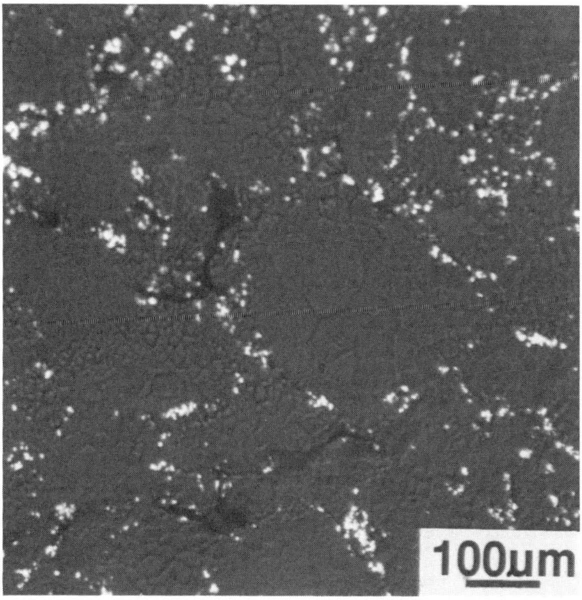

(b)

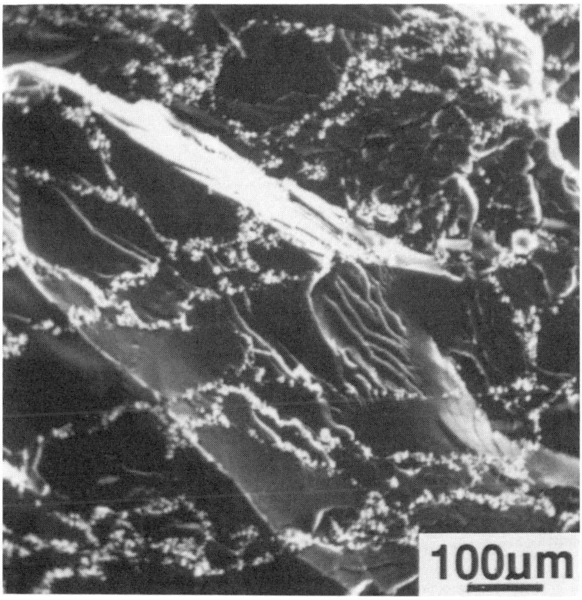

(c)

FIGURE 12 (Continued)

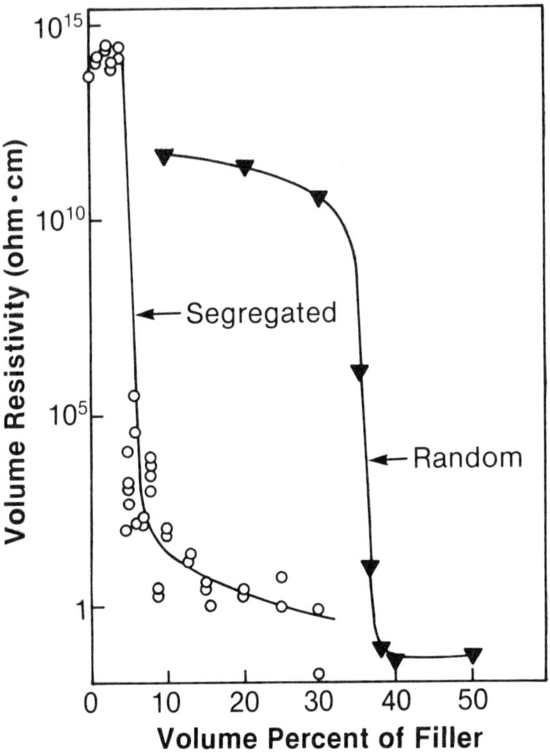

(a)

FIGURE 13 (a) Influence of filler content on electrical resistivity of a random and a segregated particulate distribution. (b) Photomicrographs illustrate the actual geometries of the bakelite/50% silver (random distribution, ▼) and polyvinyl chloride/7% nickel (segregated distribution, o) composites at V_c (cf Figure 10). Similar results were obtained for a segregated distribution formed from polyvinyl chloride and copper (cf Figure 14). (Adopted from (a) Kusy and Turner, (Ref. 54); (a and b, top) Gurland (Ref. 50); (b, bottom) Malliaris and Turner (Ref. 55).)

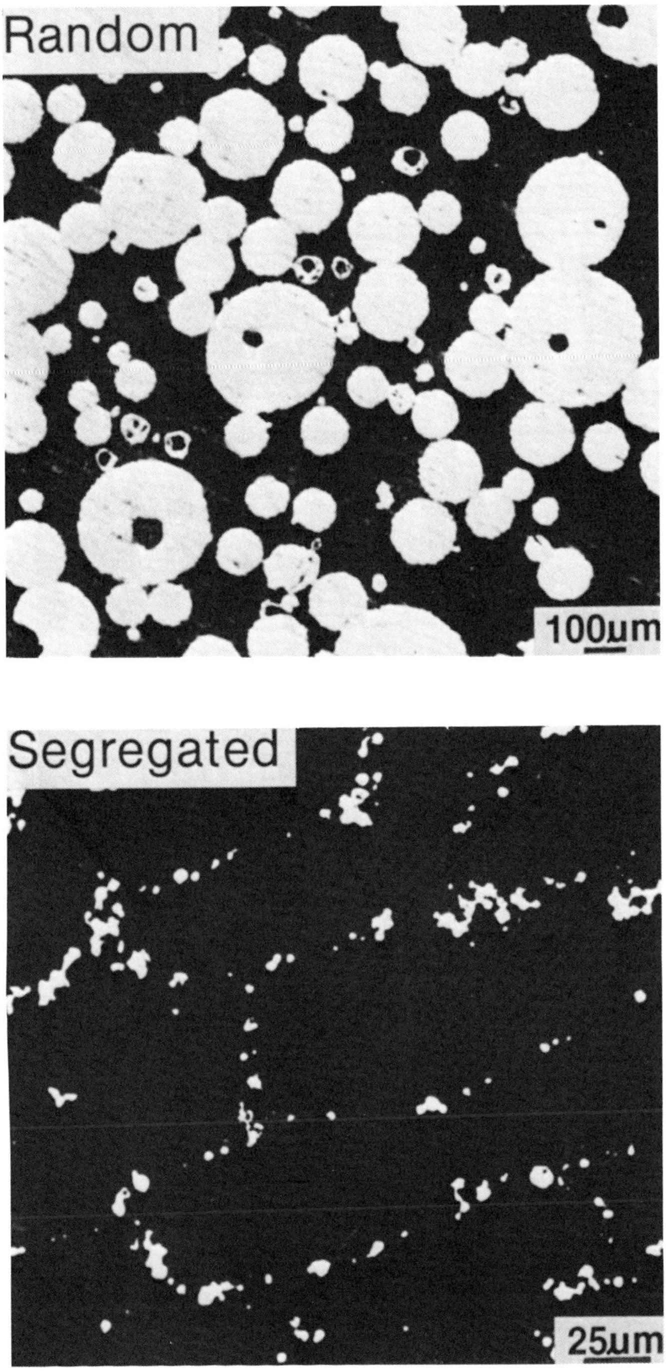

(b)

FIGURE 13 (Continued)

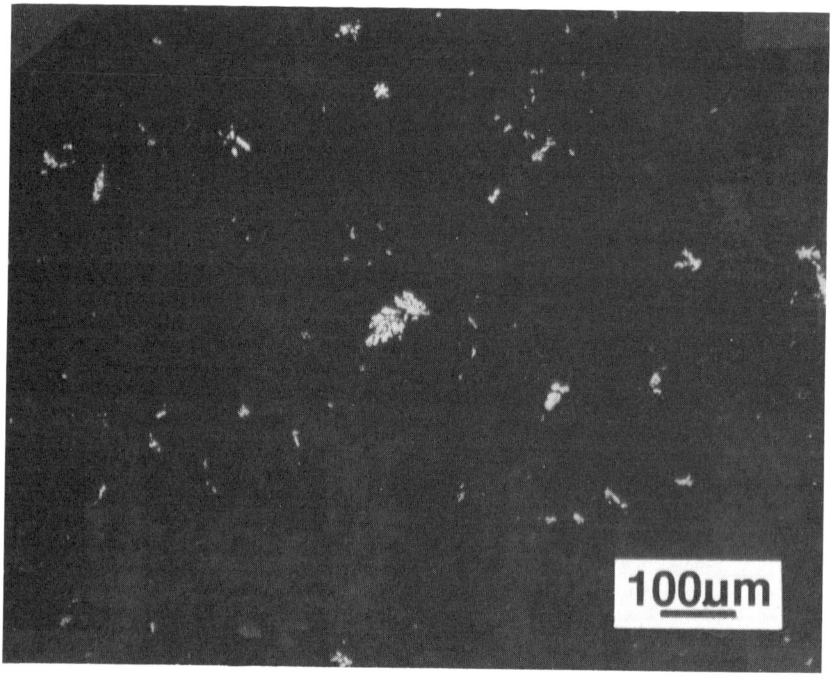

(a)

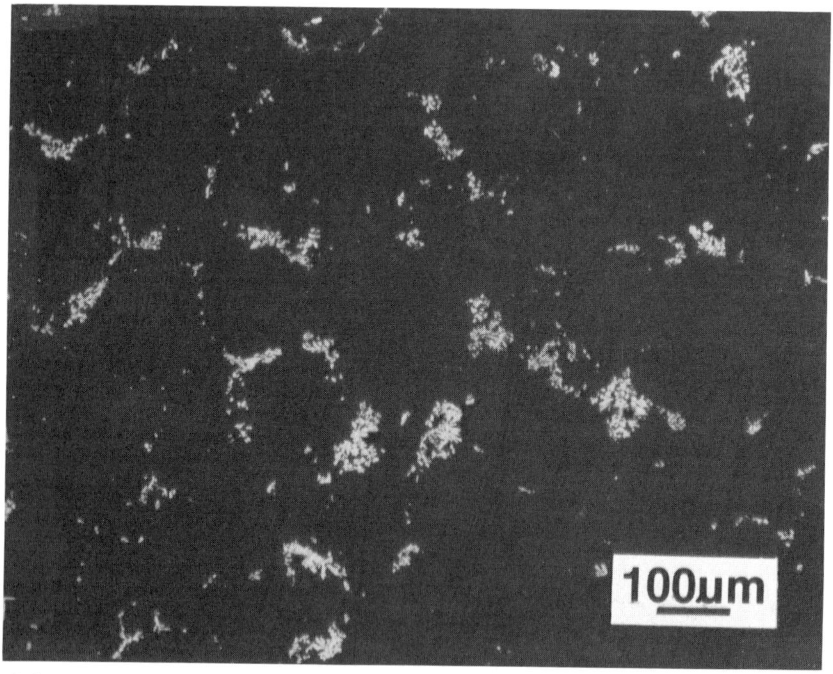

(b)

FIGURE 14 Photomicrographs of sectioned, ground, and polished polyvinyl chloride composites as a function of dendritic copper loading: (a) 1%; (b) 5%; (c) 10%; (d) 16%. Note that long after three-

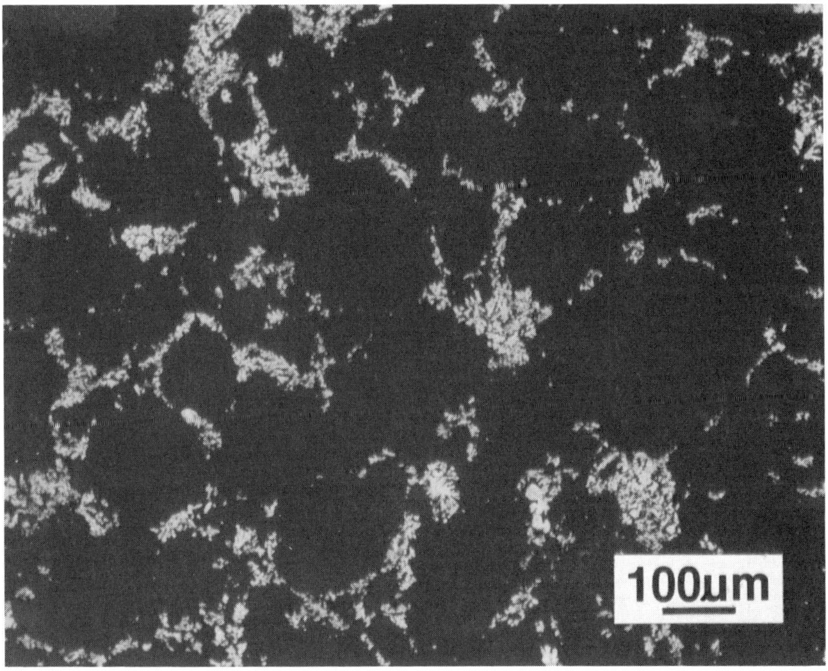

(c)

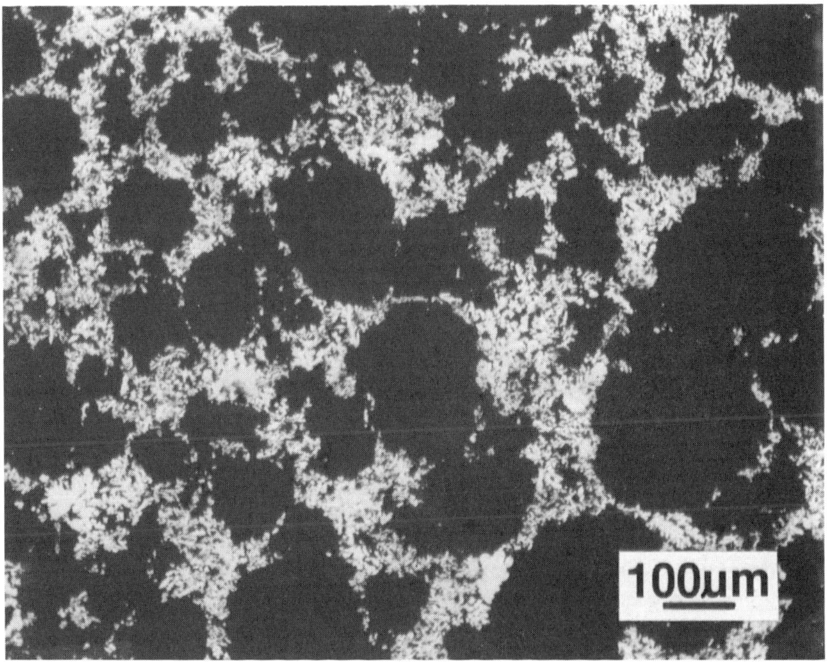

(d)

dimensional continuity has been established ($V > V_c$), the metallic particles form neither a monolayer nor any continuous paths in two dimensions.

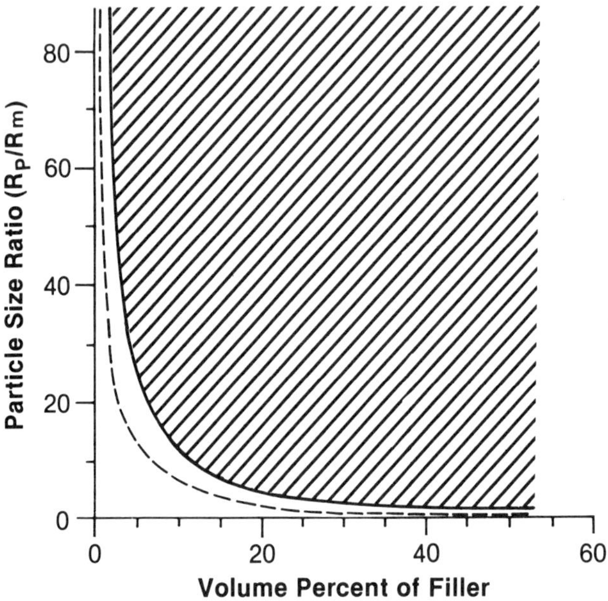

FIGURE 15 Theoretical (– – –) and practical (——) minimum requirements for continuity. The hyperbola-like functions show that large particle size ratios achieve continuity at small filler loadings and that large filler loadings are required when the particle size is nearly equivalent. The hatched region represents the zone of continuity. (Adopted from Kusy, Ref. 48.)

Using this spherical model, segregated mixtures with high particle size ratios (e.g., in the limiting case the polystyrene/silver composite of Hochberg, Ref. 46) and random mixtures with low particle size ratios (e.g., the phenol formaldehyde/silver composite of Gurland, Refs. 49, 50) indicate the extent to which particle size ratio can vary.

Particle Shape

Generally speaking, the formulator should strive to maximize both the volume/surface area occupied by the polymeric phase and the surface area/volume possessed by the metallic phase. Based on this criterion, spherical polymeric particles would minimize the potential contact surface area while maximizing the volume displaced; and flake-shaped, needlelike, or fibrous metallic particles would increase the surface/volume ratios (perhaps even providing some cold welding of metallic particles as a result of high contact stresses), thereby enhancing charge mobility. If all other factors are satisfactory, the critical volume loading for continuity decreases as the aspect ratio increases (see Chapter 3 or Refs. 59 and 61). In this regard the formulator should avoid excessive shear of the metal particles in the mixing, calendering, injection molding, or extrusion procedures, or else L/D will decrease and V_c will increase. Particle size, shape, and distribution will be reconsidered by Delmonte in Chapter 2.

Metal Penetration

Even if the aspect ratio of the metal particles is not reduced by excessive stresses, the benefits of a favorable segregated geometry can be lost by penetration of the metal into the polymer. This can occur when blending time is protracted, or when the compaction pressure-temperature is unsatisfactory. In the former case the metal particles literally become embedded in the polymer matrix, whereas in the latter case the discrete reduction of viscosity of the crystalline zones above their melting points ensures penetration into the polymer. When the proper processing temperature is selected, however, increasing the compaction pressure can actually reduce the

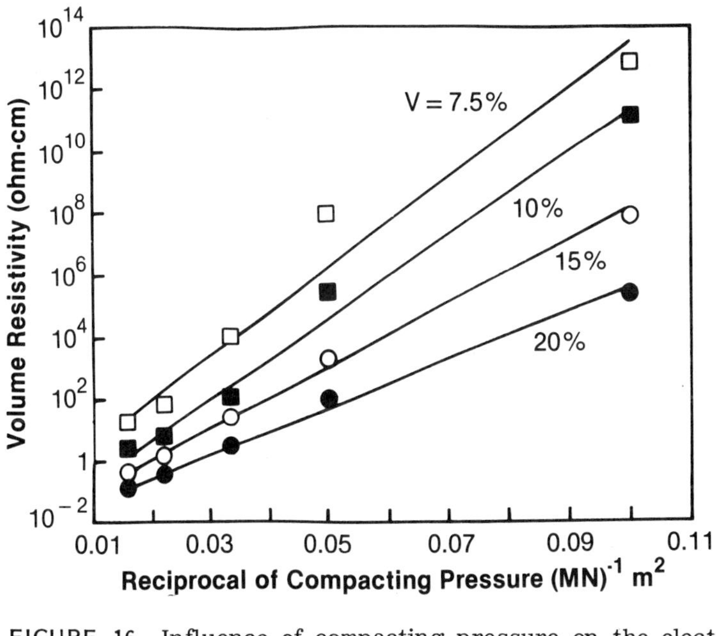

FIGURE 16 Influence of compacting pressure on the electrical resistivity of polyvinyl chloride/copper composites. (Adopted from Bhattacharya, Ref. 62.)

resistivity by reducing the occurrence of mesophase structures (cf Chapter 5). Some results are shown in Figure 16 for an amorphous polyvinyl chloride/copper composite (Ref. 62).

Prepolymer Viscosity

If ester or epoxide prepolymers are mixed with metals to form a conductive polymer/metal composite, the early set viscosity is critical. Obviously, too dense a metal relative to the fluid medium will result in rapid, differential sedimentation according to Stoke's law (Ref. 63). If some semblance of homogeneity is to be retained, the metal particles must remain in suspension until the polymerization reaction can sufficiently increase the polymer viscosity. For iron, nickel, and cobalt powders both stabilization and formation of a highly structured network can be attained by imposing a magnetic field which is adequate to counter the settling effects of gravity or, alternatively, to premagnetize the powder. Gul' and Golumbera (Ref. 64) have used magnetism to fabricate conductive composites.

Electrostatic Attraction

If the metallic particles are small enough (approximately $<5\ \mu m$), the inherent charge differential between the two primary constituents may be used to advantage. In the normal course of blending, a charge is built up so that each polymer particle is surrounded by at least a partial monolayer of metal (Ref. 48). Together with their associated polymer particle, these highly structured runs of metal form the "unit" of aggregate (cf. Figure 11). Previous experience has shown that mixtures such as polytetrafluoroethylene/copper and poly(methyl methacrylate)/lead are difficult to maintain as segregated composites because the polytetrafluoroethylene and the lead tend to coalesce upon themselves (Ref. 65). Only by using smaller metal particle sizes or by preferentially softening the surface of the poly-(methylmethacrylate) beads with an acetone solution could favorable results be obtained (Ref. 53).

Oxide Layer Thickness

Presuming that the metal particles are free from extraneous contaminants, the oxide layers which naturally form on their surfaces significantly contribute to the loss of electrical and thermal conductivity. Electrical and thermal loss factors ranging from 10 to 1000 have been associated with the formation of these oxide layers (Refs. 55, 66, 67), each of which acts as a tiny resistor which requires electron hopping to negotiate. Although interphase reactions may cause such layers to reform, Figure 17 (Ref. 58) shows that every attempt should be made to admix the metal into the polymer in its purest state.

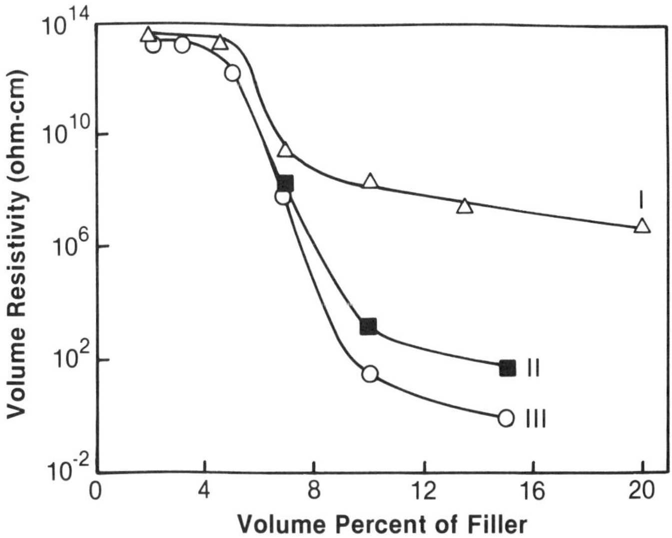

FIGURE 17 Influence of filler loading and oxide content on the electrical resistivity of polyvinyl chloride composites using (I) 100% copper oxide, (II) copper containing 47% by weight of oxide, and (III) reduced copper with negligible oxide. In all cases, $R_p/R_m \approx 25$. (Adopted from Bhattacharya et al., Ref. 58.)

Polymer Bead Shear

Analogous to the decreasing aspect ratio which occurs in metallic fibers subjected to increasing shear forces, the degeneration of the polymer beads by shear can increase the V_c required, too. In a segregated conductive composite this occurs if special precautions are not taken to retain the identity of each polymer bead during processing. For this reason compression molding is most often used. In contrast, in a random conductive composite, polymer bead shear is an expected consequence of the processing methodology (typically calendering, injection molding, or extrusion) and is not of primary importance here.

Properties of Experimental Metal-Filled Polymers

It has been shown that metal-filled composites literally switch from an insulator of electricity to a conductor of charge at a specific volume fraction of loading (e.g., Figure 13). Only in a rather narrow transition region does the number of (distance between) gaps become prevalent (small) enough to permit charge transfer processes similar to those found in a n- and p-type junctions to dominate.

Compared to electrical conductivity, thermal conductivity is a simple curvilinear function whose energy transport mechanism is oblivious to the onset of electrical continuity (Figure 18). Several models have been recently reviewed (Ref. 68), one of which considers fibers as well as spheres (Figure 18a) (Ref. 61). According to this model, increasing the L/D ratio from 1 to 35 will increase the thermal conductivity some eightfold at 30% volume loading of metal. Early work in polyvinyl chloride/nickel and polyvinyl chloride/copper composites (Figure 18b) (Ref. 69) showed that the inherent thermal conductivity of the metal filler influences the rate of heat transfer through the composite. In these two segregated networks, addition of 15 vol% nickel or copper increased the thermal conductivity of the composite relative to the thermal conductivity of the polymer (K_c/K_p) by factors of 4 and 20, respectively. De Araujo et al., (Ref. 70) showed as a function of testing temperature that the thermal

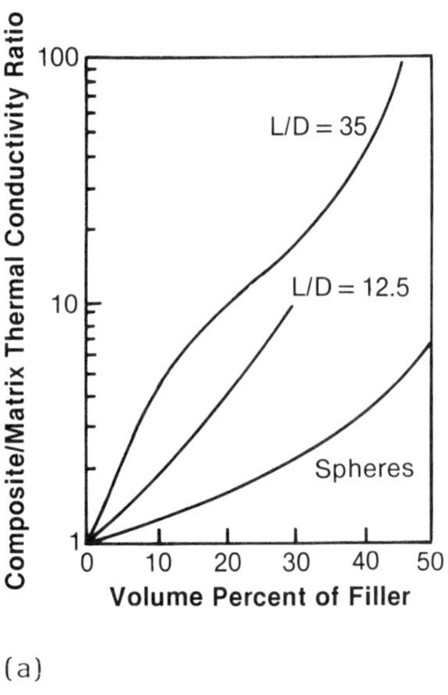

(a)

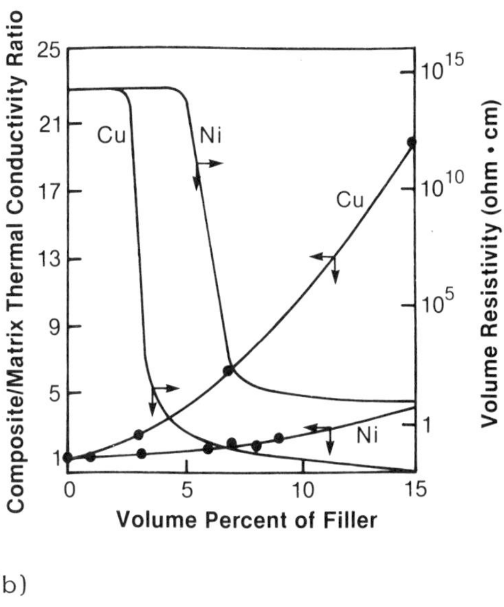

(b)

FIGURE 18 Relative increase in thermal conductivity as a function of filler loading: (a) influence of aspect ratio via one theoretical approach; (b) comparison of the thermal and electrical conductivity curves of copper- and nickel-filled polyvinyl chloride composites; (c) influence of test temperature on the absolute and relative thermal conductivities of copper-filled polymers (theoretical lines of Meredith and Tobias are shown for comparison); (d) influence of filler type and polymer matrix on the conductivity ratio. (Adopted from Battelle-Columbus Laboratories, Ref. 61, Kusy and Corneliussen, Ref. 69, de Araujo et al., Ref. 70, Meredith and Tobias, Ref. 71, and Bigg and Bradbury, Ref. 72, respectively.)

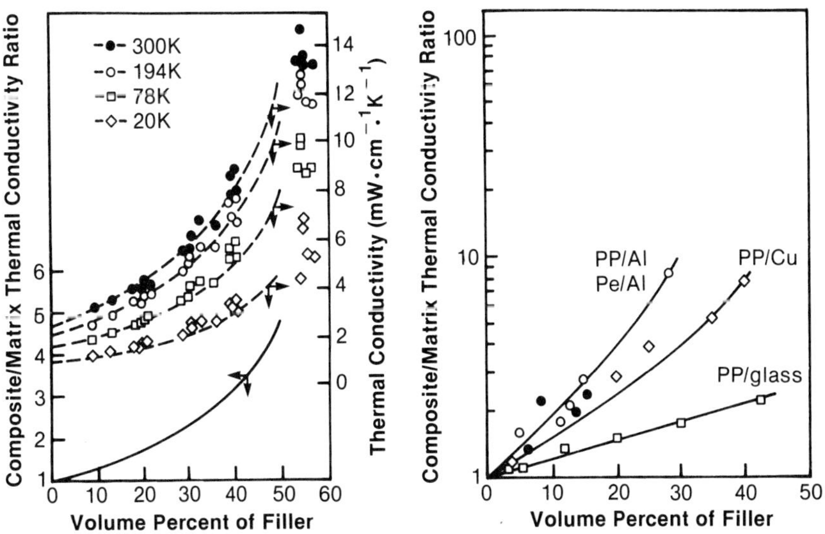

(c) (d)

FIGURE 18 (Continued)

conductivity of an epoxy/copper composite increased at all levels of loading (Figure 18c). No effect was observed, however, when the mean sphere diameters were varied from 11 to 100 μm. Moreover, although differences in the absolute values of thermal conductivity were observed as a function of temperature (dashed lines) (Ref. 71), no differences were noted in the normalized values (solid line). Comparison of this random network at 15% volume loading and 300 K with the polyvinyl chloride/copper data cited previously (Ref. 69), suggests that the segregated geometry conducts heat over an order of magnitude better than the random geometry. The overall improvement in heat transfer from milled glass fibers to copper particles and aluminum fibers is shown in Figure 18d (Ref. 72). With only 15% loading, as much as a twofold increase in K_c is obtained in the polypropylene-based composites. As expected, changing the matrix from polypropylene (PP) to polyester (Pe) had no significant effect. The solid lines drawn in for each frame in Figure 18 represent different theoretical models which individually look quite reasonable but collectively agree only qualitatively. More information concerning thermal conductivity is presented in Chapter 4.

Compared to the electrical and thermal properties, the mechanical properties of conductive polymer/metal composites are inconclusive. Gordon (Ref. 73) recently compiled some of the various theoretical models which have been used to describe the elastic modulus, ultimate tensile strength, and elongation at break of composite materials, and these are shown in Figures 19a, 20a, and 21a (for a theoretical discussion of various mechanical property theoretical models see Chapter 4). As reviewed earlier in this chapter, the elastic moduli should increase regardless of the nature of the filler. Figure 19a supports that general statement, but the three limited laboratory studies project a somewhat clouded picture. In Figure 19b (Ref. 74), both the secant (□) and the tangent (o) modulus of polyvinyl chloride/nickel composites increase steadily until V_c is achieved. At this point a discrete shift in the elastic moduli occurs, and the slope either increases or decreases slightly. The data presented in Figure 19c (Ref. 75) not only confirm that a change in slope occurs at V_c, but also suggest that V_c corresponds to the maximum of a continuous function. There is even an indication that the elastic moduli at V_c may increase as R_p/R_m increases. Finally, the curves reproduced in Figure 19d [76] have no maxima, show no discontinuities, and are concave downward. Thus it is fair to say that all three experiments differ with one another and with the theories which are presented here. A systematic, parametric laboratory study of random and segregated conductive composites should be quite informative.

Unlike particle fillers, fiber additions have been shown to enhance the tensile strength of polymeric materials (compare Figure 2b, with Figure 3 and 4). In two of the three laboratory investigations reported here [Figure 20b (Ref. 74) and c (Ref. 75)], an initial strengthening is followed by a general loss in strength as $V > V_c$. In the third case (Figure 20d) (Ref. 76), a monotonous decrease in strength is reported which is indicative of poor adhesion between filler and matrix. In Figure 20b the strengths reported for $V < V_c$ are quite low [10.3 versus 55.2 MPa via conventional processing (Ref. 77)], suggesting that incomplete coalescence had occurred despite the fact that the specimens were optically clear (Ref. 78). In Figure 20c it was also not clear whether the particle size ratio at $V = V_c$ had any systematic effect on the ultimate tensile strength. None of these laboratory experiments followed any of the theories shown in Figure 20a (Ref. 73).

When tested in tension, sharp metal particles or fibers tend to create stress concentrations from which cracks initiate and propagate. As a result, matrix deformation is reduced in some pseudologarithmic fashion with increasing filler loading (Figure 21a) (Ref. 73). When milled glass and aluminum fibers are admixed into polypropylene, a precipitous loss equal to 95% of the unfilled polypropylene, occurs at 7 to 8% volume loading (Figure 21b) (Ref. 72). This relative decrease is greater than that shown by any of the theories and indicates that good matrix and filler adhesion exists. The discontinuity coincides quite closely to the precipitous drop in electrical resistivity (cf. Figure 3 of Ref. 72).

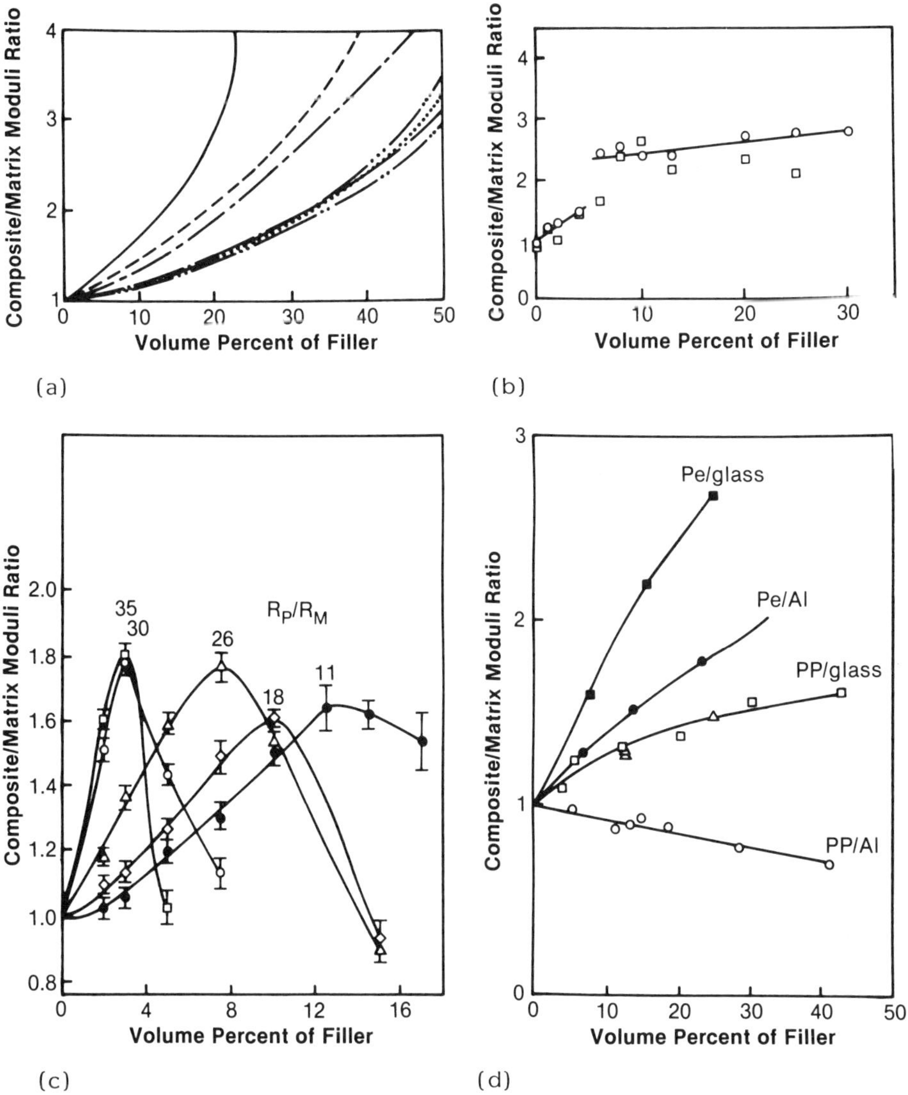

FIGURE 19 Relative changes in tensile moduli as a function of filler loading: (a) compilation of various theoretical models; (b) polyvinyl chloride/nickel mixtures; (c) polyvinyl chloride/copper blends; (d) various metal-filled polypropylene and polyester composites. (Adopted from Gordon, Ref. 73, Kusy and Turner, Ref. 74, Bhattacharya et al., Ref. 75, and Bigg, Ref. 76, respectively.)

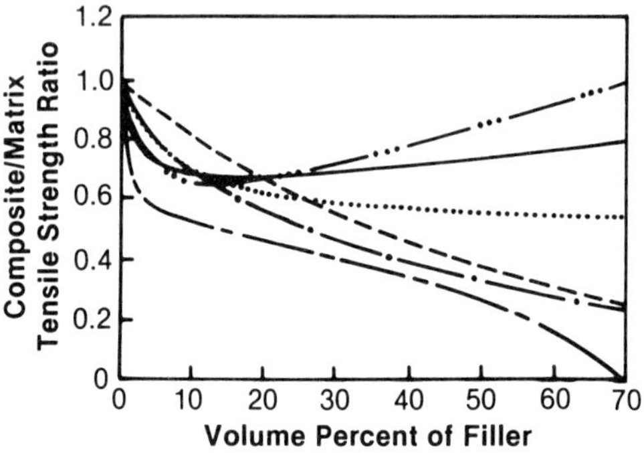

(a)

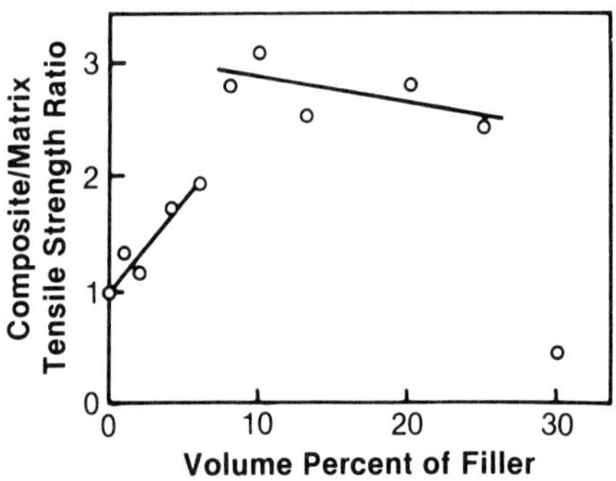

(b)

FIGURE 20 Relative changes in tensile strength as a function of filler loading: (a) compilation of various theoretical models; (b) polyvinyl chloride/nickel mixtures; (c) polyvinyl chloride/copper blends; (d) various metal-filled polypropylene and polyester composites. (Adopted from Gordon, Ref. 73, Kusy and Turner, Ref. 74, Bhattacharya et al., Ref. 75, and Bigg, Ref. 76, respectively.)

In a review article Delmonte (Ref. 79) summarized the effects of increasing the concentration of metallic particles in a polymer matrix. Table 10 is an updated version of that earlier table as it incorporates the data presented in Figures 13 and 18–21. As this field matures, further modifications of this table will undoubtedly be required.

Conductive Polymers

Conductive polymers are the latest addition to metal-filled polymers. Research in this field can be traced back to the 1960's, when the

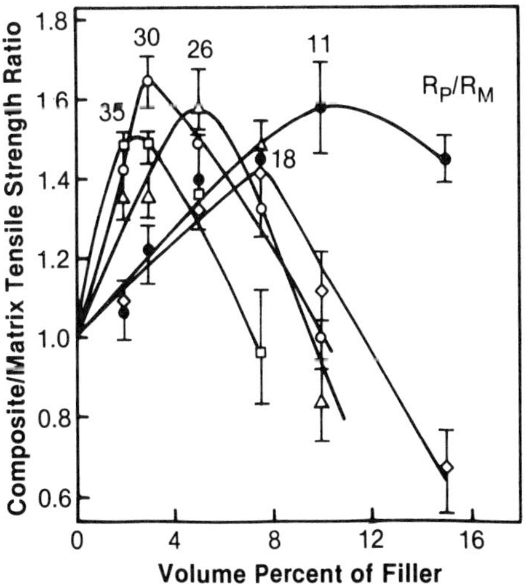

(c)

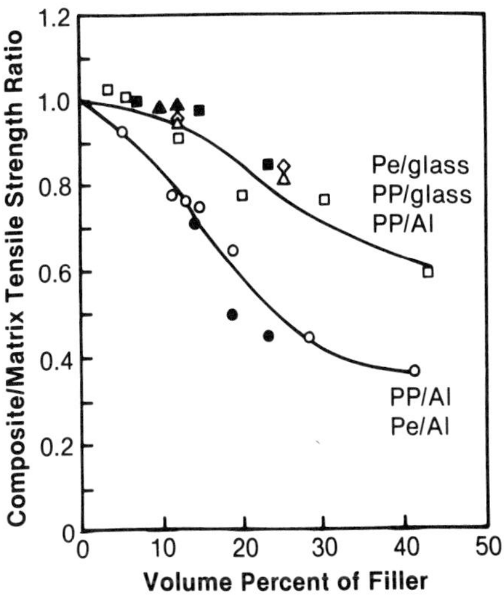

(d)

FIGURE 20 (Continued)

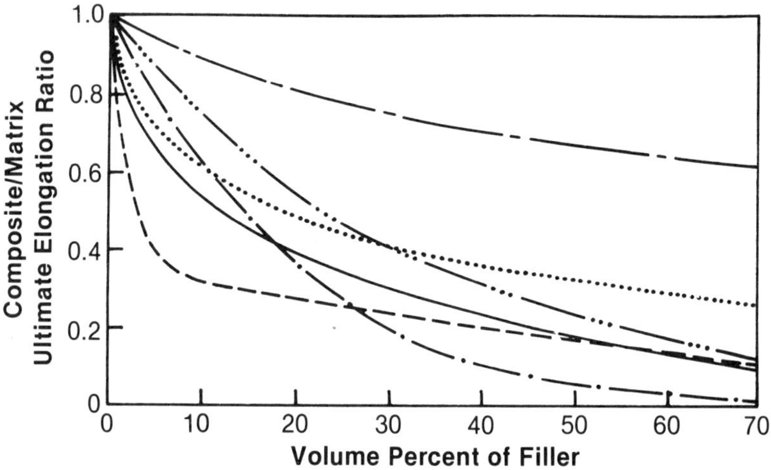

(a)

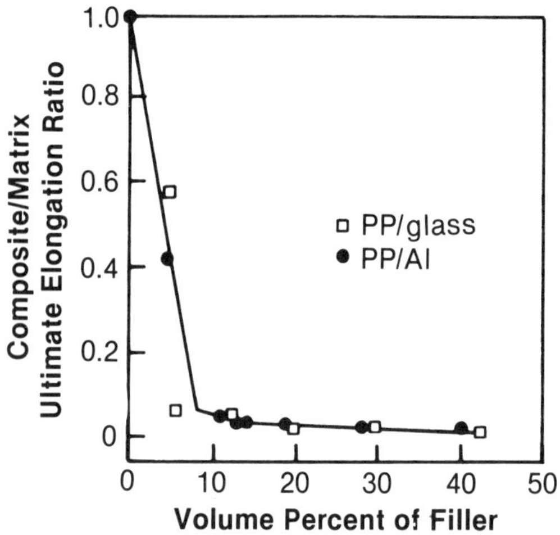

(b)

FIGURE 21 Relative decrease in ultimate elongation as a function of filler loading: (a) compilation of various theoretical models; (b) influence of filler type in polypropylene composites. (Adopted from Gordon, Ref. 73, and Bigg and Bradbury, Ref. 72, respectively.)

TABLE 10 Trends in Properties of Metal-Filled Composites with Volume Loading

Property	Effect of increasing concentration of metallic particles
Mechanical	
Flexural and compressive moduli of elasticity	For $V < V_c$, increase; but for $V > V_c$, increase or decrease
Tensile and flexural strength	For $V < V_c$, increase or decrease; but for $V > V_c$, decrease
Compressive strength	Increase
Elongation at break	Large decrease
Physical	
Density	Usually increases
Shrinkage during cure	Decrease
Thermal	
Thermal expansion coefficient	Usually decreases
Thermal conductivity	Increase
Glass transition temperature	Little or no change
Electrical	
Volumetric resistivity and dielectric strength	Decrease
Dielectric constant	Increase
Dissipation factor	Increase
Shielding effectiveness	Increase

Source: Updated from work of Delmonte (Ref. 79).

government funded National Aeronautics and Space Administration (NASA) research projects on polymers with intrinsic conductivity. Concurrently, a large number of such polymers were investigated in the Soviet Union and subsequently tabulated by Gul' (Ref. 80).

At present there are three ways in which organic polymers may be rendered conductive: (a) chemical doping, (b) electrochemical doping, and (c) ion implantation (Refs. 81–83). Regardless of the method used, transforming a disorganized collection of molecules into a conductive low-dimensional solid requires that two conditions be met (Refs. 84–89). First, sufficient intermolecular orbital overlap must exist between the component molecules so that a continuous electronic pathway is provided for carrier delocalization within similar crystallographic environments. Thereby, the associated energy barriers are reduced (that is, the so-called hills and valleys are smoothed out) so that the electron can move more readily through the solid. Second, the connected molecular entities must have fractionally occupied electron valence shells. These formal fractional oxidation states may be mixed valence, incomplete charge transfer, or partial oxidation states.

The first conductive polymer to emerge was polyacetylene, $[(-CH-)_n]$. Initially, only the two isomers were studied; the undoped conductivities were found to be 10^{-11} and 10^{-6} ohm^{-1} cm^{-1} for the *cis* and the *trans* forms, respectively (Figure 22). In the early 1970s, however, Shirakawa and co-workers (Refs. 90–94) pioneered the technology which produced highly conductive doped polyacetylene. Since that time, several other preparative procedures have been reported (Refs. 95, 96). In one, the polymer films were polymerized on the surfaces of a glass reactor by exposing acetylene (1 atm) at −78°C to titanium catalyst (0.3 mol/liter). Then the *cis*-rich polyacetylene films (100 μm thick) were doped with iodine vapor in equilibrium with its liquid phase at 25°C. Upon doping with iodine, whatever crystallinity existed was mostly destroyed and some *cis*/*trans* isomerization occurred such that the electrical conductivity for these flexible, golden-silvery films varied in accordance with Table 11 (Ref. 96). Thus the geometric configuration of the polymer backbones has a significant influence on electrical conductivity, although Chien and others (Refs. 97–99) claimed that the actual backbone length for this (M_n = 400–870,000) and other

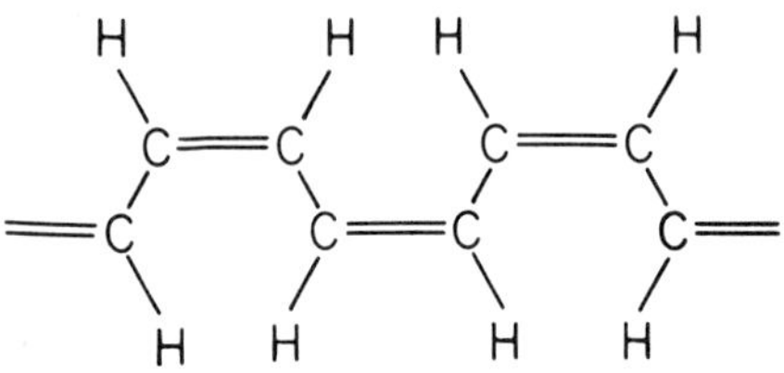

(a)

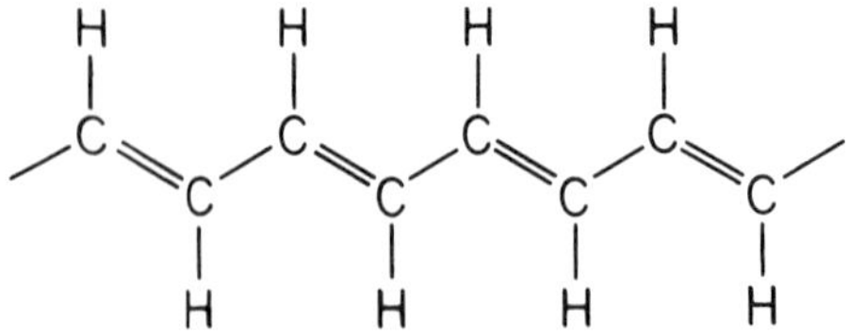

(b)

FIGURE 22 The chemical structure of polyacetylenes: (a) *cis* isomer; (b) *trans* isomer.

TABLE 11 Influence of Isomer Content on the Electrical Conductivity of Iodine-Doped Polyacetylene

Cis-isomer (%)	Conductivity (ohm^{-1} - cm^{-1})
98	500-700
74	105-250
28	50-80
24	50-80
7	3 x 10-4
3	6 x 10-3
0	6 x 10-5

Source: Adapted from Deits, Cukor, and Rubner (Ref. 96).

conductive polymers is of minor importance. The morphology (Ref. 100) and band structure (Ref. 101) of polyacetylene are shown in in Figures 23 and 24. Similar results have been reported for bromine-doped *trans*-polyacetylene, although the dopant energy levels change to -9.5 and -12.5 eV.

Two primary mechanisms of conduction have been suggested:

1. Bond alternation defects called neutral and charged solitons (Refs. 102, 103)
2. Inhomogeneous doping and the formation of small metallic regi embedded in undoped polymer (Ref. 104)

The soliton model was based originally on the concepts advar by Pople and Walmsley (Ref. 105), which suggested that discrete mobile electronic structures conduct electricity. Specifically, th structures involve the presence of topological kinks along the p meric backbones which extend over several monomer units. Co tion between the energy decrease by delocalizing the radicals o over as many carbon bonds as possible and the energy increas lengthening double bonds and shortening single bonds to achie localized bond length results in a stable soliton length of abou bonds. These defects, which introduce a localized electronic at mid-gap, move readily along the chain (if the radical type) form of a solitary wave. Using solitons and other defect str

which are beyond the scope of this book, significant electrical conduction is achieved at low concentrations of metal.

Unlike the first, the second proposed conduction mechanism does not invoke new concepts. In analogy to the segregated structure of polymer/metal composites discussed in the previous sections, metal doping is viewed as an inhomogeneous process which creates local conductive, metallic regions separated by insulative, polymer regions. The concentration of these doped and undoped regions determines whether the electrons tunnel between metallic regions (true at low dopant concentrations) or move along a metallic network which has coalesced (true at high dopant levels). Mortensen et al., (Ref. 106) proposed that a percolation process is a consequence of this inhomogeneous doping and that the dopant increases the number of carriers just as an impurity does in a classical inorganic semiconductor. The white globulelike structures observed by Rolland et al., (Ref. 107) in iodine-doped polyacetylene and the reticulated features shown by Jeszka et al., (Ref. 108) in a poly(bisphenol A carbonate) complexed

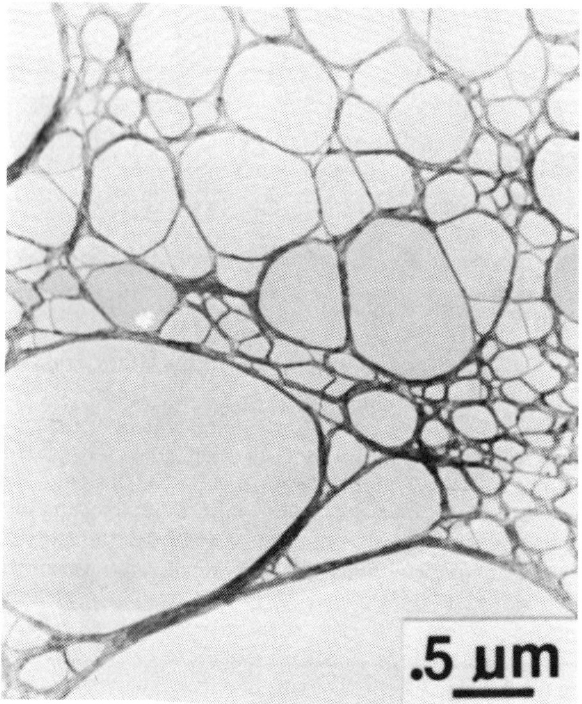

FIGURE 23 The morphology of polyacetylene. (Reproduced from Ref. 100 with permission of J.C.W. Chien, University of Massachusetts.)

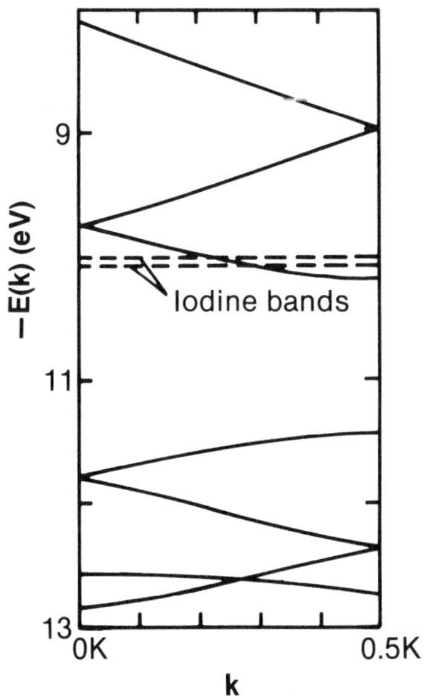

FIGURE 24 Band structure for iodine-doped *trans*-polyacetylene. (Adopted from Bhaumik and Mark, Ref. 101.)

with 1% 7,7,8,8-tetracyanoquinodimethane (TCNQ):tetrathiotetracene (TTT) suggest that the metal species is not distributed in all polymer conductors at random, but rather may form dendritelike paths which interpenetrate the polymer matrix.

Since the discovery of polyacetylene-doped compositions, many conductive polymers have evolved. A representative list of these conductive polymers and some typical dopant and metal ions which have been incorporated are given in Table 12 (Ref. 109). Most of these fall into three families of highly conducting composites: the polyenes, the poly(phenylenes), and the poly(phenylene chalcogenides).

The first group is based on the polyacetylene structure (Ref. 81) in which some of the hydrogen atoms in the $(—CH—)_n$ are replaced by one or more substituent groups. Poly(2,5-pyrrole) (Ref. 110), poly(2,5-thenylene) (Ref. 111), and poly(1,6-heptadiyne) (Ref. 112) are three examples. Conductivities usually range from 10^{-4} to 1200 ohm^{-1} cm^{-1}, although copolymers or homopolymer mixtures will yield values which are proportional to the amount of unsubstituted polymer.

Although poly(*p*-phenylene) (Refs. 113–115) is rather intractable like the three polyenes mentioned above, poly(*m*-phenylene)

TABLE 12 Some Conductive Polymers Which Have Been Successfully Complexed from Various Dopants and/or Metal Atoms

Conductive polymers	Dopants	Metals
Polyacetylenes	Iodine	Si
Polyalkadiynes	Bromine	Ni
Polyamide/palladium	Potassium	Ge
Poly(acrylaminomethyl-acrylates)	Dichlorodicyanoquinone	K
Poly(arylene vinylenes)	Chloranilic acid	Sn
Poly(azomethines)	Fluoranil	Pd
Polybenzimidazole/metal	Chloranil	
Poly(bisphenol A carbonate) complexes	Bromanil	
Poly(bithiophene)	Nitrosyl salts	
Polyimide complexes	Dihydroxybenzoquinone	
Poly(metallocenylenes)	Antimony pentafluoride	
Poly(metal phthalocyanines)	Arsenic pentafluoride	
Polyphenylenes	Tetracyanoquinodimethane	
Poly(phenylene oxide)	Tetrathiotetracene	
Poly(phenylene sulfide)	Sulfuric acid	
Polypyrrole		
Poly(sulfur nitride)		
Poly(vinyl alcohol) complexes		
Poly(vinyarylamines)		
Poly(vinylbenzene anthracenes)		
Poly(vinyl carbazole)		
Poly(vinyl phenol)		
Poly(vinyl pyridine)		
Quaternary ammonium polymers		

Source: In part, adopted from data in Reference 109.

(Ref. 116) is both fusible and soluble. Because the backbone of the chain is more flexible, both melt and solution processing are possible. Historically, it has been one of the impediments to commercialization of conductive polymers that once they are rendered conductive they are infusible and insoluble. In the last case this has also restricted the characterization (Ref. 117) and application (Ref. 118) of these materials. The electrical conductivities of this group generally range from 10^{-3} to 500 ohm^{-1} cm^{-1}.

Of the three groups mentioned, the last group is the most surprising. Whereas at one time an unlimited number of compositions of the first group were thought possible, the third group was not even considered to be feasible. Previously, it was believed that the π-electrons from conjugated carbon atoms were required for conductivity. The phenylene chalcogenide polymers disproved this preconception (see also Ref. 119), however, as both poly(*p*-phenylene sulfide) (Refs. 120, 121) and poly(*p*-phenylene oxide) (Ref. 120) have been found to be highly conductive. Not only do poly(phenylene chalcogenides) have a discontinuous carbon backbone, but they also are grossly nonplanar. Overall conductivities have been measured and range from 10^{-3} to 1 ohm^{-1} cm^{-1}.

Before the various conductive characteristics of these polymers are surveyed, the donor-acceptor complexes formed from doped organometallics should be mentioned. Figure 25 shows the regular, planar structure of Group IV metallophthalocyanines. Although Ge and Sn have been substituted for the Si metal atom and Br, K, halogenated anils, anilic acids, and quinone have been used in place of the I_2 dopant, a very successful composition has been $[(Si\ (Pc)O)\ I_{1.55}]_n$, for which the conductivity was 1.4 ohm^{-1} cm^{-1} (300 K) (Ref. 122). Indeed, donor-acceptor complexes may assume many general structural forms, six of which are shown in Figure 26 (Ref. 122):

Three forms of segregated stacking:

(1) With canted donors
(2) With D_{nh} donor stacking
(3) With zigzag donor stacking

Two forms of integrated stacking:

(4) Simple donors
(5) Donor dimers

One form of ion clusters:

(6) Without stacking

At this point the different combinations of chemical species, processing techniques, and geometric arrangements should suggest

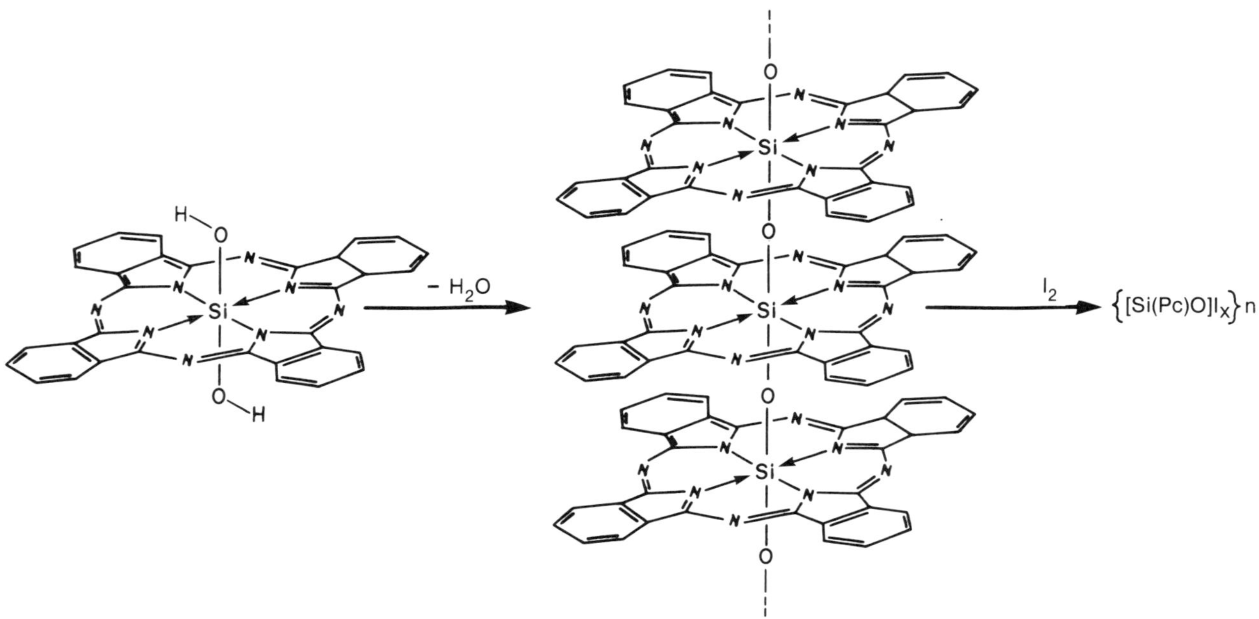

FIGURE 25 Formation of siloxane phthalocyaninate cofacial arrays via a condensation reaction and subsequent doping with iodine.

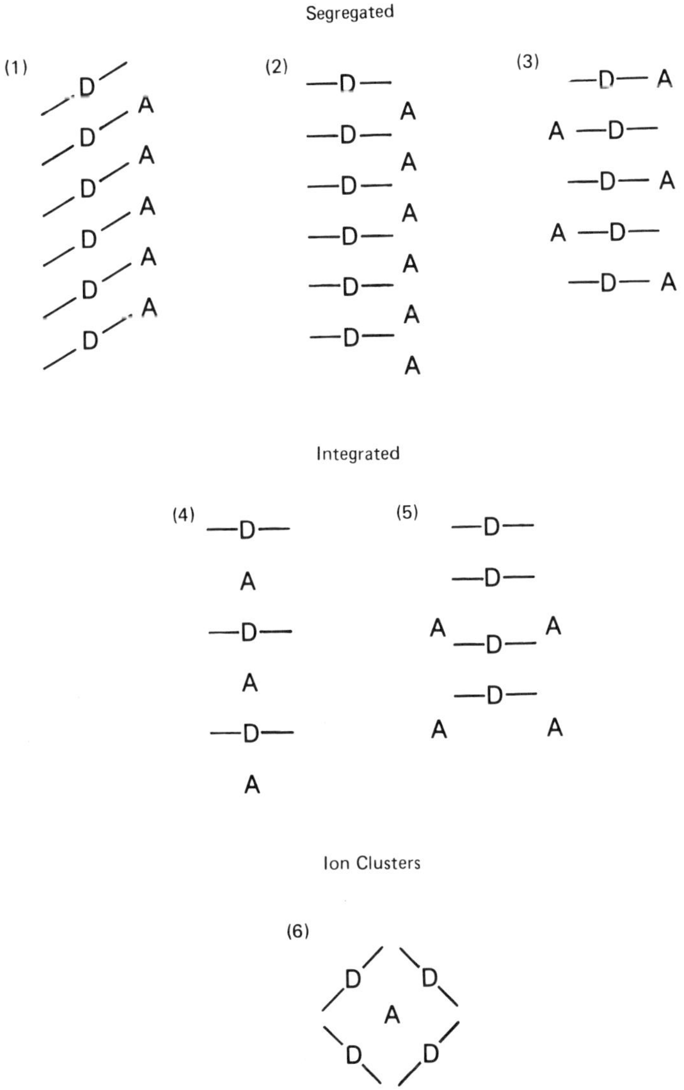

FIGURE 26 Six general structural forms of donor (D) and acceptor (A) complexes. (Adopted from Dirk et al., Ref. 122.)

to the reader that a new class of materials is being conceived and that soon many applications will be making their way into the marketplace. Some specific examples of the conductivities which are possible in these novel laboratory materials compared to conventional organic polymers, a semiconductor, and pure metals are shown in Table 13 (Ref. 123).

A summary of some of the more important characteristics of conductive polymers is detailed in Figures 27–29 and Table 14.

In order to tailor the electrical conductivity for a particular application, the reaction time may be controlled. Figure 27a shows the

TABLE 13 Relationship of the Electrical Conductivity of Conductive Polymers to Other Materials

Material	Dopant	Conductivity (ohm^{-1} cm^{-1})
Copper	—	6×10^5
Mercury	—	10^4
Polyacetylene	Arsenic pentafluoride	1200
Poly(*p*-phenylene)	Arsenic pentafluoride	500
Polypyrrole	Iodine	100
Polypyrrole	Perchlorate[a]	40
Polyaluminophthalocyanine fluoride	Iodine	4.5
Polyphthalocyaninesilorane	Iodine	1.4
Polyphenylene sulfide	Arsenic pentafluoride	1
Germanium	—	2×10^{-2}
Polyurethane	—	10^{-9}
Epoxy	—	10^{-14}
Polystrene	—	10^{-16}

[a]Polypyrrole perchlorate film from electropolymerization of pyrrole in perchlorate solutions.
Source: In part adopted from Reference 123.

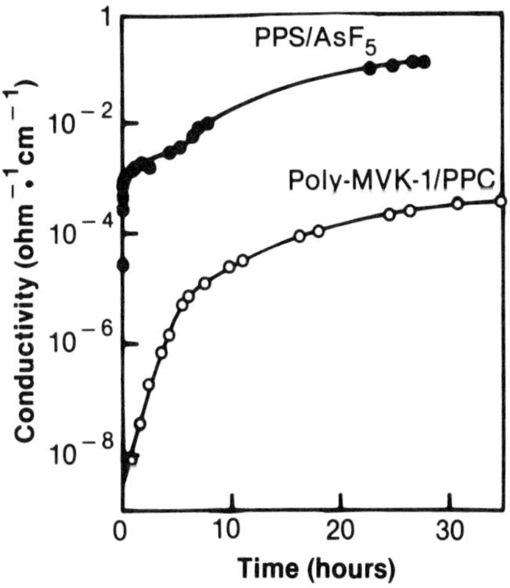

(a)

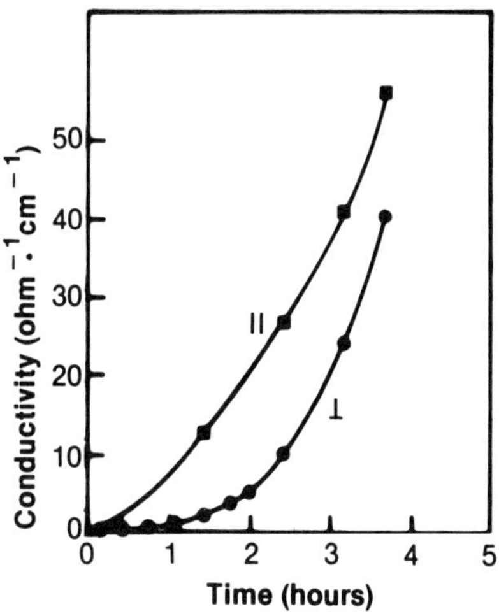

(b)

FIGURE 27 Influence of dopant/complex content on the electrical conductivity of polymers: (a) for poly(*p*-phenylene sulfide) reacted with arsenic pentafluoride and poly(vinyl ketone) complex reacted with phosphoryl chloride; (b) for *p*-terphenyl single crystal reacted with arsenic pentafluoride and tested to show anisotropy both parallel and perpendicular to the long axis of the crystal; (c) for polyacetylene reacted with iodine and for polycarbonate reacted with TTT:TCNQ complex; (d) for polyacrylonitrile and poly(*p*-phenylene oxide) films implanted with bromine ions. [(a) and (b) adopted from Ogawa et al., Ref. 124, Chance et al., Ref. 125 or Shacklette et al., Ref. 126. (c) Generated from the data of Rolland et al., Ref. 107, and Jeszka et al., Ref. 108. (d) Adopted from work of Wasserman et al., Ref. 83.]

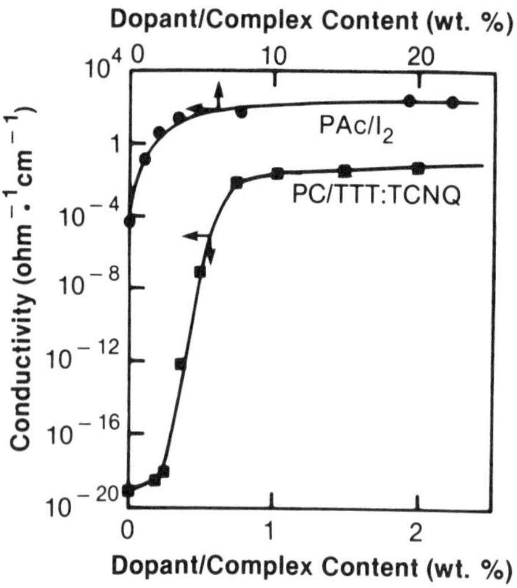

(c)

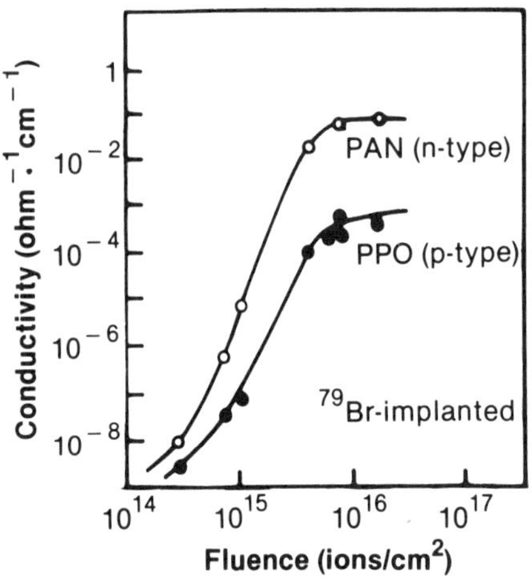

(d)

FIGURE 27 (Continued)

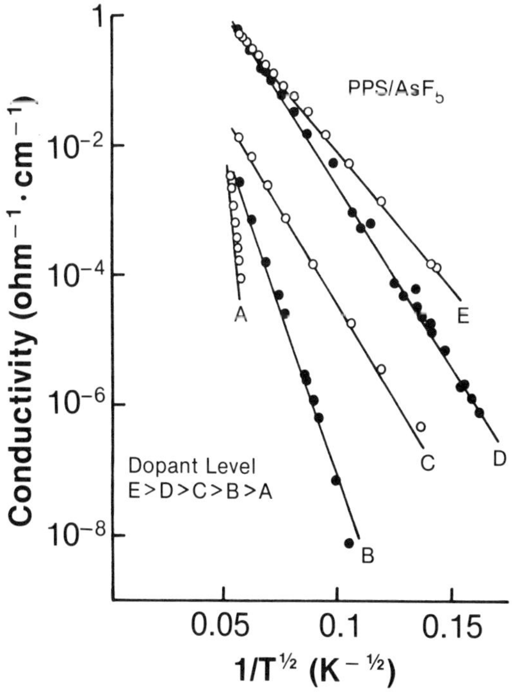

(a)

FIGURE 28 Influence of temperature on the electrical conductivity of polymers: (a) as a function of dopant level of arsenic pentafluoride in poly(*p*-phenylene sulfide); (b) as a function of complexing iodine-doped, face-to-face phthalocyanine polymers with different ion species (Sn, Ge, and Si); and (c) as a function of doping cofacial siloxane phthalocyaninates with high-potential quinones (ClA, TCNQ, and DDQ). ((a) Adopted from Chance et al., Ref. 125; (b) and (c) adopted from Dirk et al., Ref. 127.)

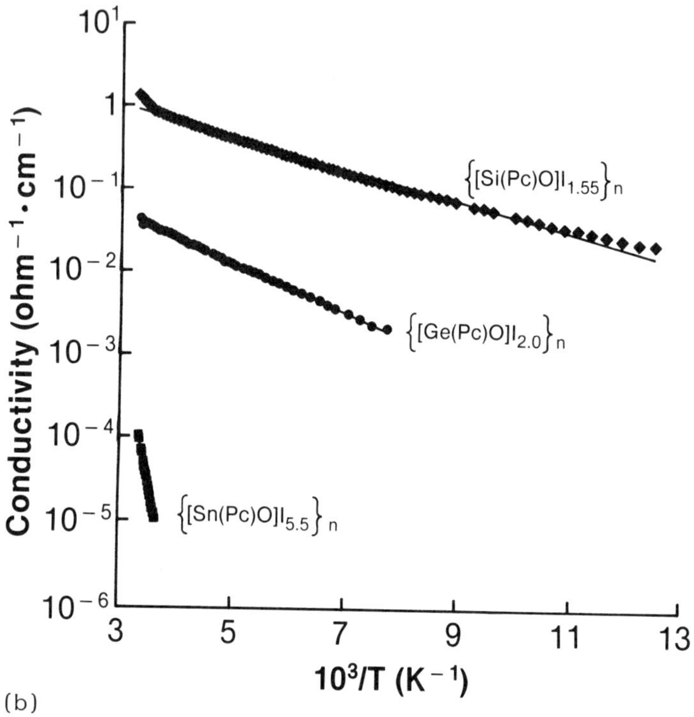

(b)

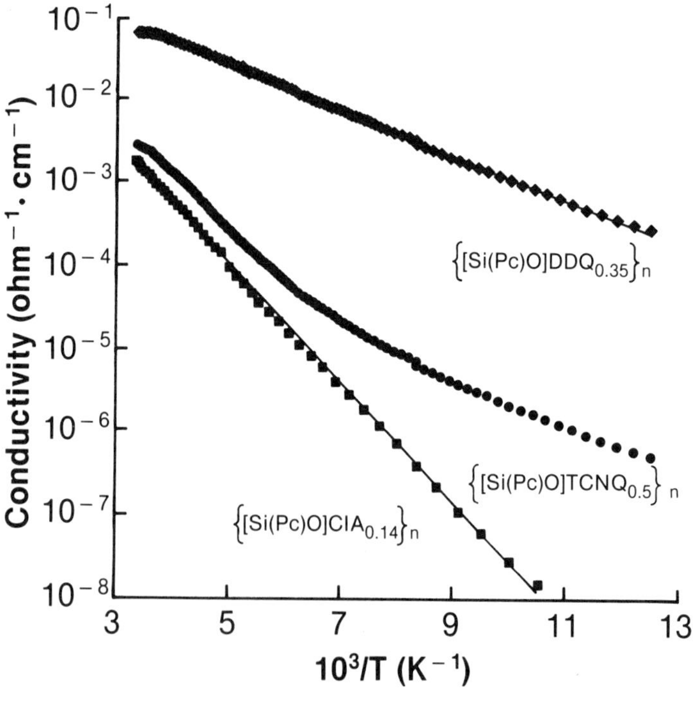

(c)

FIGURE 28 (Continued)

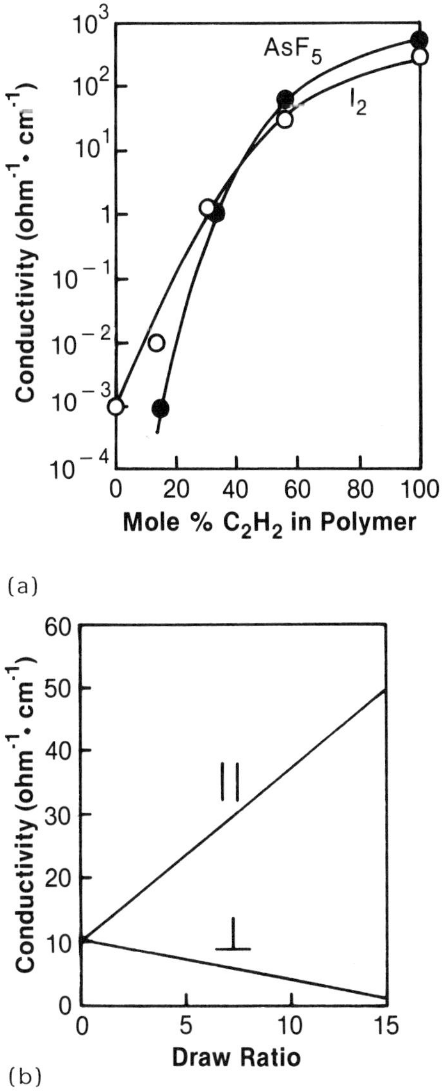

FIGURE 29 Influence of (a) copolymer composition and (b) draw ratio on the electrical conductivity of polyacetylene-methylacetylene and poly(p-phenylene vinylene) films. In the former case the effects of both arsenic pentafluoride and iodine dopants were investigated. In the latter case the effects of anisotropy both in the draw direction (∥) and transverse to it (⊥) were investigated after doping with arsenic pentafluoride. (Adopted from Wnek et al., Ref. 117, and Gagnon et al., Ref. 128.)

TABLE 14 Conductivities of Potassium-Complexed *p*-Phenylenes

Chain length (n)	Conductivity (ohm^{-1} cm^{-1})	Appearance
3	4×10^{-5}	Yellow-green
4	2×10^{-5}	Green-gold
6	5×10^{-1}	Gold
16	7	Gold

Source: Adopted from Shacklette et al. (Ref. 126).

change in conductivity for a poly(vinyl ketone) complex reacted with phosphoryl chloride (Ref. 124) and a poly(*p*-phenylene sulfide) reacted with the Lewis acid, arsenic pentafluoride (AsF5) (Ref. 125). When a single crystal is used, the conductivity may be studied both parallel to and perpendicular to the original c-axis. Such experimental results are illustrated in Figure 27b for a slab of *p*-terphenyl which was doped with AsF_5 at 600 mm (Ref. 126). Clearly, anisotropic behavior is seen which is indicative of different carrier charge mobility in different crystallographic directions. When the dopant concentration is plotted as the independent variable in Figure 27c, a steep conductivity curve is seen for the I_2-doped polyacetylene (Ref. 107), and a sigmoidal curve is seen for the same TTT:TCNQ complexed polycarbonate which was discussed earlier in this section (Ref. 108). In the first case note that the polymer exhibits a transition from a semiconductor to a metal at about 1–3%, depending on the nature of the dopant. Coincident with this transition, the activation energy drops over an order of magnitude. In the polycarbonate complex only 0.5–2.0% by weight of TTT:TCNQ is required, and conductivities as high as 3×10^{-2} ohm^{-1} cm^{-1} can be achieved at room temperature with just 1% by weight loading. If, instead of using chemical means to dope the polymer, ion implantation is used, then the ion flux or fluence can be taken as the independent variable, yielding curves similar to that for the TTT:TCNQ complex. Two curves are shown in Figure 27d (Ref. 83), one for an n-type (PAN) and the other for a p-type (PPO) conductor. Although in the limit the polyacrylonitrile-based material appears to be about 100 times better than the poly(*p*-phenylene oxide)-based material, both improved at least 1 million times after a fluence of 5×10^{15} ions/cm^2 of ^{79}Br. Ion implantation, which may be thought of as a low-temperature pyrolysis process, has been performed using F, Cl, I, and As ions as well, Conductivities decrease by as much as 14 orders of magnitude in air-stable, ion-implanted films.

In addition to investigating the influence of conductivity versus (what is in essence in each case) the dopant level, the influence of the chelated metal ion and the doping material on the conductivity of the polymer network as a function of temperature may be minimized. Besides showing that the conductivity improves with increasing dopant level at a given temperature, the temperature sensitivity decreases with increasing dopant level for poly(*p*-phenylene sulfide) doped with AsF_5 (Figure 28a) (Ref. 125). For low dopant levels this immediately suggests thermal switching applications in which current would be discretely turned on (off) only when the temperature exceeded (dropped below) a specified value. Such an effect may be controlled to some extent in iodine-doped, face-to-face phthalocyanine polymers by varying the metal ion specie from Sn to Ge to Si ((Figure 28b) Ref. 127) or in siloxane phthalocyanate cofacial polymers by varying the doping agent from chloranilic acid (ClA) to tetracyanoquinodimethane (TCNQ) to dichlorodicyanoquinone (DDQ) (Figure 28c) Ref. 127). Note that only for one of the oxiding quinones, TCNQ, was a distinctly nonlinear conductivity observed over a broad temperature range.

Other means of modifying the conductivity may include changing the chain length, the copolymer composition, or the draw ratio. Table 14 (Ref. 126) shows that poly(*p*-phenylene) doped with potassium changes conductivity by over five orders of magnitude with an increase in chain length from 3 to 16. These results clearly contradict the general statement made earlier that chain length has no effect on charge mobility. In Figure 29a (Ref. 117), the conductivity changes which result from copolymerization are shown. Originally this approach was advocated to enhance the processability of what are often highly conductive but equally intractable polymer complexes. But as is seen for I_2- and AsF_5-doped acetylene-methylacetylene copolymer films, the conductivity decreases as the mole percent of C_2H_2 decreases from polyacetylene ($>10^2$ ohm^{-1} cm^{-1}) to poly(methyl acetylene) ($<10^{-3}$ ohm^{-1} cm^{-1}). Without stabilizers these copolymers are not practical as they are prone to oxidation. Relatively smaller but significant changes in conductivity may be observed if an AsF_5-doped poly(*p*-phenylene vinylene) film is deformed to a draw ratio of 15 (Figure 29b) (Ref. 128). The anisotropy which results in the longitudinal versus the transverse direction is greater than 30-fold, changing from 10 ohm^{-1} cm^{-1} for the unoriented film to 50 and 1.5 ohm^{-1} cm^{-1} for the parallel and perpendicular draw directions of the oriented film. Other effects on conductivity have been observed, including a frequency-dopant concentration dependence, a humidity dependence of cationic and anionic microgels, and an effect of microgel coverage (Refs. 129–131).

APPLICATIONS OF METAL- OR METALLOID-FILLED COMPOSITES

Up to this point the distinctions between particulate-filled and reinforcing composites and the physical characteristics and mechanical properties of three types of metal-filled polymers have been presented. With this background the extensive range of applications of commercial metal-filled polymers can now be discussed in three parts: (a) the general properties of, (b) the various forms of, and (c) the specific applications of metal-filled polymers. In the latter case this includes such current uses as EMI/RFI shielding, electronic switching, and solid-state batteries as well as such classical applications as ESD, aesthetics, and radiation shielding. Although neither part of the discussion is exhaustive, this survey is intended to give the reader a perspective on the current level of technology, the diversity of applications, and the bright future which these types of composite materials should have.

General Properties

The addition of certain metal or metalloid atoms to polymers can enhance the mechanical, physical, thermal, electrical, and aesthetic properties of a product. Recognition of the importance of metal- or metalloid-filled polymers has lead to literally hundreds of commercial carbon-filled and metal-filled thermoplastic and thermosetting products. The list of just some of the conductive blends alone is impressive (Tables 15 and 16) (Ref. 109). Here, virtually every type of component material is represented: engineering plastics (nylons and polycarbonates), high-temperature plastics (polyetheretherketones and polyether sulfones), elastomers (silicones and urethanes), highly conductive fillers (gold, silver, and copper), and specialized fillers ("quick quenched" aluminum flakes, silver-plated beads, and nickel-plated graphite fibers). In the future more products will certainly become available as the commercialization of conductive polymers, synmetals, organometallics, low-dimensional materials, and metallopolymers becomes a reality.

The general properties of 10 commercial polymer/metal composites are surveyed in Table 17 (Ref. 132). Several of these hybrid composites (i.e., composites which utilize the advantages of two or more fillers) contribute bulk (A), flame retardancy (B), low friction (E), high temperature (G), or electrical conductivity (H). Mechanical properties of all products are adequate, the highest tensile strength and modulus being recorded for Nylon 6/6 (E) and polyphenylene sulfide (G) reinforced composites, respectively. Compared to the handbook values for each matrix alone, certain physical and thermal properties increased (specific gravity, heat distortion temperature, and thermal conductivity) while others decreased (water absorption,

TABLE 15 Some Commercial Thermoplastic and Thermosetting Carbon-Filled Polymer Products

Matrix/modifier	Examples
	Thermoplastic
ABS/carbon	Fiberfil (Absafil J-1200), RTP (685, 687); Thermofil (Emi-Gard)
FEP/carbon	Fiberfil (Fluorofil J-1400)
Modified PPO/carbon	LNP (Stat-Kon Z)
Nylon 6/carbon	Thermofil (Emi-Gard), Fiberfil (J-1700, J-3-CF)
Nylon 6/12/carbon	Thermofil (N6), LNP (IC1006), Fiberfil (Electrafil J-4)
Nylon 6/6/carbon	Fiberfil (J-1, NY-1EC); LNP (RC 1006, RC 1008, Stat-Kin R-15), RTP (285, 287), Thermofil (N3, N7, N11)
Nylon 6/6/high-impact/carbon	LNP (Stat-Kon V)
Nylon 6/6/Ni-coated carbon	Fiberfil (J-1/CN)
Olefin copolymer/carbon	Asea Kabel, Sweden (ET Semicon), Colonial Rubber (201G, 301G)
PAI/carbon	Amoco (7130)
PBT/carbon	LNP (Thermocomp C), Fiberfil (Tetrafil J-1850, 1105)
PEEK/carbon	LNP (PDX 9738)
PES (polyether sulfone)/carbon	Fiberfil (J-1100), Thermofil (K-30NF)
PET/carbon	Fiberfil (Terafil, J-1800)
Polyacetal/carbon	Fiberfil (Formaldafil); LNP (Stat-Kon K)
Polycarbonate/carbon	Fiberfil (Polycarbafil J-50, G-50); RTP (385, 387) Thermofil (Emi-Gard)
Polycarbonate/carbon film	Schweitzer (Kimflow)
Polyetherimide/carbon	LNP (Thermocomp C)

TABLE 15 (Continued)

Matrix/modifier	Examples
Polyethylene/carbon	UCC (Bakelite HFDA, HFDE), USI (Petrothane YR, XL)
Polypropylene/carbon	Fiberfil (Profil J-60); Ensign-Bickford (SX1C), LNP (Stat-Kon M, ME)
Polystrene/carbon	Fiberfil (Styrafil J-30)
PPS/carbon	Fiberfil (Sulfil J-1300); LNP (OC-1006), RTP (1387 HEC); Thermofil (Emi-Gard), Compounding Technology (SF40CF)
PTFE/carbon	Allied (Halon 4022, 2015), Custom Compounding (CC 123)
PVDF/carbon	Fiberfil (Fluorofil J-1405)
Thermosetting	
Polyester, unsaturated/carbon	Premix (BMC-EMI, TMC-EMI, SMC-EMI), Plastic Systems
Silicone/carbon elastomer	GE, Dow Corning
Urethane/foamed carbon	Toyo Gomu Kagaku (Japan)

Source: Adopted from Reference 109.

TABLE 16 Some Commercial Thermoplastic and Thermosetting Metal-Filled Polymer Products

Matrix/modifier	Examples
Thermoplastic	
ABS/Al flake	Thermofil (Emi-Gard)
ABS/steel	Thermofil (Emi-Gard)
EPDM/metal tape	3M Company
Fluoropolymers	LNP (Stat-Kon)
Nylon 6/steel	Thermofil (Emi-Gard)
Nylon 12/elastomeric polyamide	Daicel (Japan)

TABLE 16 (Continued)

Matrix/modifier	Examples
Nylon 6/6/Al flake	Thermofil (Emi-Gard); LNP (EMI-X320), Plastic Systems (FC 636, 637)
Nylon 6/6/metallized glass	Fiberfil (MG-1/40)
Nylon 6/6/steel	Thermofil (Emi-Gard)
PAN/copper	Bancroft (Thunderon)
PBT/Al flake	Diamond Shamrock; LNP (EMI-X 423)
Polycarbonate/ABS/Al flake	Mobay (Bayblend ME 2540, 6540)
Polycarbonate/Al flake	LNP (EMI-X 520)
Polycarbonate/metallized glass	Fiberfil (FMG-50, MG-50)
Polycarbonate/steel	Thermofil (Emi-Gard), Fiberfil (G-5055, Electrafil 1547)
Polypropylene/glass beads	Thermofil (Emi-Gard), RTP (199)
Polypropylene/mineral	Fiberfil (Profil PP-60); RTP (199)
PTFE/bronze	Custom Compounding (CC 146)
PVC/Al flake	Diamond Shamrock (Magnex), Thermofil (Emi-Gard)
Thermosetting	
Acrylic/nickel	Acme, Electrokinetic Systems
Epoxy/copper	Electrokinetic Systems, Key Polymer (Marpoxy)
Epoxy/gold	Acme (E-Solder 3205)
Epoxy/silver	Ablestik, Acme, Amicon, Chomerics, W.R. Grace, Transene
Phenolic/Al flake	Fiberite (FM 32175)
Polyimide-based	Rogers (Envex), Rhone Poulenc (Kinel 5508)
Silicone/silver lubricant	Cool-AMP (Conducto-Lube)
Silicone/silver paint	Acme (E-Cote)
Urethane/nickel, silver paint	Electrokinetic Systems (X-Coat 225); Chomerics

Source: Adopted from Reference 109.

		(A)	(B)	(C)	(D)	(E)	(F)	(G)	(H)	(I)	(J)
Electrical											
Volume resistivity	ohm-cm	—	10^4	10^0	10^0	10^2	10^3	10^1	10^3	10^2	10^4
Surface resistivity	ohm/sq	10^5	—	—	10^1	10^2	10^3	10^1	10^3	10^2	10^4
Shielding effectiveness[f]	dB	—	20	55	55	—	—	—	—	—	—

[a]Products evaluated:
(A) PP-62/calcium carbonate/20/electrically conductive carbon black
(B) PC foam metallized glass-50/25/flame retardant
(C) PC long fiber-50/stainless steel fiber/10
(D) Nylon 6/6 chopped fiber-1/nickel-coated carbon fiber/15
(E) Nylon 6/6 chopped fiber-1/PAN carbon fiber/30/PTFE fluorocarbon/15
(F) PS chopped fiber-30/PAN carbon fiber/20
(G) Polyphenylene sulfide (PPS) chopped fiber-1300/PAN carbon fiber/50/high modulus
(H) PPS chopped fiber-1300/PAN carbon fiber/20/molybdenum disulfide/10/PTFE fluorocarbon/15
(I) Acetal chopped fiber-80/pitch carbon fiber/30
(J) Polyurethane (PU) chopped fiber-100/pitch carbon fiber/25

[b]Unless otherwise noted, all tests conducted at 23°C using applicable ASTM procedures.

[c]Izod impact test bars 6.35 × 12.7 mm.

[d]Nonbreaking.

[e]Deflection temperature at 1.82 MPa.

[f]Attenuation at 1000 MHz, 3-mm section.

Source: Consolidated from product data of Wilson-Fiberfil International, Ref. 132.

mold shrinkage, and coefficient of expansion). Overall volume resistivities were quite satisfactory for electrostatic discharge (ESD), and the shielding effectiveness (SE) was marginal in one case (B) but excellent in the other two cases (C and D).

A detailed investigation of the influence of loading on commercial polymer/metal composites is shown in Figures 30–32 (Ref. 133). These materials were specifically formulated to provide low volume resistivities and excellent EMI/RFI shielding. In each case the reinforcing fibers are nickel-plated graphite; the matrices are Nylon 6/6, polycarbonate, acrylonitrile-butadiene-styrene terpolymer (ABS), or polyphenylene sulfide.

In Figure 30 three mechanical properties are reported: the tensile modulus, the tensile strength, and the tensile elongation. For

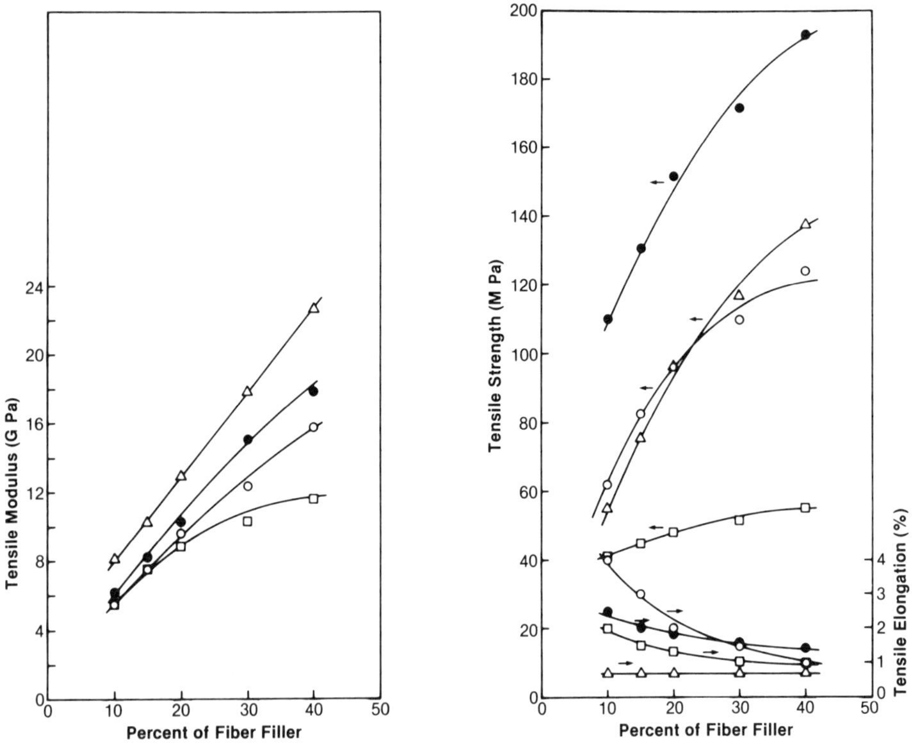

(a) (b)

FIGURE 30 Tensile properties of four commercial products: nickel-plated graphite fibers in Nylon 6/6 (●), polycarbonate (o), acrylonitrile-butadiene-styrene (□), and polyphenylene sulfide (△): (a) tensile modulus; (b) tensile strength and elongation. (Data of RTP Company's 200-, 300-, 600-FR-, and 1300-HEC series, respectively. Ref. 133.)

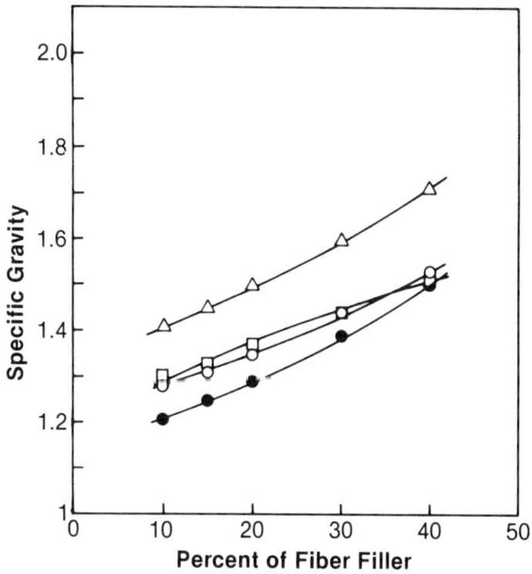

(a)

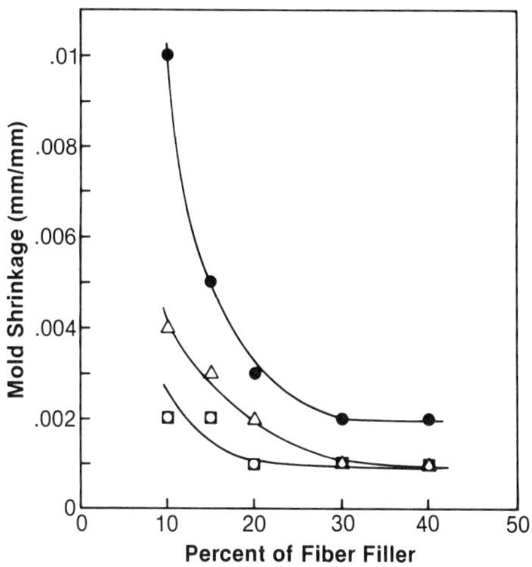

(b)

FIGURE 31 Physical properties of four commercial products: nickel-plated graphite fibers in Nylon 6/6 (●), polycarbonate (o), acrylonitrile-butadiene-styrene (□), and polyphenylene sulfide (△): (a) specific gravity; (b) mold shrinkage. (Data of RTP Company's 200-, 300-, 600-FR-, and 1300-HEC series, respectively. Ref. 133.)

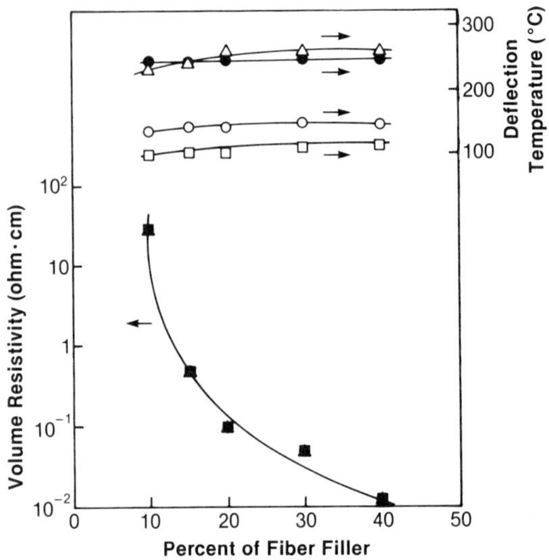

(a)

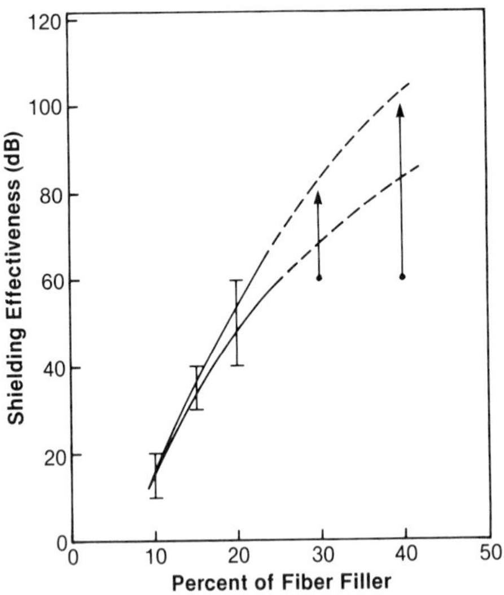

(b)

FIGURE 32 Thermal and electrical properties of four commercial products: nickel-plated graphite fibers in Nylon 6/6 (•), polycarbonate (o), acrylonitrile-butadiene-styrene (□), and polyphenylene sulfide (△): (a) volume resistivity and deflection temperature; (b) shielding effectiveness. (Data of RTP Company's 200-, 300-, 600-FR-, and 1300-HEC series, respectively. Ref. 133.)

loadings of 10, 15, 20, 30, and 40%, no discontinuity is seen for any of these commercial product curves (cf. Figures 19–21 for laboratory materials, however), although linear behavior is the exception rather than the rule. Only the polyphenylene composites appear to follow a linear relationship, ranging in stiffness from a low of 8 GPa at 10% loading to a high of 23 GPa at 40% loading. Clearly, further filler additions to the ABS matrix do little to stiffen or strengthen that composite. By comparison, the Nylon 6/6 composites are about three times stronger and 25–50% more ductile. Overall, these results are consistent with the general effects of fiber reinforcement in conventional composites (cf. Figure 2)—an attractive feature which many of the early conductive composites failed to achieve.

Two physical properties are investigated in Figure 31: the specific gravity and the mold shrinkage. Since almost all metals are denser than polymers, the specific gravity logically increases with loading content. However, because nickel is deposited onto the surface of carbon fibers, the density is lower than it would otherwise be—without any loss in electrical conductivity. Even when loaded with 40% nickel-coated carbon fibers, the specific gravity is, at its worst, only two-thirds that of aluminum. At the same time mold shrinkage decreases, initially at a rate greater than that of fiber addition. One possible explanation of this phenomenon is that the metal restricts shrinkage by forming a "bird's nest" structure. The resulting entangled fiber network limits thermally induced changes by increasing the viscosity and reducing the flow. This may involve the cold welding of fibers to various degrees.

One thermal property and two electrical properties are illustrated in Figure 32: the deflection temperature at 1.82 MPa, the volume resistivity, and the shielding effectiveness at 1000 MHz. As expected, the deflection temperature is greatest for polyphenylene sulfide (260°C) and Nylon 6/6 (250°C) and least for ABS (115°C). In Figure 32, the volume resistivity is shown to decrease four orders of magnitude from 10 to 40% loading after having precipitously dropped some 12 to 14 orders of magnitude in the first 10% of loading. That the electrical resistivity of all four polymer/metal composites superpose is not surprising, since the amount and type of filler control the electrical properties, and they were the same. Returning for a moment to Figure 30 and the observation that no discontinuity in either the tensile modulus or tensile strength was seen, that observation is no longer decisive since the critical loading occurred at $<10\%$. Figure 32 also shows that the shielding effectiveness is adequate for 50% of the commercial applications after about 14% loading and for 95% of the applications after about 17% loading (Refs. 134, 135). For this reason the exact value for the shielding of the 30 and 40% loaded compositions was not measured.

Properties of aluminum- or iron-filled epoxies (Ref. 79) and carbon black- or carbon fiber-filled hybrid polypropylenes (Ref. 133) have been reported in the product literature as well.

Various Forms

The applications of metal-filled polymers are so diversified that the end product may be found in all forms including bulk materials, laminations, foams, caulking compounds, sealants, adhesives, solders, tapes, paints, coatings, lacquers, greases, and powders.

Bulk Materials

Bulk materials represent the largest portion of sales, being compression-, injection-, or extrusion-molded, cured from monomer, cast from the melt, or cast from solution (Refs. 136–138). Resultant products may be standard sheet, rod, or bar form, which are later die-cut or machined (metal-filled polymers tend to be free-machining, Ref. 74). They may be complex die parts which require no further preparation (e.g., drilling or painting) or simple potting materials which dissipate electrical and/or thermal energy. In the last regard, specific materials are produced today to fulfill not only charge and heat dissipation requirements (cf. Figure 18b) but also heat dissipation without charge transfer (so-called semi-insulators) (Refs. 139, 140). In fact, some materials have been produced which will "switch" from an electrical insulator to an electrical conductor while dissipating heat (Refs. 141, 142). Several applications of bulk materials are listed in Table 18.

Laminations

Laminations are a special type of bulk material which give the final product additional characteristics. The plywood and fiberglass industries have capitalized on anisotropic properties. Whereas these industries typically ply several layers of the same composite at various orientations to one another (e.g., see Table 6), the layers of a metal-filled composite are often quite dissimilar. They may be filled with different metals to give one surface of the product metallic appearance without losing function, or they may be filled with some other pigment or dye to mask the functional nature of the bulk material. One manufacturer has used a coinjection molding process to produce a teleprinter housing in which both the inner and outer surfaces are aesthetic, fine-textured, light-stable, colored layers and the core is a fiber-reinforced, shielding layer (Ref. 143). Such laminations reduce the number of manufacturing steps and thereby cut costs while providing various types of radiation shielding protection.

Foams

Metal-filled foams may be rigid or flexible and have open- or closed-cell pores. In addition to conductive carbon/vinyl foams, silver-filled silicone, polyurethane, and plasticized vinyl foams are

TABLE 18 Some Uses of Metal-Filled Polymers in Bulk Form

Bearings
Braking devices and clutches
Cargo pallets
Decor (utensils, table tops)
Density controls
Drop hammers and dies
Electrical components (resistors, thermistors)
Electrodes
Engine mufflers
Ground apparatus (electronics, conveyors)
Heating elements (pipes, flooring)
Housings for EMI/RFI shielding
Magnetic tapes, disks, and cards
Magnets (motors and appliances)
Maintenance and repair products
Potting and encapsulation
Radiation shielding (γ, thermal n)
Sheathing
Small connectors and fasteners
Structural components (furniture, door knobs)
Switching power supplies and switch contacts
Tooling (molds, holding fixtures)
Tote boxes for sensitive components

available. Typical properties of a conformable, conductive, open-cell (8 pores/cm) foam are shown in Table 19 (Ref. 144). When foams are used as a gasket material, both the surface and volume resistivity decrease in proportion to the compression ratio. Other advantages of foams are that they come in large sizes (40.6 × 40.6 × 0.32 cm), can be readily cut to size, fit into tight locations, and can be reused. Most applications capitalize on some combination of the inherent conductivity, the lossiness, and the transmission of heat and/or air. Some applications are summarized in Table 20.

Caulking Compounds

Caulking compounds are used to fill large gaps where potential differences must be minimized, thermal continuity must be maintained, and/or some type of shielding is important. In the last case caulking can close potential air gaps, reduce the transmission of radioactive nucleides, or attenuate electromagnetic radiation. As Figure 33 shows, caulking may be in the audible region, the microwave frequencies, the radio-frequency (RF) wave bands, or the high-energy x- or γ-ray wavelengths. Caulking compounds are relatively inexpensive ($0.15 to $0.45 per lineal foot per 0.32 cm bead diameter) and have low volume resistivities ($\approx 10^{-3}$ ohm-cm) and high attenuation capabilities (50–100 dB). Densities are moderate (1.8–4.9 g/cm^3), and the useful temperature ranges are outstanding (-55 to +205°C) (Ref. 145). Caulking compounds may be applied by using conventional equipment—either a spatula, putty knife, plastic syringe, or caulking gun. Since they remain pliable, caulking compounds may be used again.

TABLE 19 Typical Properties of Foams[a]

Property	Value
Elongation (%)	>100
Compression set	Very low
Durometer	Very low
Specific gravity	0.1
Service temperature (°C)	−65 to +125
Volume resistivity (ohm-cm)	<0.15 uncompressed
Surface resistivity (ohm/sq)	<0.5

[a]Applicable ASTM procedures were used.
Source: Adopted from Reference 144.

TABLE 20 Some Uses of Metal-Filled Polymers as Foams

Acoustical noise suppressor
Buoyant materials for hydrospace (oil well riser modules, deep submergence vehicles)
Corrosion-resistant/high-temperature gaskets
Diode washers
Grounding ("living hinges")
Heat escape mechanism
Low closing force shielding gasket
Microwave absorbers
O-rings
Radiation shielding pads
RF filter for air vents
Transistor cases
Vibration insulator and shock resistance pads
"Wraparound" materials (to control emission in use or to protect components in storage)

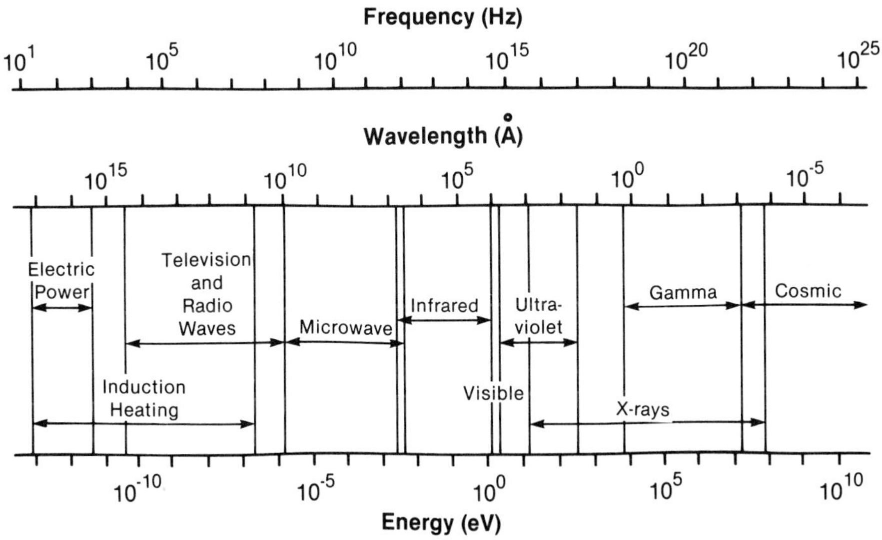

FIGURE 33 Classifications and characteristics of electromagnetic radiations.

Sealants

When smaller gaps must be filled and some bonding is required, sealants may be used as an alternative. Like conventional silicone sealants, these materials cure on exposure to moist air. Metal-filled sealants can form an impervious, corrosion-resistant, conductive bond.

Adhesives and Solders

For closing still smaller gaps and creating a conductive structural bond a metal-filled adhesive may be employed. Both one- and two-part high-performance composites exist, some silicone and epoxy types of which set at room temperature (sometimes denoted as *cold solders*), while other epoxies must be heat-cured. Silver, gold, palladium, copper, nickel, and carbon fillers have produced adhesives with typical properties as shown in Table 21. Moreover, Delmonte has reported (see Table 11-3 in Ref. 79) the change in volume resistivity and lap shear strength as a function of weight percent of silver loading. Results show that the volume resistivity and lap shear strength decrease from about 10^{14} ohm-cm and 20.0 MPa at 0% loading down to 640–0.023 ohm-cm and 11.7–10.2 MPa at 70–75% loading, respectively. Because these materials are heavily loaded, their costs are high, the silver type exceeding $175/kg. Metal-filled adhesives have been used in electronics component and device applications where the heat of brazing or silver soldering would be prohibitive; as substitutes for spot-welding, rivets, or other mechanical fasteners; to cement aircraft and aerospace components; and to bond propane torches (as was done for the torches used in the XXIII Olympiad, Ref. 146). Metal-filled adhesives may be applied to bond almost any material, including nichrome wire, conductive gaskets to metal flanges during assembly, or electrical components

TABLE 21 Typical Properties of Adhesives[a]

Property	Value
Lap shear strength (MPa)	5.49 – 14.5
Specific gravity	2.9 – 3.4
Thermal conductivity (W/m-K)	1.44 – 7.21
Thermal expansion (m/m-°C)	45×10^{-6}
Volume resistivity (ohm-cm)	$4 \times 10^{-4} - 5 \times 10^{-2}$

[a]Applicable ASTM procedures were used.

TABLE 22 Typical Properties of Solders[a]

Property	Value
Bond strength (MPa)	2.06
Flexural strength (MPa)	66.6 – 84.0
Lap shear strength (MPa)	4.81 – 5.49
Thermal conductivity (W/(m-K))	2.16 – 11.5
Thermal expansion (m/m/°C)	$36 \times 10^{-6} - 63 \times 10^{-6}$
Volume resistivity (ohm-cm)	$2 \times 10^{-4} - 6 \times 10^{-3}$

[a]Applicable ASTM procedures were used.

to heat-dissipating surfaces. Entire conductive parts may be assembled from components, chips may be bonded to boards, circuit integrity may be restored to damaged equipment, or threaded fasteners may be rendered RF-tight. In the latter case the outstanding shielding effectiveness which is attainable (>95 dB from 1 to 10^3 MHz) has been detailed in the product literature (Ref. 147). A variety of adhesives are available, ranging from creamy thixotropic pastes to hot-melt types.

Solders may be metal-filled epoxy or polyester composites. They are selected for the assembly of exterior shells and body components (Ref. 79). Ideally, they should be one-component pastes capable of curing under a variety of manufacturing circumstances in 1–2 minutes at 150°C. Good sandability, grindability, and compatibility are essential for overcoats without seams, and gap-filling capabilities of tolerances are required. Some solders have been manufactured which may be applied by rolling, brushing, or even spraying. Representative properties of solders are shown in Table 22.

Tapes

Conductive pressure-sensitive tapes are used for four principal purposes. As a production aid tapes provide mechanical, thermal, and/or electrical continuity between working parts. Taping a wire to a part that is to be plated, providing a thermal dissipation path during the welding of components, and fabricating electroformed joints by electroplating a thick layer of metal on top of a layer of conductive tape are just a few examples. Interconnects for cables or as grounds provide temporary and reusable conductive paths.

Self-heating tapes consisting of carbon black-filled, irradiation-cross-linked polyethylene are marketed (Ref. 148). By far the largest market on a volume basis is that for tapes used in electrical transmission lines. As Figure 34 shows, two layers of thin metal tapes are generally incorporated for service lines above 5000 volts in order to equalize the electric field and prevent corona damage (Ref. 149). Metal-filled tapes could function in the same manner.

Paints, Coatings, and Lacquers

Metal-filled paints, coatings, and lacquers have many applications, primarily when electrical conductivity, corrosion resistance, and/or shielding are important (Table 23). Specific applications have included corrosion-resistant antifouling paints (CuO, Cu, or Sn loaded) for ship hulls, nickel-filled coatings for frame shielding, and graphite- or silver-filled lacquers to provide scanning electron microscopy (SEM) specimens with satisfactory electron bleed-off. Several types of materials are marketed today. Some are one-part, air-dryable, solvent-based, graphite, silver, nickel, copper, or copper-graphite materials. Others are abrasion-resistant, one-component, nickel-filled, water-based, solventless, acrylic materials. The resulting flexible or rigid products may be applied by dipping, spraying, silk-screening, roll-coating, or brushing onto such metals as steel and aluminum and such bulk polymers as ABS, nylon, polycarbonate, and polystyrene as well as structural foams. Typical surface resistivities and service temperatures of conductive acrylic paints, elastomeric conductive coatings, and a silver lacquer are shown in Table 24. For the case of nickel-filled acrylic paints, the shielding effectiveness can be quite good (32–80 dB for 30–1000 MHz) because the solids content is very high (56–67%) (Refs. 145, 150).

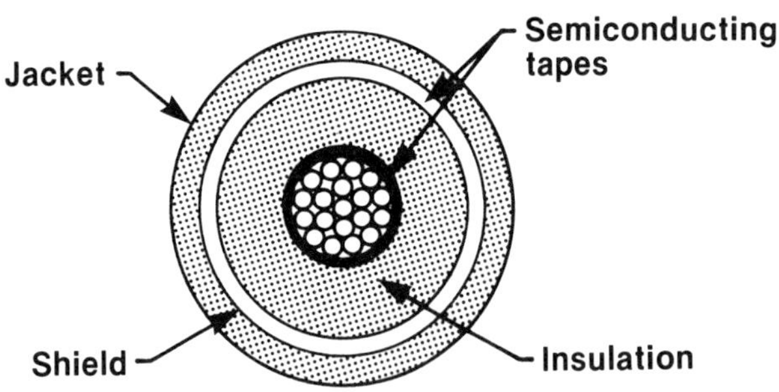

FIGURE 34 Schematic illustration of a medium-voltage (10^3 – 10^5 volts) power line, showing its layered structure. These cables are used in industrial plants and large buildings to power 4,160- and 13,800-volt motors.

TABLE 23 Some Uses of Metal-Filled Paints, Coatings, and Lacquers

Aesthetic coatings for wood, plastic, ceramics, metals, and concrete
Circuit repair
Conductive patterns on dielectric surfaces
Corrosion-resistant marine coatings
EMI/RFI shielding
Fill minute voids in porous castings
Formation of capacitor plates
Formation of ground planes
Heat sinks during soldering operation
Reduce contact resistance of metal enclosure flanges
Transducers
Waveguides, RF reflectors

Greases

Conductive greases are generally carbon- or silver-filled materials. Using brightfield, reflected light microscopy the centers of the coarse and fine particles may be highlighted—the larger diameter particles displaying a greater bright spot than the smaller diameter particles (Figure 35a) (Ref. 53). Evidence of the actual dimensions

TABLE 24 Typical Conductive Paint, Coating, and Lacquer Service Temperatures and Surface Resistivities

Conductive material	Service temperature (°C)	Surface resistivity (ohm/cm^2)
Acrylic paint	-54 to +120	0.05 – 2.0[a]
Elastomeric coating	-54 to +120	0.001 – 0.05[b]
Silver lacquer	-54 to +177	0.16[b]

[a]Thickness = 5.0×10^{-3} cm.

[b]Thickness = 2.5×10^{-3} cm.

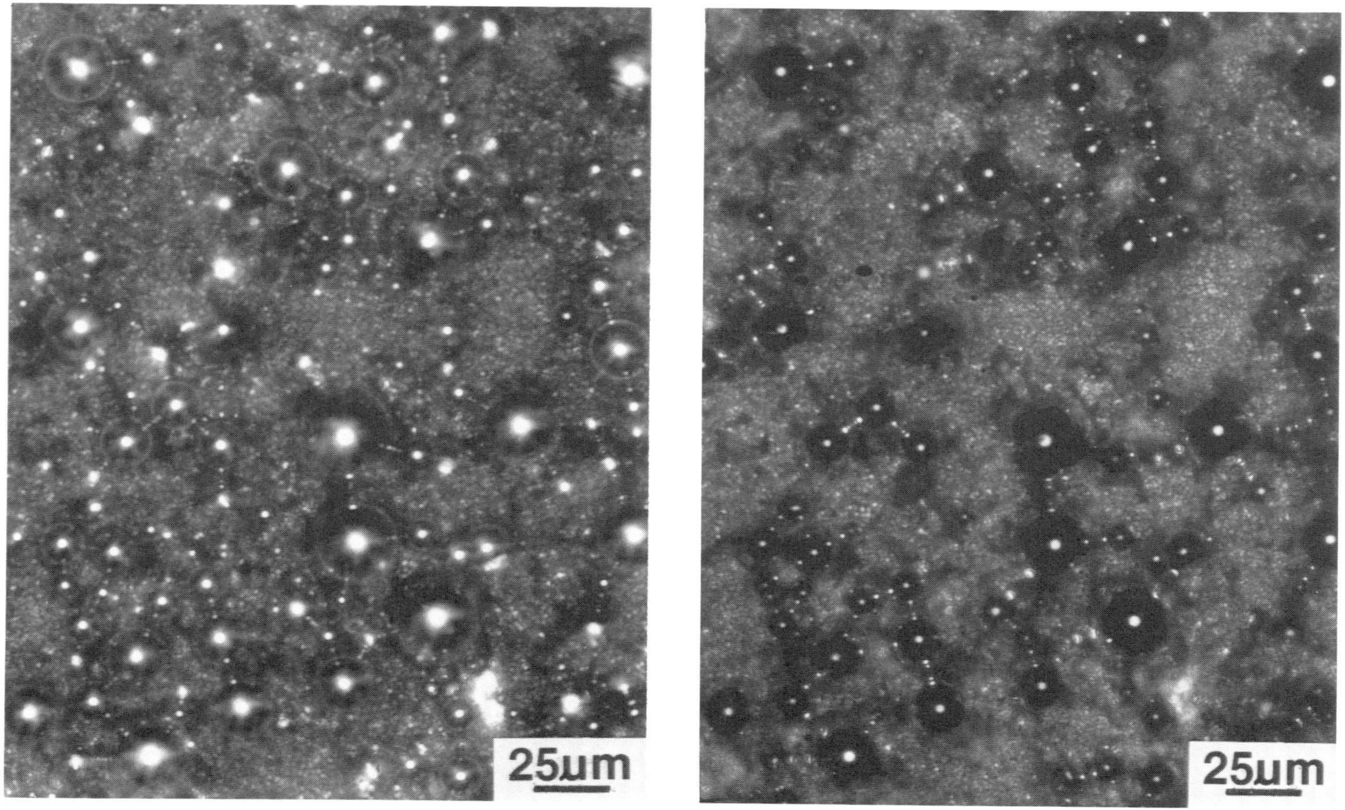

(a) (b)

FIGURE 35 Photomicrographs of a commercial conductive grease. Using (a) brightfield and (b) sensitive tint illumination, large silver particles may be seen dispersed within a fine flour of the same.

of the particles may be deduced by using sensitive tint, reflected light microscopy (Figure 35b). The latter photomicrograph also suggests that short-range runs exist within the composite which could transfer potential charge differences if continuity were achieved in three dimensions or if some form of mechanical agitation were applied to the lubricant. Gross particle migration and charge transfer naturally occur as the lubricant is subjected to shear forces between opposing moving parts. Mating surfaces are thereby lubricated, and potential differences are minimized as short-range runs are continually created and destroyed. Other applications include prevention of arcing and switch freezing, inhibition of corrosion in hostile triboelectric environments, and implementation of a luting agent for RF noise reduction along seams of opposing metal components.

Powders

As an end product, polymer-metalloid powders have been utilized for one major purpose, as dry inks in the copier industry (Ref. 151). These *toners* are fabricated by melt-mixing 10% carbon black into a polymer matrix followed by micronization to about 10 μm diameter. Using the principles of contact- or triboelectrification, a photoconductive surface is charged by corona discharge. When the image of the original document is focused onto this charged drum or belt, charge transfer occurs: black areas remain charged while white areas are discharged. To develop the latent image, the polymer-metalloid powder is charged oppositely from the electrostatic image. When presented to this image, the dry ink adheres only to the black areas. The resulting black-and-white image of the original document which is on the photoconductive material is captured by electrostatically transferring the dry ink to a sheet of paper. The polymer-carbon particles are fused onto the page and the copy of the original document is complete.

Specific Applications

In this last major section 10 specific applications of metal-filled polymers are considered:

- Heating elements
- Radiation shielding
- Aesthetics
- Magnetic materials
- Electrostatic discharge (ESD)
- Eleectromagnetic/radio-frequency interference (EMI/RFI)
- Switches
- Transducers
- Transparent composites
- Batteries

The first five applications listed represent the more classical uses of this class of materials, while the last five are either more recent or other potential areas of development. Undoubtedly this list could have been made longer, but these applications will give the reader an appreciation of this class of materials and their vast potential.

Heating Elements

Conductive piping has been an attractive use of metal- or metalloid-filled polymers for some time. These self-heating pipes are particularly important in the petrochemical and mining industries where heated hoses, tubes, and ducts can facilitate the transport of viscous products or prevent the freezing of liquid feedstocks, water,

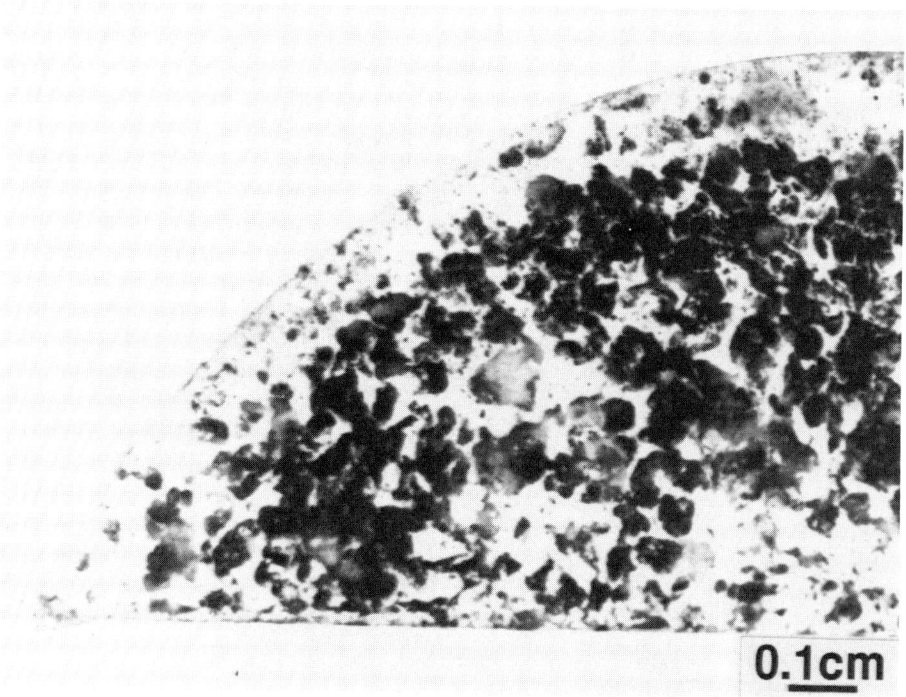

(a)

FIGURE 36 Corrosion resistance of polyethylene/10% nickel composites: (a) macroscopic appearance of the surface and (b) its interior after long-term exposure to acid; (c) microscopic appearance of the interior showing the retention of a conductive metallic network (cf Figure 12 and text). (Reproduced from Kusy, Ref. 65.)

or wastes. Immediate advantages of such composites over alternative approaches are low current densities, absence of a high-temperature source, reduction of thermal hot-spots, inherent corrosion resistance, and long term thermal stability.

As in conductive flooring or walkways and portable heating panels such as those used at oil rig drilling sites and at open-air installations of chemical plants, low current densities exist in metal- and metalloid-filled polymer heating elements. The potential danger of electrical shock or electrocution is thereby reduced. The ever-present risk of a fire is diminished, too, since there are no hot-spots generated by heating elements radiating at temperatures of thousands of degrees, nor are any potential charges built up between the fluid within the pipe and its external environment.

Laboratory work on polyethylene/nickel composites 15 years ago illustrated the potential corrosion resistance of such materials (Figure 36) (Ref. 65). After exposure to 50% nitric or 100% sulfuric

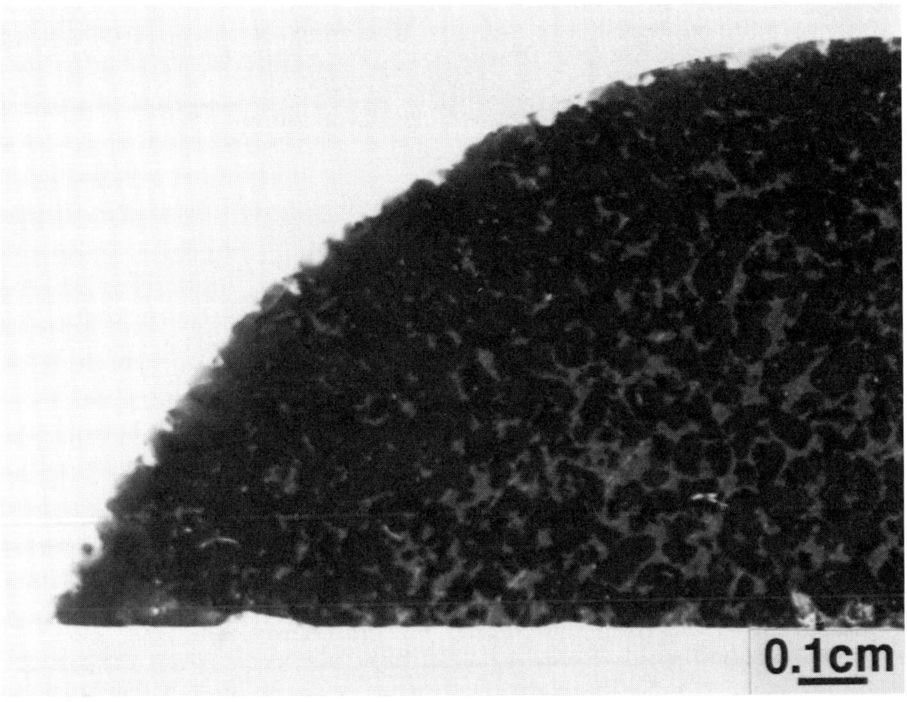

(b)

FIGURE 36 (Continued)

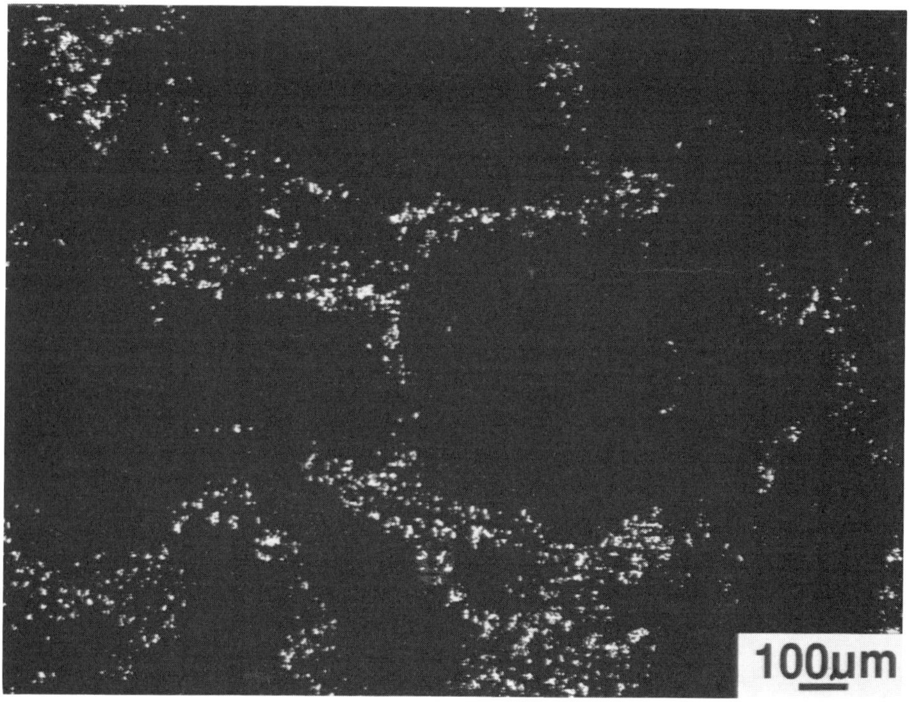

(c)

FIGURE 36 (Continued)

acid at, in the first case, both ambient room temperature and the boiling point, only about 5.5% by weight of the composite (that is, 11.1% by weight of the metal) was lost. An examination of the external (Figure 36a) and internal (Figure 36b) macrostructure revealed that, while the surface layer of metal had been dissolved, the internal microstructure remained intact (Figure 36c). Indeed when the metal-depleted surface of a test specimen was ground off and the specimen repolished, the original electrical resistivity was restored (that is, about 10^2 versus 10^{11} ohm-cm). Insofar as long-term stability of metal-filled polyethylene is concerned, Turner et al. (Ref. 152) demonstrated that, in principle, the resistivity could be stabilized by exposing the polymer to high-energy radiation. After a dose of 16.5 Mrad (^{60}Co, $\sim$3 Mrad/hr), the cross-linked composites could be heated (150°C) to temperatures in excess of the thermodynamic melting point (110–140°C) without appreciable loss of resistivity (Figure 37). Later, Narkis (Ref. 153) reported that, after

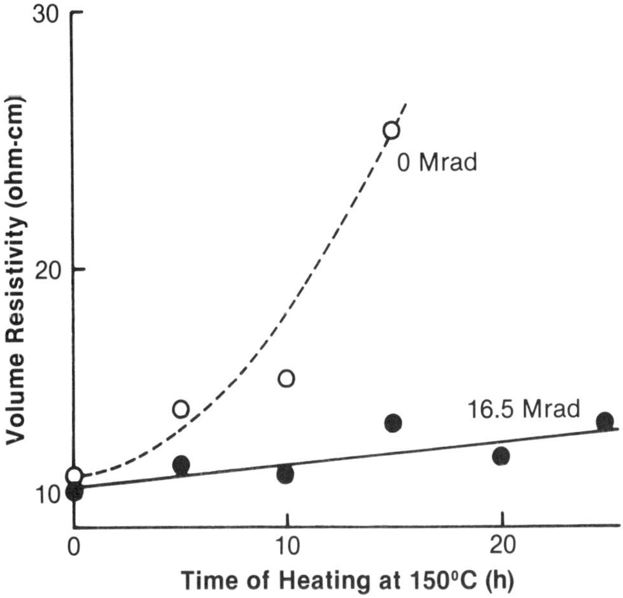

FIGURE 37 Influence of heating period on electrical resistivity of as-received and radiation-stabilized polyethylene/10% nickel composites. (Adopted from Turner et al., Ref. 152.)

an initial voltage drop in the first 200 hr, a carbon black-filled polyethylene is stable for 4000+ hr in air at 115°C.

Other properties of conductive carbon black-filled polyethylene should be mentioned (Ref. 153). First, when the voltage-current characteristics are measured, the pipe exhibits distinctly ohmic behavior which is independent of the distance between electrodes. As a function of surface temperature the resistance does vary, but not until the temperature rises to more than 110°C. Figure 38 shows these similar trends for two different pipe materials, independent of the voltage gradient. As Figure 39 illustrates, however, the voltage gradients do influence the self-heating behavior of conductive pipes in air. Both the rate of temperature rise and the ultimate surface temperature increase as the voltage gradient is changed from 2.7 to 5.2 volt/cm. For a 10 ohm-cm conductive pipe, typical power ratings of 30, 60, and 120 watt/m are reported for voltage gradients of 150, 220, and 330 volt/m. Corresponding equilibrium surface temperatures on a pipe of 20 mm diameter and 2 mm wall thickness are 65, 90, and 130°C. Over the temperature range 23–100°C tensile properties

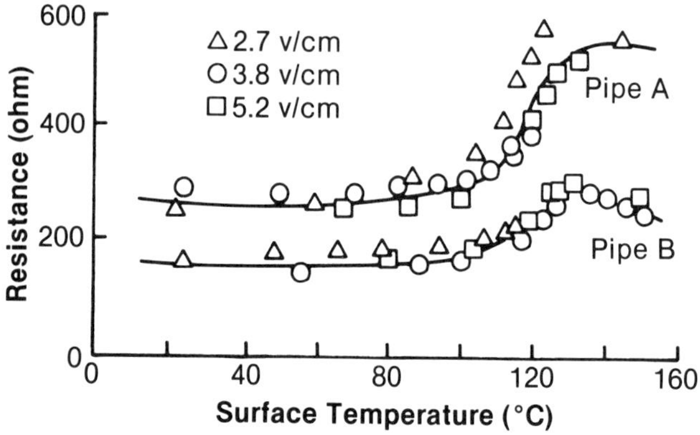

FIGURE 38 Resistance of two different carbon black-filled polyethylene pipes as a function of surface temperature and voltage gradients. (Adopted from Narkis, Ref. 153.)

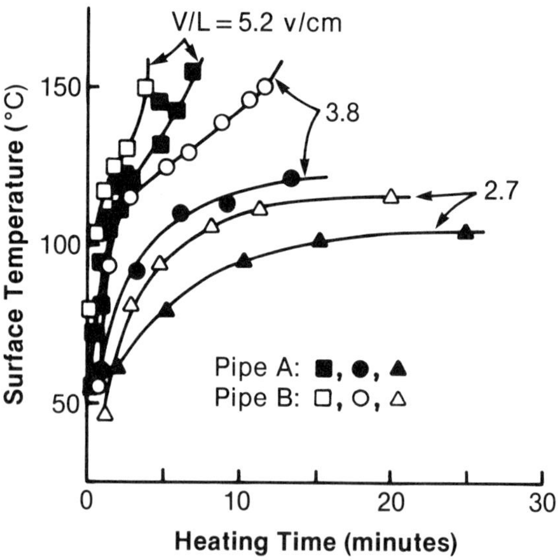

FIGURE 39 Influence of various voltage gradients on the self-heating behavior (surface temperature) of two different carbon black-filled polyethylene pipes. (Adapted from Narkis, Ref. 153.)

decrease from 20.6 to 12.3 MPa, while the elongation at break increases from 125 to 267%. Varying the inherent resistivity of the pipe from 10 to 110 ohm-cm has little influence on the ultimate tensile strength but improves the elongation at break because the filler content (and the number of stress raisers) decreases.

In addition to direct resistance heating of pipe, flooring, or wall panels, RF heating and passive transmission of energy should be considered. Figure 40 (Ref. 72) demonstrates the influence of a small amount of aluminum filler (3.8% by volume of fibers, L/D = 35) compared to the RF heating rate of other polyethylene materials. Clearly, neither the unmodified polyethylene nor the glass fiber-loaded composite (L/D = 250) had the same effect on the heating rate: a slab temperature of 65°C could be attained in 10 versus 40 min. Collection of solar energy, transmission of that energy into storage beds, and later dissipation of it into a home are three applications in which passive, corrosion-resistant heat exchanges are useful. In the transportation and desalination of seawater, metal-filled polymers have been considered (Ref. 154) as an alternative to corrosion-resistant titanium alloys (Ref. 155).

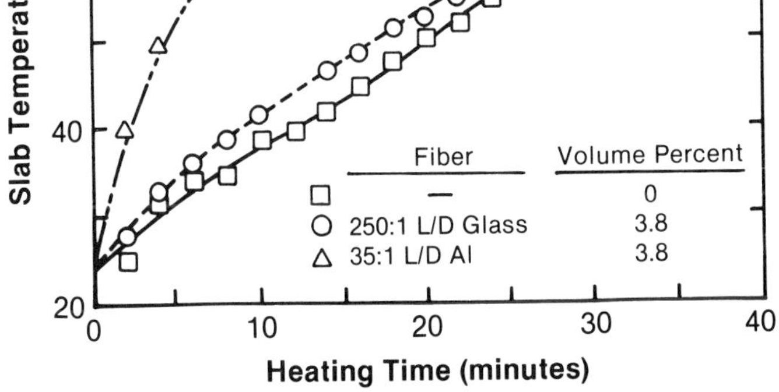

FIGURE 40 Influence of fillers on the RF heating rate of polyethylene composites. (Adopted from Bigg and Bradbury, Ref. 72.)

Radiation Shielding

When the energy levels of electromagnetic radiation become great enough, the structure of materials may be radically altered. Ultimately the properties are affected to the extent that function is impaired or curtailed. Figure 41 (Ref. 156) shows that those electronic components which rely on doping are the most susceptible to damage. In contrast, metals are the most radio-resistant group, while most organic polymers are intermediate. The two photons which the radiation safety officer is most concerned with are x- and γ-rays (Figure

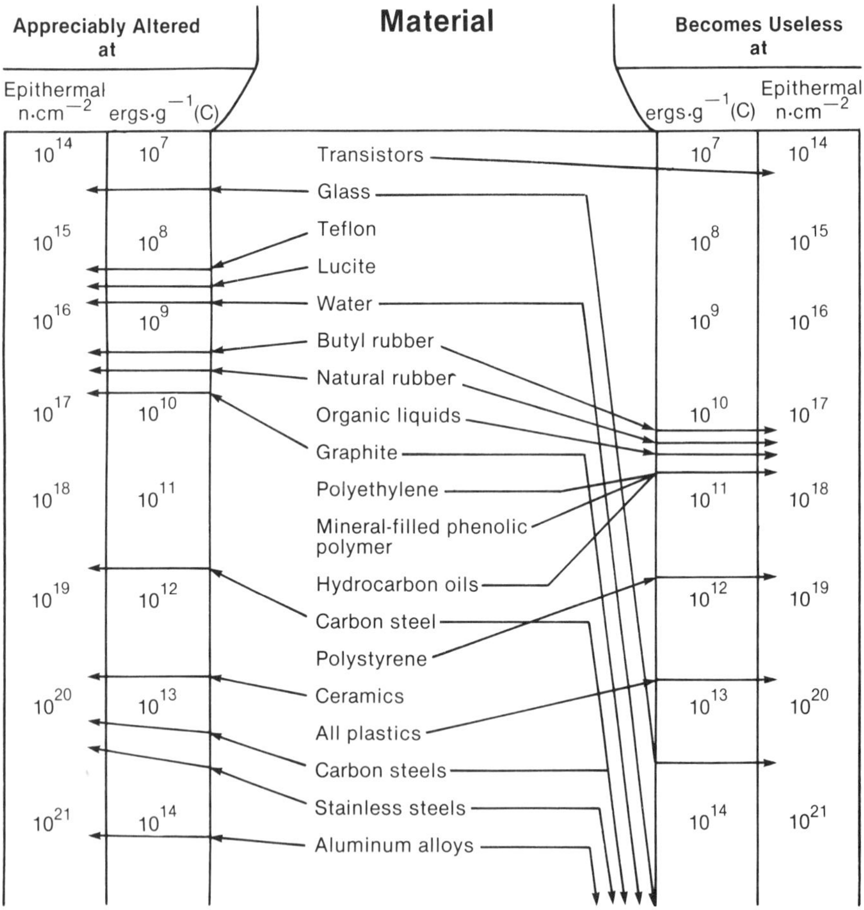

FIGURE 41 Susceptibility of various materials to radiation damage. (Adopted from Hennig, Ref. 156.)

33). Because electromagnetic radiation follows an exponential law in which the logarithm of the ratio of the residual to the original intensity (I_x/I_0) equals the product of the absorption coefficient ($-\mu$) times the thickness (x) (Ref. 157), these photons do not have a maximum range and cannot be completely stopped by any shielding material. Consequently, half-life thicknesses are used to ascertain the thickness required to reduce exposure rate by one-half (Ref. 158). When relative comparisons are made as a function of either photon energy or material (Ref. 159), concrete and aluminum behave similarly at all energy levels (Table 25). At the energy extremes, however, both lead and iron are superior to either aluminum or concrete because their higher atomic numbers (Z) result in greater photoelectric and pair production absorption. Except when poly-(methyl methacrylate)-packaged mercury is used as the innermost shield of powerful γ-ray units, most x- and γ-ray shielding is constructed of some type of lead, concrete, or iron-based material.

Metal-filled polymers play a role in primary and secondary protection. When constructing a partition to reduce the emissions from an unshielded source, lead bricks or concrete blocks are often used. To protect workers in the same room from leakage along ceiling lines, plumbing, and electrical fixtures, lead-filled epoxy blocks may be custom-poured and shaped. Similarly, when a flexible shield is required to protect personnel against the radiation scattered by a material which may be in the way of the photon source, a protective

TABLE 25 Comparison of Various Materials for Radiation Shielding at Different Energies

Photon energy (MeV)	Lead	Concrete	Iron	Aluminum
	Relative absorption with change in energy			
0.1	100	100	100	100
1.0	1.29	37.6	17.3	38.1
10.0	0.93	13.5	8.5	14.2
	Relative absorption with different materials			
0.1	3120	100	203	95.0
1.0	108	100	93.8	96.8
10.0	213	100	128	100

Source: Adopted from Blatz, Ref. 159.

vest or apron may be made from lead- or lead salt-filled rubber or plastic sheeting. Such shields have been used by physicians in fluoroscopy and by dentists when taking x-radiographs. In both cases the effectiveness of such products is directly proportional to the quantity of the lead present.

Unlike electromagnetic radiation, the radiation for particulate sources (such as alpha- and beta-emitting radioactive nucleides), accelerators, neutron sources, and reactors may be attenuated. Among the common atomic particles, beta particles and neutrons are of particular interest.

Beta particles are electrons traveling at relativistic speeds. Thus, although their penetration is a function of their energy, they follow a relativistic energy-mass relationship rather than a classical kinetic energy relationship (Ref. 159). Because β-particles cover a broad spectrum of energy levels, their maximum range must be determined. Without knowing any further information, Figure 42 (Ref. 160) would suggest that lead, iron, and copper make the best shields. Unfortunately, placing a beta-emitting source in a high-Z material can result in bremsstrahlung, or "breaking radiation." This radiation occurs as the rapidly decelerating electron transforms into an high-energy photon, an x-ray. To avoid creating this additional hazard, materials from the second group [for example, poly(methyl methacrylate) or glass] are used to attenuate the β-particles over substantially longer distances. Having done this, the beta absorber is then placed in a second, high-Z material to attenuate the photon emission.

One way to make effective beta shields from polymer/metal composites is to add aluminum flake to thermosetting epoxies. Alternatively, a coinjection process may be used in which a high-impact polymer on the inside of a containment can is simultaneously injected with a mixture of lead, copper, or iron with the same or a compatible polymer on the outside. In either case an opaque radiation containment can results which also serves as a tough structural component.

Neutron shielding presents special problems, too, since these uncharged atomic particles, with a mass similar to that of a proton, pass readily through electron-dense materials such as lead. It is critical to quickly moderate fast neutrons to thermal neutrons so that they can be captured in materials with high absorption cross sections (Ref. 157). Following the principles of momentum transfer, neutrons are essentially unaffected by collisions with electrons and elastically recoil when they collide with a heavy nucleus such as lead. Only when collisions involve particles of similar mass (for example, a hydrogen atom or a proton) will maximum transfer of kinetic energy and momentum occur. Consequently, water, paraffin, and plastics have been used to moderate neutron sources. Once moderated, the thermal neutrons are absorbed either by mixing a material with a

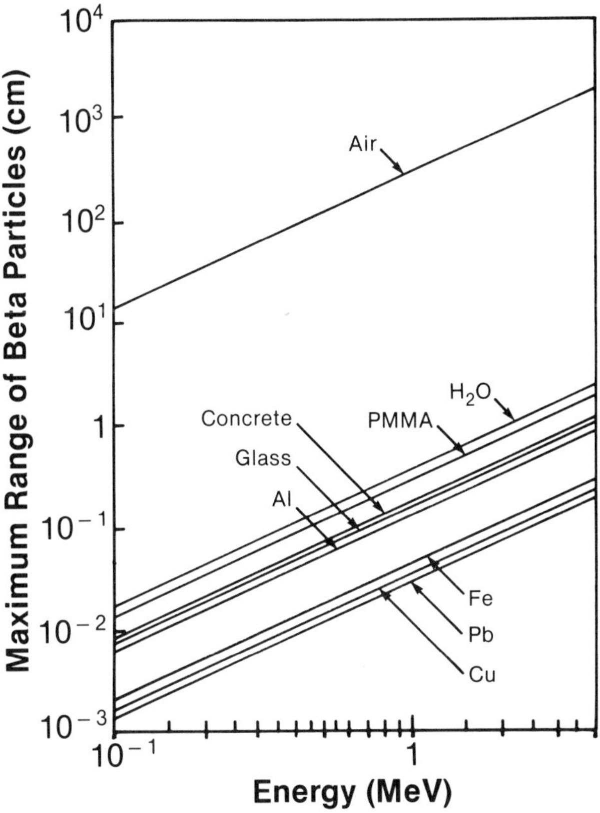

FIGURE 42 Maximum range which beta particles can travel in a shielding material as a function of accelerating energy. (Adopted from Ref. 160.)

high neutron capture cross section uniformly into a moderator or via a separate layer. Lithium, boron, cadmium, and indium can efficiently absorb thermal neutrons.

The relative effectiveness of lithium- and boron-filled polyethylene composites is illustrated in Figure 43 (Ref. 161). On a logarithmic scale, these two composites are much more efficient neutron absorbers per unit of thickness than the two moderators alone (graphite and paraffin). High-Z materials which effectively absorb γ-radiation (iron and lead) are not better than paraffin whether they are employed alone or admixed with composites having a high neutron capture cross section. Once again, coinjection molding techniques provide a means of building a unitized layer structure which moderates fast neutrons and absorbs thermal neutrons.

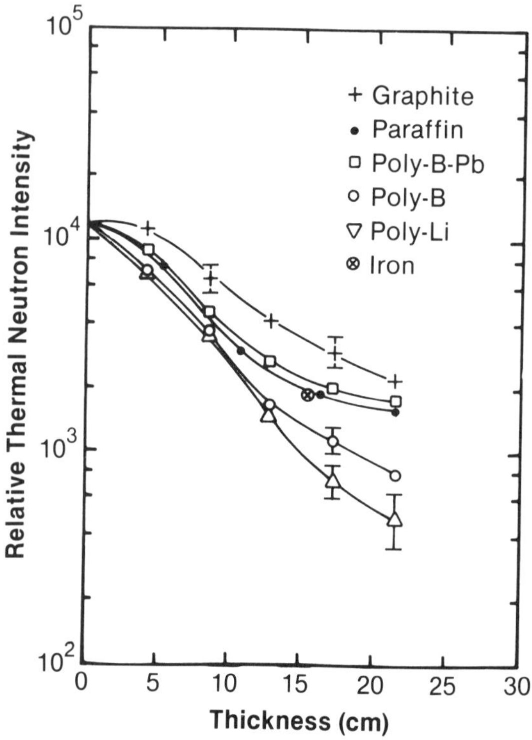

FIGURE 43 Relative effectiveness of different neutron moderating and absorbing materials as a function of shielding thickness in a 2.6-MeV neutron source. (Adopted from Gujrathi and D'Auria, Ref. 161.)

Aesthetics

Aesthetics is an important consideration in marketing any product. To the average consumer, functional appearance ("looking as if it will work") is a critical factor in making a purchase. Not too long ago plastics were considered inferior to metal alloys; they lacked the massiveness which was associated with serviceability. Today, the number of structural plastics and the improvement of plastic part design have provided the consumer with reliable products having a high ratio of strength to weight. The marketplace education was not accomplished overnight, however, and metal-filled plastics played a role in this transition.

A historical example of manufacturer redesign and consumer acceptance may be gleaned from the writing industry. About 25 years ago sturdy, reliable, and inexpensive ballpoint pens were introduced which featured a clear barrel and a brass tip (A in Figure 44a) (Ref. 162). Undoubtedly because the raw material and the processing steps were more expensive than those for an alternative molded plastic part, a major portion of the tip was modified. To avoid giving the impression of cheapening the product with a plastic part at a time when consumer acceptance was low, however, a brass-filled plastic was selected (B in Figure 44a and Figure 44b). Since that skepticism has all but disappeared, the same part is today made of a pigmented plastic (C in Figure 44a).

In 1960 a patent was granted for the colloidal dispersion of aluminum and copper in plastics and the compression molding of samples for decorative purposes (Ref. 163). Since that time both decorative and aesthetic metal-filled composites have grown, particularly in the automotive and building and furnishing industries. The opulent appearance of gold leaf paints on windows, wall hangings, or inlaid tables; the glittering finish of metallic-flake gel coat on an automotive body or of a metallic-flake elastomer in the floor mats; and the simulated metallic luster of front-end grilles, door panels, and bumper guards are a few examples. The most recent products include the gold and silver colored aircraft wings of Transformer toys, the silver-flecked lavender bottles of Silkience shampoo, and the glittering barrels of Pilot razor point pens. Indeed, one airline provides with its meals metal-filled compliant plastic dinnerware which is both aesthetically pleasing (gold colored) and functionally sound (Figure 45a) (Ref. 162). The longitudinal (L) and transverse (T) microstructures of such composites typically reveal a sparse ($<0.5\%$ by volume), fine distribution of metallic particles (Figure 45b). Here, the transparent character of the polymer suggests that greater volume fractions of particles are present. Moreover, for this application the formation of a polymer skin, which occurs from molding in a hot die and often proves troublesome in conductive applications (Ref. 53), provides a transparent protective coating. Not only alloys such as yellow and red brass but also pure metals such as copper and aluminum will continue to find decorative applications as filings and flakes in metal-filled polymer composites.

Magnetic Materials

Magnetic composites are found in every household today. They may be random polymer-metal mixtures of pure metals (iron, nickel, or cobalt), alloy powders (Alnico, Cunico, or Vicalloy), or heavily loaded ceramics (spinels such as barium, nickel, or manganese-zirconium ferrite). Typical properties of three barium ferrite-filled nylon composites are summarized in Table 26 (Ref. 6). Note first that of all the materials discussed in this chapter, the filler content is greatest in these composites—in effect reducing the role of the polymer to that of a binder. As expected for particulate-filled

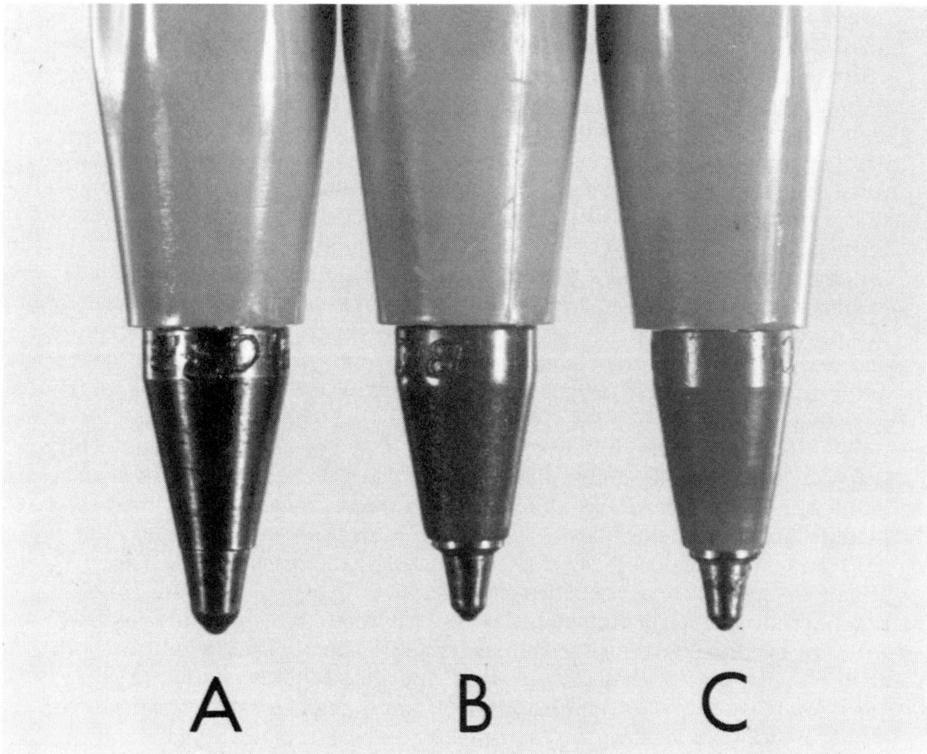

(a)

FIGURE 44 Use of metal-filled polymers to simulate metal: the BIC pen. (a) At one time a metal-filled polymer B had been substituted for the brass shoulder A. However, today a pigmented polymer C is used. (b) Energy-dispersive analysis (KEVEX) showed that brass was the filler metal.

systems, magnetic composites have a higher modulus but a lower tensile and impact strength than the unreinforced base resin. With regard to physical and thermal properties, the specific gravity increases three to four times, while the coefficient of thermal expansion decreases. Two important characteristics of a magnetic composite, its remanence and its coercive force, increase linearly with volume loading as the maximum volume fraction is achieved. The energy product increases proportionally with the square of the filler volume. The magnetic properties of these and other magnetic composites can be tailored through fabrication and orientation techniques.

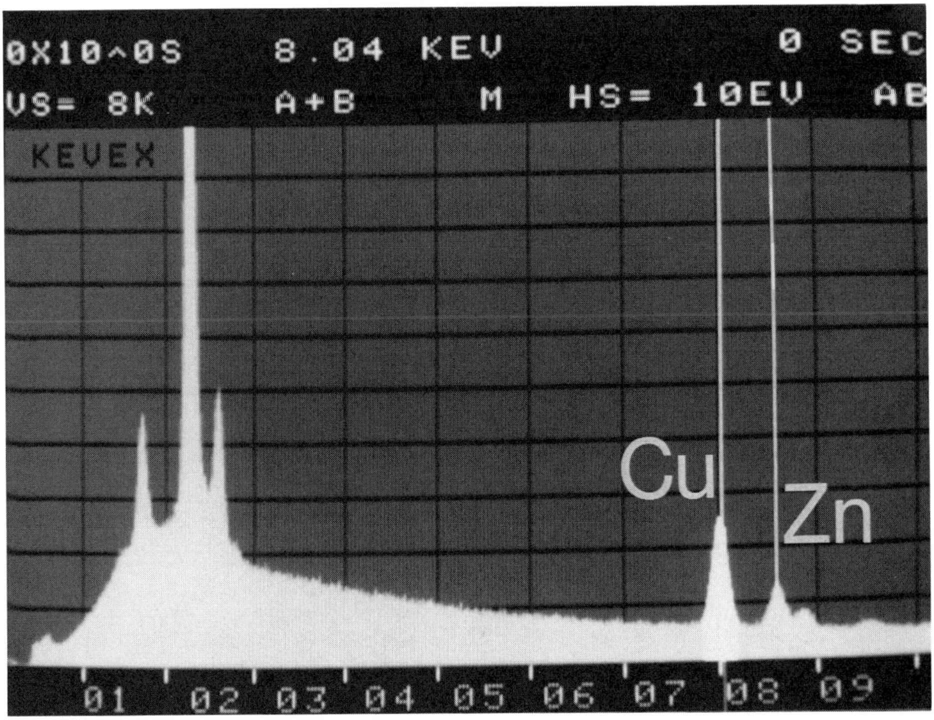

(b)

FIGURE 44 (Continued)

For example, when magnetic stability is important, maximum saturation may be followed by demagnetization to a specific remanence level.

The magnetic strips within the silicone gaskets which seal refrigerator doors and the message holders of all types which dot refrigerator doors are two common household uses of magnetic composites. Unlike harder magnetic materials, which are difficult to machine and to grind, magnetic composites can be injection-molded into any number of intricate shapes and then selectively magnetized. Predictably, applications have increased beyond the simple clamps and latches to include molded ferromagnetic cores (choke coils, clutches for servomechanisms, intermediate-frequency (IF) transformers, oscillator cores, threaded cores, toroids, transformer cores), bearing sleeves, timing-motor rotors, indicators and sensors, and magnetic tapes, inks, disks, and cards. In the first example the composite may serve as a structural member, whereas in the last

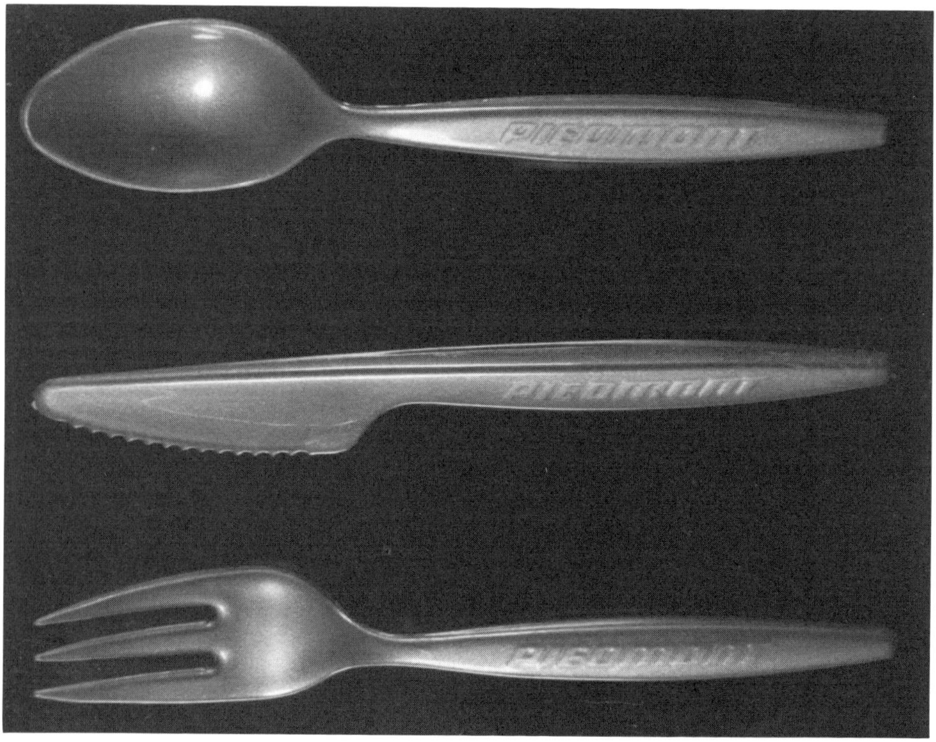

(a)

FIGURE 45 Use of metal-filled polymers to improve aesthetics without sacrificing function: (a) Piedmont Airline's dinner place setting. Here, the gold luster of the metal-filled utensils is shown. (b) Using a SEM, the backscattered electron images of a longitudinal (L) and a transverse (T) section show that very little metal is present and that some subsurface pores exist. Computer generated x-ray maps (not shown) indicate that copper is present.

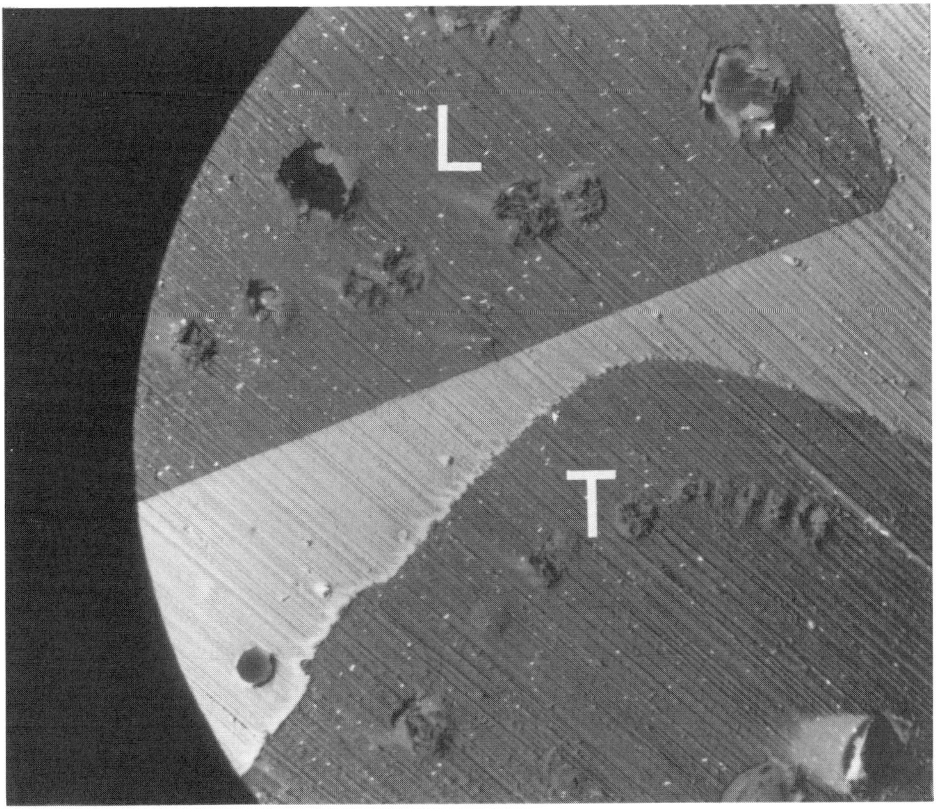

(b)

FIGURE 45 (Continued)

example powdered iron or its oxides are often combined with a polyester matrix to form a magnetic tape.

Electrostatic Discharge

Electrostatic discharge has long been recognized as a hazardous and occasionally a life-threatening problem. The potential danger of static discharge occurs whenever electrons can be added to or removed from the surface of a material without equilibrium being reestablished with its surroundings. Because charge mobility is severely restricted, polymers are prone to have potential differences with their environment. In the past the plastics formulator modified the

TABLE 26 Typical Properties of Three Magnetic Composites[a,b]

Properties	Units	(A)	(B)	(C)
Mechanical				
Tensile strength	MPa	48.2	58.6	55.1
Elongation	%	1.0	1.0	1.0
Flexural strength	MPa	82.7	86.1	89.6
Flexural modulus	GPa	8.61	8.96	9.30
Izod impact strength, notched[c]	J/m^2	3.2	3.2	3.2
Physical				
Specific gravity	—	3.20	3.45	3.60
Thermal				
Deflection temperature under load[d]	°C	110	110	110
Coefficient of linear thermal expansion	10^{-5}/°C	3.6	3.33	3.24
Thermal conductivity	W/(m-K)	0.92	0.94	0.97
Magnetic				
Residual induction[e]	T	0.184	0.196	0.230
Coercive force[e]	kA/m	130	139	155
Intrinsic coercive force[e]	kA/m	191	223	215
Energy product[e]	kTA/m	4.77	6.76	9.55

[a]Products evaluated:
(A) 83% barium ferrite/17% nylon
(B) 86% barium ferrite/14% nylon
(C) 88% barium ferrite/12% nylon
[b]Unless otherwise noted, all tests conducted at 23°C using applicable ASTM procedures.
[c]Izod impact test bars 6.35 mm.
[d]Deflection temperature at 1.82 MPa.
[e]Oriented.
Source: Adopted from Theberge (Ref. 6).

compounds in order to provide a path for dissipation of charge. Sometimes plasticizers were used, but these leached out over time or settled via gravitational forces, rendering the composite prone to charge buildup. Other solutions for ESD have been humidity-active additives, air ionization, and copper foil.

Among carbon black-filled, metal-filled, and conductive polymers the first two have been readily used as low-cost, lightweight ESD alternatives (Table 27). Static decay rates of a carbon- and metal-filled versus a nickel-painted and unmodified polycarbonate are shown in Table 28 (Ref. 164). All of the modified polycarbonates performed similarly, a charge of ± 5 kV decaying in <0.06 sec compared to >100 sec for the unmodified polymer (metals decay in <0.01 sec). In practice, potentials may easily exceed 30,000 volts.

TABLE 27 Some Antistatic Carbon Black- and Metal-Filled Composites Marketed Today

Adhesive tapes
Boots and bushings
Caulking compounds
Conductive paints
Electromedical body contacts
Flange gaskets
Floor and worktable grounding mats
Fluid gaskets
Foams
Heat-shrinkable tubing
High-voltage cable conductors
Hot-melt conductors
Intrusion detection sensors
Keyboard switch contact elements
Low-current switch elements
Plating powders for hinge joint switches
Rolls, belts, sheets, and tubing
Static discharge wicks

TABLE 28 Static Decay Rate of Selected Filled Plastics

Material	Applied voltage (%)	Decay rate (sec) At +5 kV	At -5 kV
Polycarbonate	50	>100	>100
	10	>100	>100
	0	>100	>100
Nickel-painted polycarbonate	50	0.02	0.02
	10	0.03	0.04
	0	0.05	0.05
Polycarbonate/ 30% metal	50	0.02	0.02
	10	0.04	0.03
	0	0.06	0.05
Polycarbonate/ 40% carbon	50	0.02	0.02
	10	0.04	0.03
	0	0.05	0.05

Source: Adopted from Litman and Fowler (Ref. 164).

The properties of a series of Nylon 6/6, polycarbonate, polyphenylene oxide-based, and polypropylene antistatic composites are shown in Table 29 (Refs. 6, 132). Surface resistivities for these four commercial products vary from 10^5 to 10^3 ohm/sq. Overall, surface resistivities of statically conductive composites may range from 10^8 to 10^2 ohm/sq.

Antistatic materials are used in three general situations: when fine powders, flammable liquids, or explosive gases are present, when relative motion between dissimilar materials occurs, and when electrical circuitry is involved.

Because grain or coal dust and fine metal or explosive compound particles can spontaneously ignite as a result of electrical discharge, the environment must be maintained as much as possible at the same potential. Therefore, grain chutes and conveyors are formulated with metal- or metalloid-filled plastics or elastomers. Aviation fueling and truck depots are potentially hazardous if fuel intake components and fuel tanks are not engineered from conductive materials. Hospital operating rooms must also eliminate potential differences because of the risk of explosion in the oxygen-enriched environment. Consequently, the oxygen gas is transported by using conductive Y and T joints (Figure 46) (Ref. 165), the floors, walls, doors, and hinges are made of antistatic materials, and the personnel wear shoes with statically conductive soles.

TABLE 29 Statically Conductive Thermoplastics[a,b]

Properties	Units	(A)	(B)	(C)	(D)
Mechanical					
Tensile strength	MPa	68.9	117	41.3	20.7
Elongation	%	2.0	2.4	2.0	11.0
Flexural strength	MPa	137.8	179	62.0	24.8
Flexural modulus	GPa	5.83	7.58	2.89	0.76
Izod impact strength, notched[c]	J/m^2	5.3	6.8–8.9	1.6	16.8
Physical					
Specific gravity	—	1.29	1.28	1.12	0.94
Linear mold shrinkage, 3 mm	%	0.6–0.8	0.05–0.3	0.4–0.6	1.0–2.0
Thermal					
Deflection temperature under load[d]	°C	252	143	129	82
Electrical					
Surface resistivity	ohm/sq	10^3	10^3-10^4	10^4	10^5

[a]Products evaluated:
(A) 85% Nylon 6/6/15% fiberglass
(B) 85% polycarbonate/10% PAN carbon fiber/5% GR fiber
(C) Modified polyphenylene oxide
(D) Polypropylene-based
[b]Unless otherwise noted, all tests conducted at 23°C using applicable ASTM procedures.
[c]Izod impact test bars 6.35 mm.
[d]Deflection temperature at 1.82 MPa.
Source: Product data abstracted from Theberge (Ref. 6) and Wilson-Fiberfil International (Ref. 132).

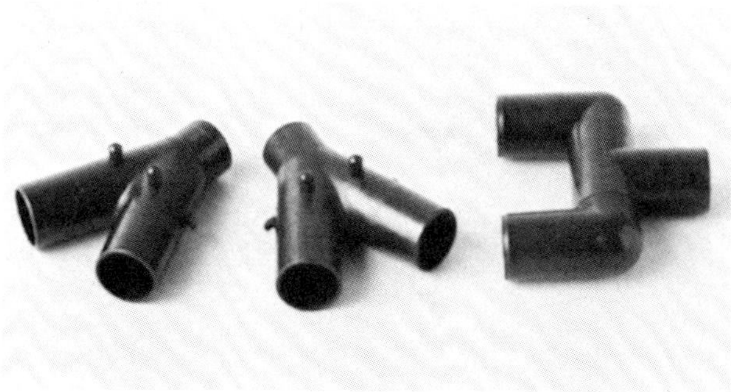

FIGURE 46 The Y and T joints in oxygen lines are made of a conductive composite in order to reduce the risk of a fire or an explosion in an operating room environment. (Reproduced from product literature of RTP Co., Ref. 165.)

Airplanes slicing through space at supersonic speeds, high-speed factory equipment, and personnel all create charge buildup and require static dissipation. Airplane wings, rotor blades, and tires have been fabricated from conductive materials, since the leading edges of the wings which cut through the air, the blades which push the air, and the tires which first touch the ground must all bleed off the charges accumulated by the plane during flight. High-speed print rollers, conveyor belts, or weaving machines (Figure 47)

FIGURE 47 Component parts such as this weaving machine arm are made of a conductive composite in order to prevent electrostatic discharge from damaging textiles or injuring employees. (Reproduced from product literature of RTP Co., Ref. 165.)

(Ref. 165) all require attention. Even the action of employees walking across a floor can cause problems with diagnostic, photographic, or computer equipment. More than one floppy disk has been destroyed by static discharge as the information was being handed to a co-worker or transferred to another disk drive.

Unlike the previous applications, in which grounding was necessary on passive devices, the electronics industry is concerned with dissipating the charges on both active and passive devices. In a personal computer, for example, there may be not only two belt-driven disk drives and a high-speed fan for cooling but also low- and high-voltage sources. Isolation of the chips (which toggle somewhere between 0 and +5 volts) from the cathode-ray tube (which typically operates at between 10,000 and 50,000 volts) is essential. Appropriate grounding of internal components from one another is the solution. In addition to computers, television sets and capacitive video disks (Ref. 166) require effective grounding protection.

Before leaving this topic, an important antistatic application of conductive polymers should be noted. Because water-soluble polyionomers mixed with hydrophilic binder polymers have poor abrasion resistance and tackiness at high temperatures and relative humidities, an alternative solution to static dissipation from high-speed rollers and sprockets which are found in the photographic industry has been sought. One approach which does not depend on hydrophilic binders for water retention and does not affect the physical properties of the products adversely is to use highly charged polymeric microgels (Ref. 131). These are cross-linked, spherical, polyelectrolyte particles (≈0.01–1.0 μm) that will disperse in solvents in which non-cross-linked polymers of similar composition would be soluble. Chemically they consist of a copolymerizable monomer, a cross-linking tetrafunctional monomer, and a charged ionic group (20–99 mole % ionic). Both cationic and anionic microgel polymers have been prepared (Figure 48). The resistivities of both types decrease with increasing coverage and relative humidity. In service, typical surface resistivities are $<10^9$ ohm/sq, with the cationic microgels having substantially lower resistivities than their anionic counterparts. Because anionic polymers contribute toward negative triboelectric charging, however, a more neutral charging film will result when compounded with a hydrophobic binder. In this way, anionic microgel films will not tend to acquire as much static charge on contact and separation with high-speed apparatus as the more conductive cationic microgel polymers under the same circumstances.

Electromagnetic/Radio-Frequency Interference

The previous five specific topics represent more traditional applications of metal-filled polymers. The closing five topics represent relatively new or perhaps future areas of product development:

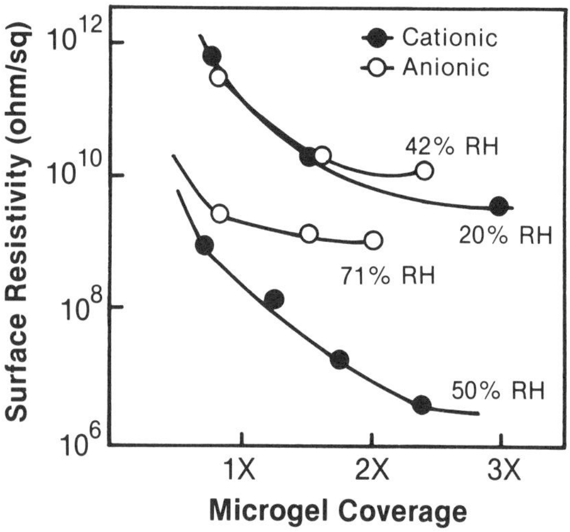

FIGURE 48 Electrical resistivity of cationic and anionic microgel poymers as a function of relative humidity and coverage. Such conductive coatings are useful to the photographic industry in order to eliminate arcing from the rollers or sprockets to the film or prints during manufacture, development, or printing. (Adapted from Upson and Campbell, Ref. 131.)

EMI/RFI, switches, transducers, transparent composites, and batteries.

Electromagnetic wave pollution has become a growing concern in recent years. It affects the reception of television and radio broadcasts, the operation of medical instrumentation and aircraft equipment, and, some suspect, even the health (stress level) of individuals (Refs. 135,167). The cause of the problem is twofold: the overall increase in the use of plastics and the expanding use of electronic devices by consumers. Higher strength-to-weight ratios, part simplification, ease of fabrication, corrosion resistance, and cost have facilitated the displacement of inherently RF-absorbent metals by intrinsically RF-transparent plastics, while the presence of microwave ovens, power tools, electric shavers, and hair dryers (household appliances), high-frequency clocks (computers), and relays, motors, and switching power supplies (automobiles) has increased transmission in the 0.1–1000 MHz range (see Figure 33). Since the elimination or redesign of filters, capacitors, special cable connectors, etc., is unrealistic, shielding is the alternative. Using the precedent set by a 1938 rule affecting restricted radiation devices, the Federal

Communications Commission (FCC) stepped in in 1979 and established the specifications for radiated emission unintentionally transmitted by hardware, the limits of conducted feedback into the main power line, and the requirements for labels and instruction manuals (Ref. 168). The first two of these are summarized in Table 30 (Ref. 169) for business (class A) and personal (class B) devices.

Some devices which either emit or absorb EMI/RFI energies are listed in Table 31. Shielding of these devices may be accomplished at either the individual component level or the device enclosure level.

TABLE 30 EMI Requirements for Different Classes of Equipment

For class A (industrial) equipment

Radiated limits:

Frequency (MHz)	Distance (m)	Field strength (μV/m)
30 – 88	3	300
88 – 216	3	500
216 – 1000	3	700

Conducted (power line) limits:

Frequency (MHz)	Maximum voltage (μV)
0.45 – 1.6	1000
1.6 – 30	3000

For class B (mass-market) equipment

Radiated limits:

Frequency (MHz)	Distance (m)	Field strength (μV/m)
30 – 88	3	100
88 – 216	3	150
216 – 1000	3	200

Conducted (power line) limits:

Frequency (MHz)	Maximum voltage (μV)
0.45 – 30	250

Source: Adopted from Reference 169.

TABLE 31 Some Devices Which Can Emit and/or Are Susceptible to EMI/RFI Energies

Alarm systems	Graphic systems
Black & White terminals	Mainframe computers
Blood analyzers	Microcomputers
CATVs	Microwave ovens
Central processors	Minicomputers
Color terminals	Miniterminals
Controllers	Modems
CRTs	Modular composing systems
Dashboard radar detection unit	PBXs
Data communications equipment	Pocket calculators
Digital reverberators	Portable computers
Digital watches and clocks	Printer buffers
Digital weighing scales	Printers
Disk drives	Statistical multiplexers
Display terminals	Temperature monitors
Electronic cash registers	Touch-sensitive CRTs
Electronic games	TV cameras
Electronic hear monitors	Typesetting systems
Graphics processors	Word processors

In either case the incident electromagnetic wave must be reflected at the shield interfaces or absorbed within the shielding material if acceptable attenuation is to be achieved (Figure 49). Regan (see Table 3 in Ref. 135) evaluated seven shielding alternatives—pressure-sensitive foils/tapes, vacuum metallizing, flame/arc spraying, cathode sputtering, silver reduction, metallic paints, and metal-filled plastics—and found that only conductive, metal-filled plastics did not require secondary processes. Carbon fibers, quick-quenched aluminum flakes, copper powders, stainless steel filaments, and nickel-metallized glass fibers have been mixed with ABS, polycarbonate, polypropylene, nylons, or phenolics to produce commercial EMI/RFI products. Although some shielding will be afforded in the high-

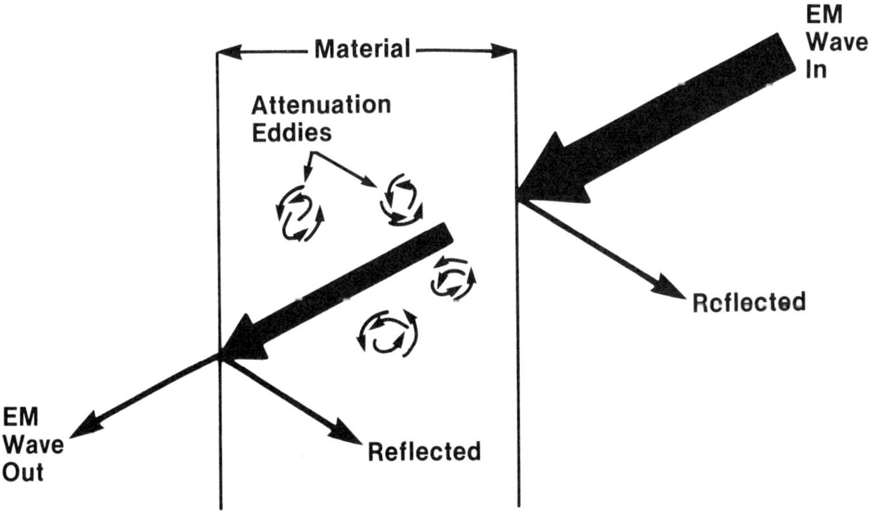

FIGURE 49 Schematic illustration showing how an incident electromagnetic wave can be attenuated by a shielding material.

frequency region when $V < V_c$, effective EMI shielding requires that $V > V_c$ (Ref. 167). This statement implies something which has been verified both theoretically and experimentally, namely that electrical conductivity and EMI shielding are interrelated. Therefore, to manufacture a polymer with satisfactory EMI protection, 25–40% by weight of aluminum flake or carbon fibers may be necessary, whereas only 5–10% by weight of stainless steel fibers may be required to obtain a shielding effectiveness (SE) of 30–55 dB (Ref. 170). [The definition of SE, the description of testing methods (shielded box or room versus transmission line method), and the detailed investigation of the associated parameters are deferred to Chapter 3.] Although the mechanical properties of the stainless steel fiber-loaded composite are not affected until about 10%, the aluminum flake composites behave in the classical manner: the modulus increases as the tensile and and impact strength decrease. Moreover, because of the high metal loading, the aluminum composite requires more power for injection, which results in more die wear than with the stainless steel fiber alternative.

The EMI/RFI shielding housing for a dashboard radar detection unit may be used to illustrate some fundamental design criteria. This housing, shown in Figure 50 (Ref. 165), was molded using a nickel-plated, graphite fiber-loaded Nylon 6/6. Figure 51 (Ref. 61, 72, 134) illustrates the SE of this composite compared to some of the other

FIGURE 50 Housings like the one shown for a dashboard radar detection unit, are manufactured from a conductive metal-filled polymer to attenuate EMI/RFI emissions to acceptable levels and/or to protect internal components from external noise. (Reproduced from product literature of RTP Co., Ref. 165.)

alternatives: the all-metal enclosure (copper or steel), other conductive composite alternatives (carbon black composite, aluminum flake- and fiber-filled polyesters), and the unfilled polymer (see also Ref. 72 and 167). Using the earlier guideline that 30–40 dB attenuation would meet 50–95% of industry needs (Ref. 134, 135), all materials meet or exceed this bench mark except the unfilled polymer. Although the SE eventually diminishes to <20 dB within the frequency band 100–1000 MHz, the four composites make adequate enclosure materials for most devices with clocks operating at ⩽50 MHz. Moreover, conductive plastics such as the aluminum flake and fiber systems have demonstrated cost effectiveness and improved reliability. Indeed, significant savings may be had for complex parts since the secondary costs associated with shielding unfilled plastics are quite labor intensive. Once a complex mold is made, part weight principally determines the cost of metal-filled EMI/RFI polymers.

Despite the advantages of injection-molded metal-filled plastics, sometimes metallic coatings, paints, or lacquers are used to provide satisfactory EMI shielding. Such silver and nickel paints may be applied as a secondary process, either as a primary coating or for supplemental protection. This top coat may either provide unshielded auxiliary equipment with some protection in "noisy" environments or improve existing shielding in critical applications.

Figure 52 (Ref. 171) illustrates the SE of six test panels coated with 2, 4, and 6 mils (±0.05 mil) of silver or nickel (top frame), and

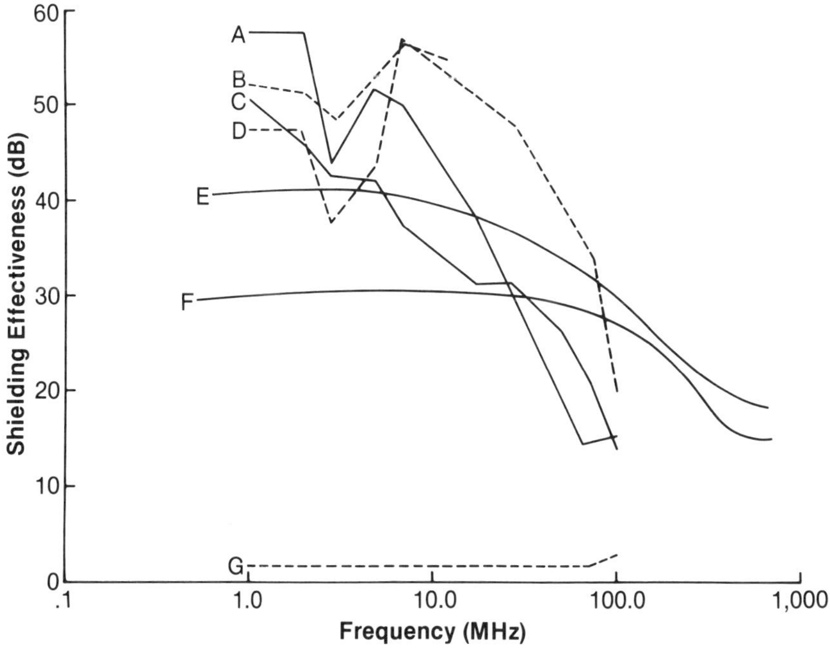

FIGURE 51 Shielding effectiveness of different materials as a function of frequency. A, Nickel-coated glass fiber composite; B, copper plate; C, carbon black composite; D, steel; E, 20% aluminum fiber in polyester; F, 30% aluminum flake in polyester; G, unfilled polymer. (Data obtained from Bigg and Bradbury, Ref. 72, Bigg and Stutz, Ref. 134, and Ref. 61.)

an actual service test of a 2 mil top coated polyphenylene oxide box (bottom frame). Figure 52a shows that the nickel coatings meet and the silver coatings exceed the 30–40 dB bench mark and that 2 mils of silver is more effective (but more expensive) than any of the nickel coating thicknesses reported. The SE improves little when coating thicknesses greater than 2 mils are applied—particularly for the cases of 4 versus 6 mils. When tested under actual service conditions (Figure 52b), the SE values of the 2-mil silver and nickel coatings are more comparable to those of the 6-mil-thick test panels at low frequencies. As the frequency increases the SE actually decreases, however, because the mating surfaces of the covers and the front and back plates were poorly bonded.

Recently, the Gesellschaft fur Oberflachentechnik (GFO), a subsidiary of Degussa AG, introduced a proprietary EMI shielding process

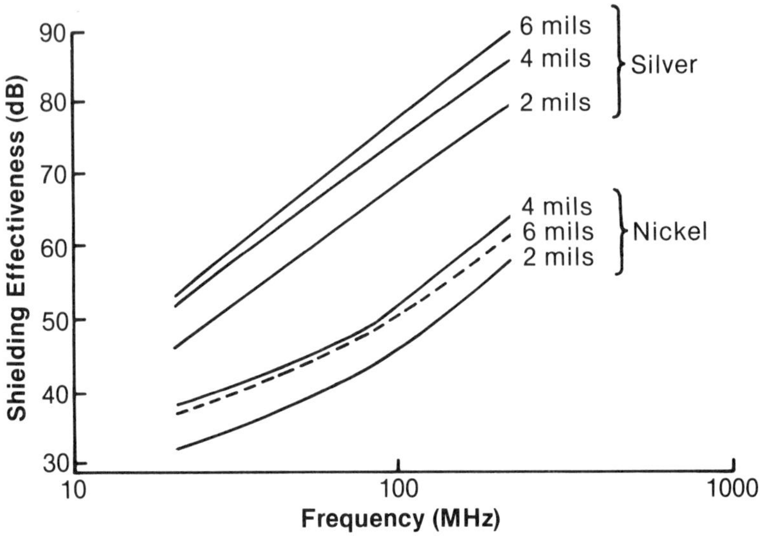

(a)

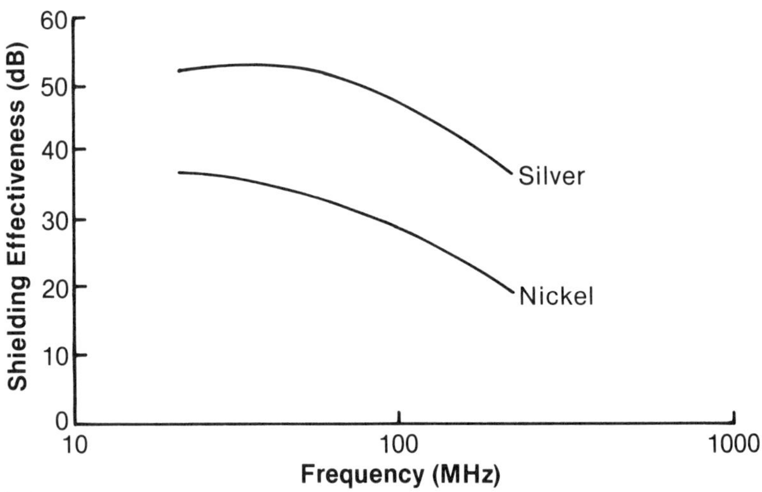

(b)

FIGURE 52 Shielding effectiveness of silver and nickel paints as a function of frequency and coating thickness (a) for test panels (b) under actual service conditions. (Adopted from Genova, Ref. 171.)

(Elamet) which can vacuum-metallize columnar crystals of aluminum onto a variety of polymer substrates. A ductile coating results which can retain SE over a component's lifetime without a basecoat or topcoat. A coating thickness of only 2.5 μm attains a SE of 45 to 75 dB over the frequency range 1 to 4 × 10^3 MHz. These homogeneous, adhesive coatings are currently being utilized for the manufacture of computer and business machine enclosures (Ref. 172).

To ensure that a nickel-coated enclosure will provide at least an SE of 30 dB from 30 to 200 MHz (e.g., the dashboard radar detection unit discussed previously), the final product must be electrically tight. In both these examples the inherently high SE of component materials may be negated if narrow slots in mating surfaces can radiate energy similarly to an antenna of comparable length (Ref. 167). Having overlapping contiguous components is one solution; using conductive gaskets is another (Ref. 173).

Metal-filled gaskets have improved heat transfer, resistance to compression set, and some lubricity. One particularly attractive product line is highlighted in Table 32 (Ref. 174). These conductive composites are comprised of silver-plated copper powders (alternatively, silver-plated aluminum or silver-plated glass) dispersed in silicone or, for more chemical resistance, fluorosilicone elastomers. Mechanical properties and service temperature range are adequate for the intended application, and the SE is outstanding. For magnetic, electric, and plane-wave fields the attenuation ranges from 65 to >140 dB over various frequency bands which span 10^{-2}–10^4 MHz (Figure 53) (Ref. 174). The gaskets used in these tests were 6.35 × 1.57 mm flat gaskets of 254 mm inner diameter. For products A and B the closure pressure was 0.69 MPa.

Switches

In Figure 13 the resistivity was decreased at least 12 orders of magnitude for $V > V_c$, and V_c was influenced by parameters such as polymer/metal size ratios, metallic aspect ratios, metallic oxide contents, and processing conditions. This precipitous drop may be utilized as a switch in self-controlled heaters or as a power limiter to peripheral devices. Two specific examples are cited in this section, the electrochemical and the thermoelectric switch.

An electrochemical switch may be used as a safety device to control an on-line process or to monitor the subsequent storage of an organic liquid. Using a conductive polymer/metal composite with $V \gtrsim V_c$, a bar of material may be situated, for example, in the interspace of a double-walled storage container. If the primary wall develops a leak from either corrosion or mechanical abuse, the imbibition of the organic liquid into the specially formulated conductive composite causes solution swelling. As the polymer expands, the gaps between the metal particles increase, and direct conduction and

TABLE 32 Typical Properties of EMI/RFI Gasket Materials[a,b]

Properties	Units	(A)	(B)	(C)
Mechanical				
Tensile strength	MPa	2.76	1.38	1.38
Elongation	%	55	150	120
Tear strength	N/m	511	511	511
Hardness, Shore A	—	80	65	80
Physical				
Specific gravity	—	3.4	3.6	4.0
Thermal				
Service temperature	°C	-55 to 125	-55 to 125	-40 to 125
Thermal conductivity	W/(m-K)	0.12	0.19	0.14
Electrical				
Volume resistivity	ohm-cm	0.005	0.004	0.008
Shielding effectiveness	dB	>78	>71	>70

[a]Products evaluated:
(A) High-durometer, silver-plated copper/silicone
(B) Medium-durometer, silver-plated copper/silicone
(C) High-durometer, silver-plated copper/fluorosilicone
[b]Unless otherwise noted, all tests conducted at 23°C using applicable ASTM procedures.
Source: Consolidated product data from Chomenics, Inc. (Ref. 174).

tunneling mechanisms are hampered. The result is a substantial increase in resistivity and the actuation of a pump, ventilator fan, valve, and/or alarm.

The principle of an electrochemical switch is illustrated in Figure 54 (Refs. 65, 175). By loading 10% nickel into a polyurethane matrix, a semiconductive composite was fabricated having a resistivity of 10^2 ohm-cm. Upon exposure to allyl methacrylate, significant diffusion occurred which resulted in an over 100-fold increase in resistivity in <1 min. Within 25 min the equilibrium resistivity of the conductive composite was comparable to that of a polyurethane swollen with allyl methacrylate after 72 hr. This result was within one order of magnitude of the resistivity of an unmodified polyurethane specimen, a relatively conductive polymer compared to others.

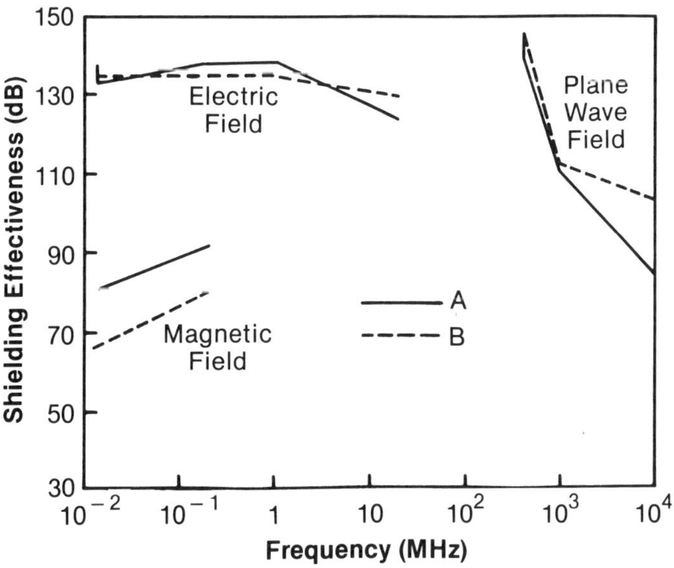

FIGURE 53 Shielding effectiveness of a high (A) and medium (B) durometer silicone/silver-plated copper composites as a function of frequency and field type. (Adopted from Ref. 173.)

A thermoelectric switch may be used for automatic control of the heat flow from one area or object to another. It may act as a thermal control over temperature-sensitive semiconductive circuitry, in cryogenic refrigeration units, and as a substitute for louvers or cover plates to protect sensitive instruments in spacecraft from the damaging rays of the sun. Because the insulator phase of conductive composites has a high thermal expansion coefficient and the conductor is loaded to the critical volume percent for low resistivity ($V \approx V_c$), a positive temperature coefficient (PTC) occurs. Thus, as the temperature is raised through the thermodynamic melting region, the resistivity increases several orders of magnitude. In part, the conduction mechanism parallels that of electrochemical switching: as the gaps increase in number and size on melting, the tunneling process is impeded. Moreover, as the composite continues to melt, additional resistance occurs in the gaps because the ratio of the amorphous to crystalline material increases. With subsequent cooling, the filler particles are excluded from the growing spherulites into the interphase regions.

In the past, Bueche (Refs. 176, 177) fabricated a 7% carbon black (HAF) loaded hydrocarbon wax, whereas Meyer (Refs. 178,

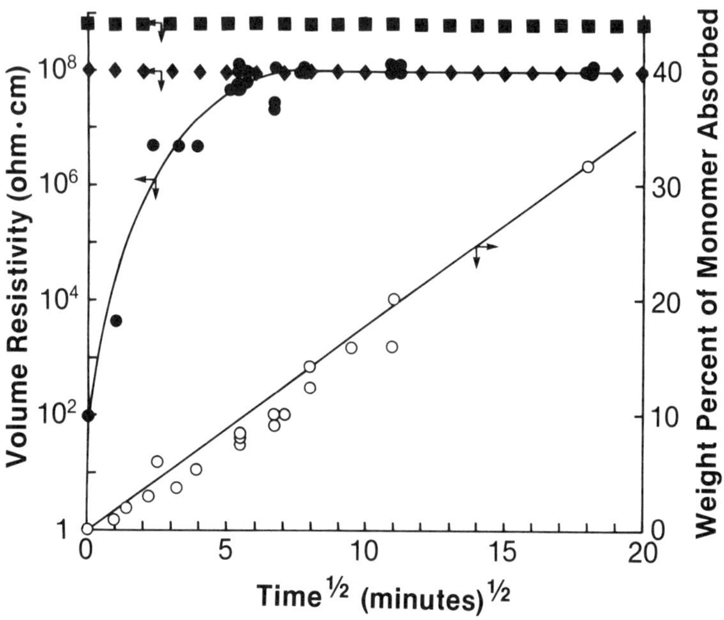

FIGURE 54 Example of an electrochemical switch: polyurethane/10% nickel. As solvent uptake increases (in this case a monomer, allyl methacrylate), volume resistivity increases a millionfold. The resistivities of both an as-received (■) and equilibrium swollen (♦) polyurethane specimen are shown for comparison. (Adopted from Kusy, Ref. 65, and Turner et al., Ref. 175.)

179) mixed 20% carbon black (Vulcan V3) in polyethylene. Today, conductive fillers may be not only conventional carbon blacks but also carbon fibers or metallic powders (for example, copper, nickel, silver, and zinc). Unlike the previous low-resistivity applications in which a furnace black (Ketjen) was preferred, optimal switching properties are obtained when the less conductive thermal blacks (e.g., MT) predominate in a formulation (Ref. 141).

The principle of thermoelectric switching is illustrated for a composite of high-density polyethylene with 2–6% Ketjenblack (Figure 55) (Ref. 180). As a melting point is approached in the unmodified thermoplastic composite (Figure 55a), a steep PTC is observed from 120 to 140°C followed by an even steeper negative temperature coefficient (NTC). On cooling, a similar but much smaller effect is observed at a somewhat undercooled temperature. When cycled through the melting point four times (Figure 55b), not only the PTC and

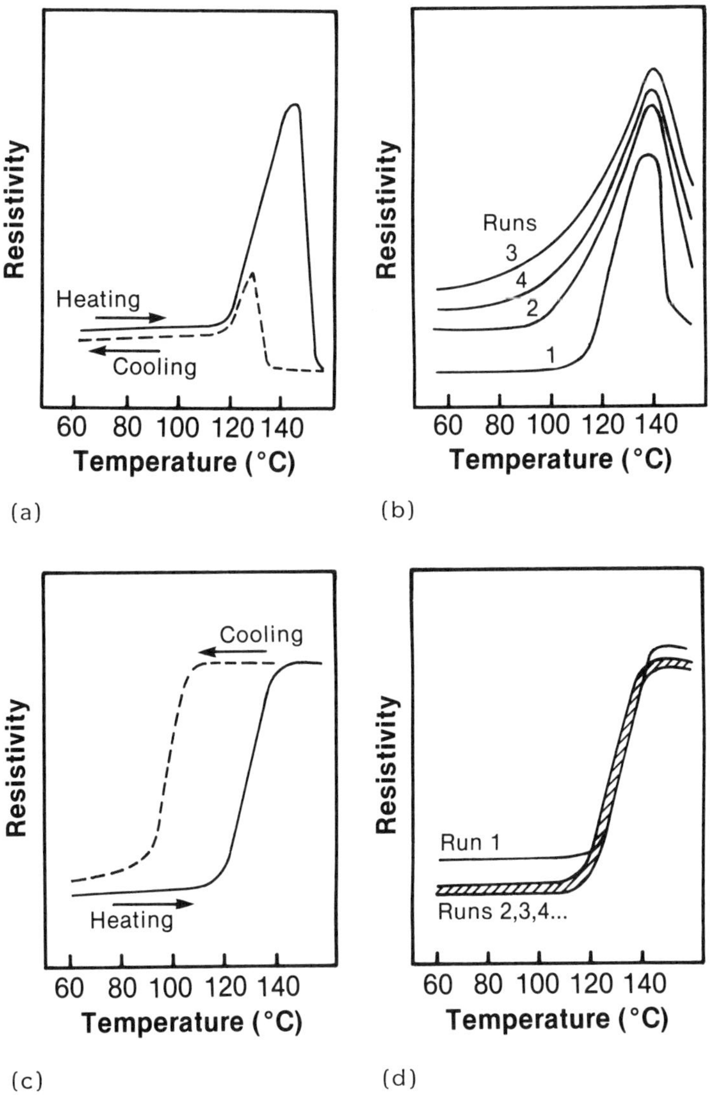

FIGURE 55 Example of a thermoelectric switch: high-density polyethylene/2–6% Ketjenblack. Without any modification, these composites have a negative temperature coefficient (NTC) on heating or cooling (a) as well as a baseline shift on cycling (b). After crosslinking, however, both the NTC (c) and the electrical stability (d) improved. (Adapted from Narkis et al., Ref. 180.)

the NTC but also the baseline resistivity (below 100–120°C) changed. Both this lack of reproducibility and the presence of an NTC had to be eliminated. Although significant improvements in reproducibility were made by formulating mixtures having two carbon blacks (Ref. 39) or by adding rubber as a "mechanical stabilizer" (Ref. 176), the products were still impractical. Ultimately, stability was attained by cross-linking the matrix (Ref. 180). Whether this was done by radiation processing (^{60}CO γ-rays) at room temperature or with chemical cross-linking agents (dicumyl peroxide) at elevated temperatures, the resulting product had high resistance to environmental stress cracking, low thermal and creep resistance, and the capability of accommodating high filler volumes without detriment to good mechanical properties (Ref 141). In the present context, cross-linking eliminated the NTC characteristics (Figure 55c) and stabilized the composites after an initial conditioning (Figure 55d). As the degree of cross-linking was increased, the baseline resistivity decreased, and the NTC progressively became less steep until it had all but disappeared (see Figure 1 in Ref. 180). Indeed, highly cross-linking these thermoplastic materials resulted in an elastic memory effect in which the filler particles were returned to the same locales upon cooling. A commercial manifestation of such an elastic or shape memory effect is heat-shrinkable tape and tubing.

Tranducers

A transducer is an energy converter, or a device that converts a nonelectrical quantity into an electrical signal. It may be a passive device which relies on a battery or other external power supply for its operation or an active element which functions as an independent voltage source. Transducers may be classified as acoustic, electrical, hydraulic, magnetic, mechanical, and optical. A microphone, speaker, piezoelectric crystal, venturi, propeller, differential transformer, reluctance transformer, Bourdon tube, gyro, and photoelectric cell are all tranducers. Two other types of tranducers are considered here: the thermistor and the strain gage.

The thermistor is a temperature-sensitive resistor. Of the three temperature transducers (the thermocouple, the resistance temperature detector, and the thermistor), it is the most sensitive to changes of temperature. Generally thermistors are fabricated from semiconductor materials, have an NTC, and are extremely nonlinear devices. As long ago as 1950 a patent describing a thermistor was issued (Ref. 181). In the claims, a ball-milled Bakelite was blended with ferrous-ferric oxide, silver sulphide, or a mixture of uranous and uranic oxides. After molding and heat curing the composite, a high NTC was reported. More recently, Narkis et al. (Ref. 180) have reported a PTC in a thermal black (MT)-loaded high density polyethylene which spanned more than six orders of magnitude of resistivity in a 10 C° region. In terms of Figure 55d, the PTC spans a 20 C° region over what is undoubtedly a large but unspecified resistivity range. If favorable PTCs were reproducible, then thermistors could be manufactured from different polymer/metal composites which could monitor temperatures over very narrow ranges.

The strain gage is a deformation-sensitive device. Three common types of gages exist: the resistive, the inductive, and the capacitive gage. In the present context, the first may be made from fine wire filament or thin-etched foil which have a high ratio of resistance change to strain change. High "gage factors" of 2.0–2.2 are reported for copper-nickel, iron-nickel, and nickel-chrome strain gage alloys. In constructing and evaluating strain gages, the temperature sensitivity, potential hysteresis, and zero-shift effects should be considered along with the strain sensitivity itself. Only then can the reliability of the strain gage be assessed.

Within the past few years, Yamaki et al. (Ref. 59) and Kost et al. (Ref. 38) have illustrated that metal- or metalloid-filled composites can exhibit significant resistivity changes when subjected to either compressive or tensile stresses. In the first experiment composites were fabricated by hot-pressing carbon fibers (length, 100 μm) with thermoplastic polyester. Depending on the compaction pressure, the specific gravity of these elastomeric composites ranged from 0.65 to 0.85. When deformed under compressive loads (Figure 56) (Ref. 59), the electrical resistivity decreased by as much as 100-fold as the number and quality of fiber contacts increased, new tunneling events occurred, and the distance of existing tunneling events decreased. Along with the experimental results, a theory based on the probability of contact between the conductive fillers was presented. Here, assuming the same fiber length, a 10 μm fiber diameter, and a 5×10^{-3} ohm-cm fiber resistivity, the theoretical curves for 12.8 and 20.1% carbon fiber by volume were determined and plotted on either side of the experimental results. When the theoretical and experimental curves were compared, neither the gage factor (i.e., the slope of the curves) nor the critical strain at which the resistivity saturates was similar. Interference of the matrix with the fiber contacts and the inherent idealization of the model contributed to the disparity.

In the second experiment, silicone rubber, Ketjenblack EC, and dicumyl peroxide were mixed on a cold two-roll mill. Thereafter, sheets (15 × 15 × 0.18 cm) were compression-molded at 170°C and air-cured at 200°C for 6 hr. As a function of carbon black concentration, the volume resistivity precipitously decreased at 6% by weight, attaining values less than 10^1 ohm-cm by 12% loading. Four compositions were evaluated near this transition region: 8, 10, 12, and 15%. Voltage-current dependence for these different concentrations of EC black in silicone was ohmic over a potential drop of 6

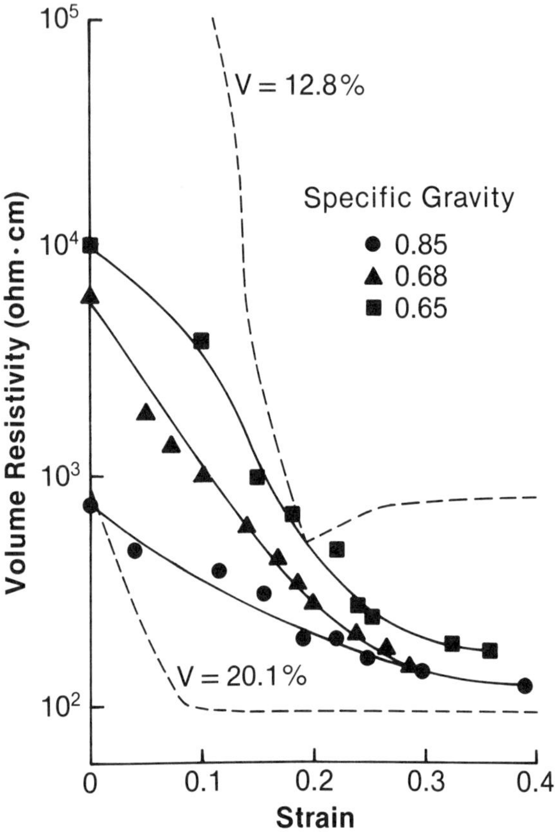

FIGURE 56 Dependence of electrical resistivity on compressive strain. By hot-pressing polyester/carbon fiber composites to different compaction pressures, the volume resistivity was decreased with increasing specific gravity. The experimental results lie between the theoretical curves for a composite loaded with 12.8 and 20.1% carbon fibers, respectively. (Adapted from Yamaki et al., Ref. 59.)

volts and 20 milliamperes. When specimens (15 × 1 × 0.18 cm) were axially stretched to a strain of 0.8 and the electrical properties were normalized with respect to the unstrained specimens, the higher ratios of resistivity were most pronounced for the volume loading in the transition region, about 8% (cf. Figure 57a) (Ref. 38). This result is sensible since continuity is just being established, and any perturbation which would increase contact-to-contact distances would drastically inhibit electron transport. When the geometric changes and the changing particle array were considered via a reduced resistance plot, a monotonically increasing linear function was observed for 8 and 10% (Figure 57b). Similar plots for stress instead of strain were markedly nonlinear, suggesting that, for transducers, the resistance or resistivity versus strain plot is a better choice. Unfortunately, varying the strain rate from 0.01 to 2.0 min^{-1} for a 10% by weight composite nearly doubled the resistance at every strain. For low strain rates, recovery was complete; for high strain rates, dewetting and chain tearing prevented total realignment. Thus, not only the concentration of the carbon black but also its morphology and structure influence the strain and time dependence of the electrical resistivity.

Transparent Composites

Because transparent conductive composites offer a unique combination of properties, their applications can be somewhat specialized, too. Transparent composites have been investigated by the telecommunications industry and the photocopying industry.

Often in televised sports broadcasts and preschool children's programs an overlay image is desired. The movements of the players, the location of a ball, or the association of two objects may be highlighted by passing a special pen over the TV monitor equipped with an osmium-coated screen. Unfortunately these screens are not only expensive but also not very durable, sometimes lasting only a single session. If a volume resistivity of 10^0-10^1 ohm-cm could be attained while maintaining a light transmission of 20–30%, a more serviceable overlay screen might be had. One design which showed promise was formed from poly(methyl methacrylate) (PMMA) beads (589–833 μm) coated with 8–10% by volume nickel powder (3–7 μm). Using a 2.5/mm/diameter die, a monolayer of PMMA beads was compression molded at 3.5 MPa for 1/2 hr at 210°C. After both the glassy surface layer (⩽1% of the thickness) and the metal-rich subsurface layer (<10%) were polished off, the bulk structure was revealed (Figure 58a) (Ref. 182). The macrostructure resembled the polygonal, crystalline microstructure of a metal—only with the PMMA beads representing the metallic grains and the nickel powder decorating the grain boundaries. At higher magnification, optical microscopy revealed that significant light could be transmitted through

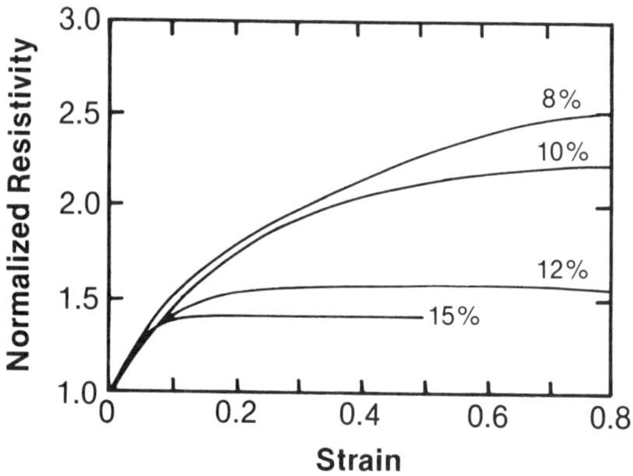

(a)

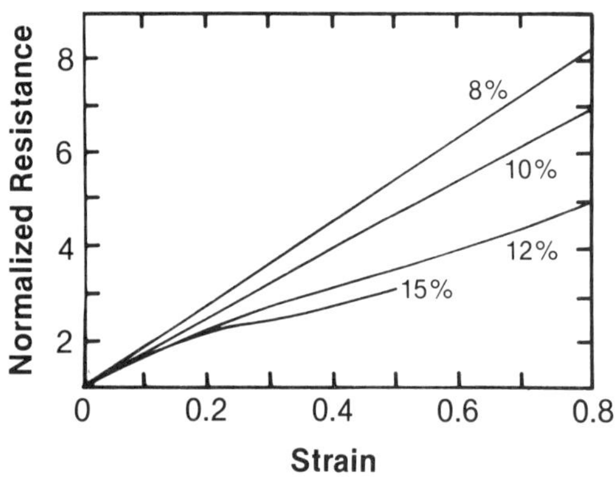

(b)

FIGURE 57 Dependence of normalized electrical properties on tensile strain. By compression-molding silicone rubber compacts having 6, 8, 10, 12, and 15% by weight of Ketjenblack EC, both (a) the greatest change in resistivity and (b) the most linear resistance behavior were found for the 8 and 10% loaded composites. (Adopted from Kost et al., Ref. 38.)

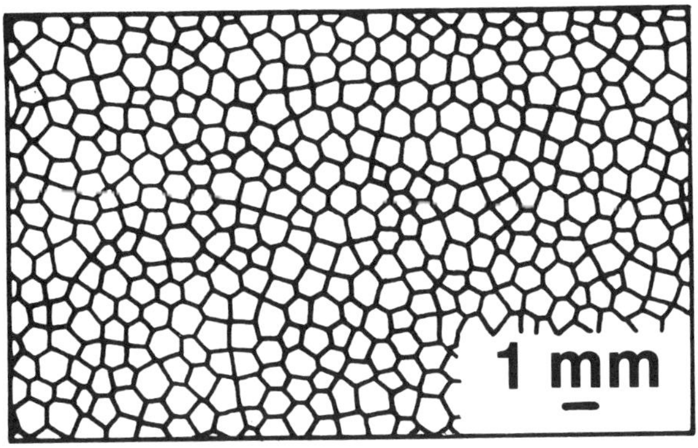

(a)

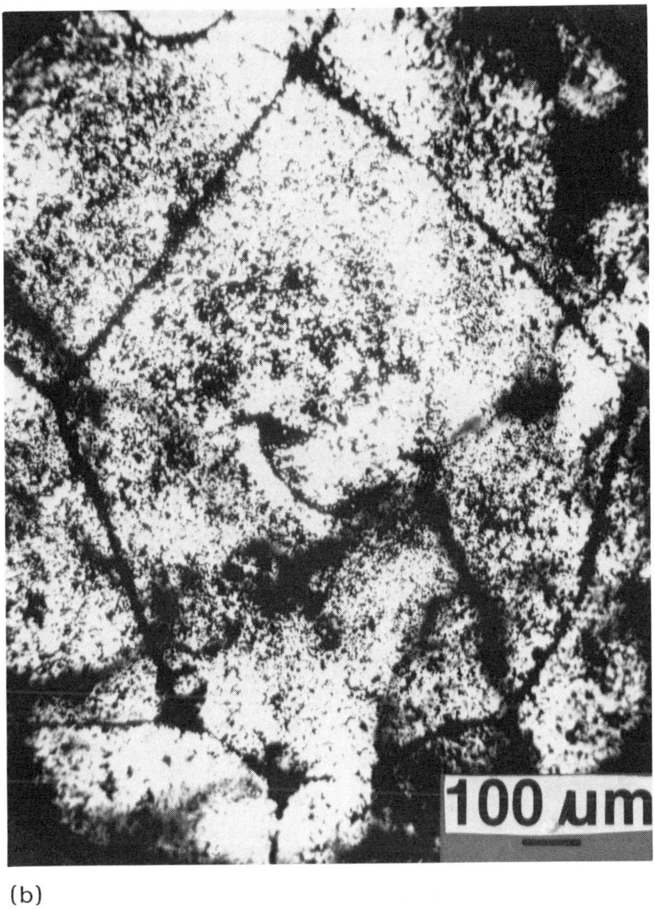

(b)

FIGURE 58 Morphology of a transparent poly(methyl methacrylate/8-10% nickel composite. (a) The transmission photomacrograph shows the bulk structure after polishing off both surfaces; (b) the transmission photomicrograph details the appearance of a single polymer bead. (Macrograph in (a) reproduced from Corneliussen and Kusy, Ref. 182.)

the cores of the polymer particles (Figure 58b) (Ref. 53). This was later quantified using a light meter. Thereby a more durable and economical overlay screen was feasible.

Transparent metal-filled composites have been proposed as photoconductors, too (Ref. 183). The photocopying industry has a particular interest in such a material for use as a particle migration imaging microfilm. The film (Figure 59) starts with a transparent metal electrode (e.g., aluminum) on top of a suitable substrate [e.g., poly(ethylene terephthalate)]. To this is added a thin polymer matrix [e.g., 1.5 μm-thick film of poly(styrene hexylmethacrylate)] and a monolayer of metal spheres (e.g., 0.3-μm selenium), which initially reside near the surface of the microfilm. With the transparent metal electrode connected to ground, the microfilm is sensitized by corona discharge. Exposure of the film causes photogeneration of electron-hole pairs in the metal spheres which repel under the influence of the applied electric field. Because charge transfer occurs within the metal-filled polymer, the metal particles acquire a net charge. The resulting *charged image* is developed by heating the microfilm above its glass transition temperature. As the intrinsic viscosity of the polymer matrix decreases, the charged metal spheres which were exposed to light are attracted toward the metal electrode while those not exposed to light remain stationary. The image is complete after the neutral metal spheres remaining on the surface of the microfilm are removed by solvation.

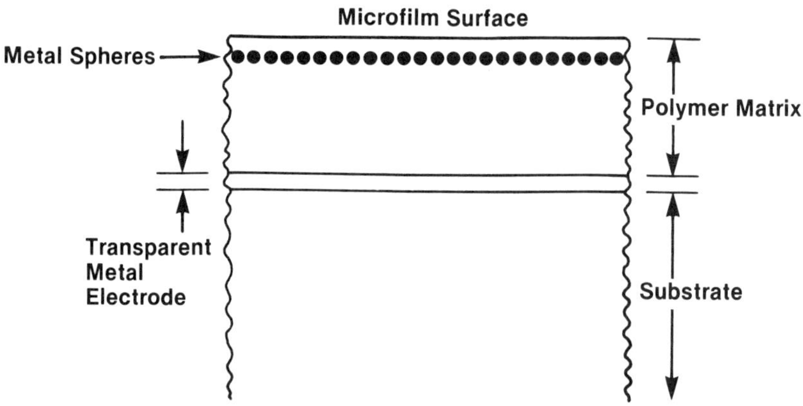

FIGURE 59 Schematic illustration of a transparent photoconductor: a particle micrography imaging microfilm. Selenium 0.3 μm in diameter and poly(styrene hexylmethacrylate) 1.5 μm thick have been successfully used as the metal spheres and the polymer matrix, respectively.

Although only two examples have been cited here, the use of transparent conductive composites will become more prevalent as the commercialization of conductive polymers grows. One application which is already being investigated is the polypyrrole-doped photovoltaic cell (Ref. 123). Some of the advantages of such a product over existing solar cells would be its greater durability and its ability to be manufactured in any configuration.

Batteries

Among the many applications of conductive polymers which look promising (Table 33), batteries are one of the most exciting. Today only about 1000 kg of poly-2-vinylpyridine complexed with iodine is used for this purpose, primarily in pacemaker batteries (Ref. 184). Within the next 5–10 years, however, a rechargeable battery should become available, perhaps based upon either polyacetylene or poly(*p*-phenylene) (Ref. 185). Such cells could come in any conceivable shape and would occupy only 33% of the volume and be only 10% of the weight of conventional lead-acid batteries. Moreover, because each conductive polymer cell can store three times more energy than a conventional cell, a conductive polymer battery of comparable size

TABLE 33 Some Applications of Conductive Polymers

Batteries
Charged microgel polymers to dissipate static electricity
Chemoselective electrodes
Durable synthetic materials to replace metals
Electrophotographic devices
Fuel cell components
High-temperature superconductivity
Photoresist elements
Rectifiers
Schottky barrier diodes
Sensors
Shielding
Solar energy conversion elements
Switching devices

would store 10 times more charge and could discharge at 10 times the present rate. Undoped polyacetylene is available in flexible sheets, but is unstable in air. In constrast, virgin poly(*p*-phenylene) is stable to 400°C but can only be fabricated at high pressures. When doped with cations (e.g., Li^+) or anions (e.g., AsF_6^-), the resisivity drops by at least 12 orders of magnitude, and both become unstable in air. When used within the confines of a battery case, instability is a less worrisome problem.

At present the successful development of "plastic batteries" requires that a compatible electrode/electrolyte combination be found (Figure 60). Two approaches are available: to use the same polymer for both electrodes and find an electrolyte that is stable to both the oxidized and reduced forms of the polymer, or to find different electrode combinations which match the oxidation and reduction potentials of the cell to the solvent's "voltage window"—that is, the difference between the solvent's oxidation and reduction reaction potentials.

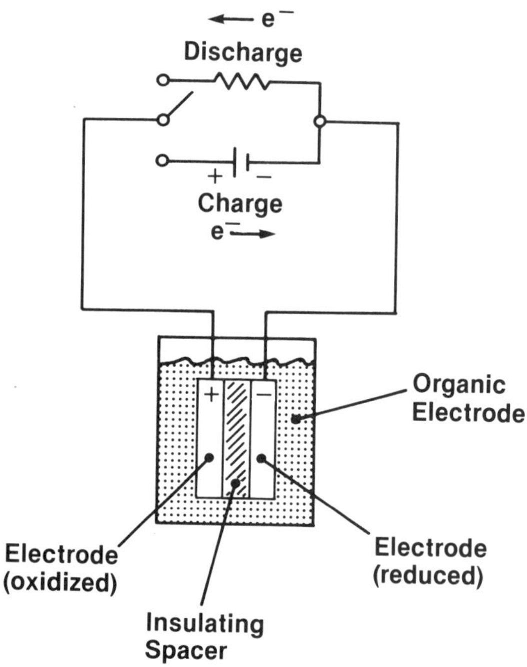

FIGURE 60 Schematic illustration of a battery. The future development of practical plastic batteries requires that compatible electrode/electrolyte combinations be found.

In the first case both electrodes can be polyacetylene or poly(*p*-phenylene). However, the presence of impurities and the instability of the reduced polymer in the electrolyte solvent must be eliminated if the shelf life of the battery is not to be curtailed by self-discharging. Alternatively, substituting another conductive polymer for one of the electrodes or a lithium or lithium/aluminum amalgam as the anode may satisfy the electrolyte solvent's voltage window but foster recharging problems in hybrid electrode pairs. Nevertheless, two hybrid pairs have been investigated, lithium/polyacetylene (Refs. 186, 187) and lithium/poly(*p*-phenylene) (Refs. 123, 188).

Using an 80% *cis*-rich polyacetylene, a 0.1-mm-thick film was doped in a dry, inert atmosphere containing lithium perchlorate (electrolyte) and propylene carbonate (solvent). Doping of the polyacetylene with the perchlorate ion was accomplished by imposing a 1-volt potential difference. After 30 min, the battery was simultaneously doped and charged to the composition $[CH(ClO_4)_{0.06}]_n$. The open-circuit voltage was 3.7 volts; the short-circuit current was 25 milliamperes. Because the mass of the film was small, the current dropped to 4 milliamperes in about 30 sec.

In the other hybrid battery, a hexafluorophosphate-doped poly(*p*-phenylene) served as the cathode and lithium dissolved in aluminum foil functioned as the anode. Once again the solvent was propylene carbonate, but the electrolyte was lithium hexafluorophosphate. An open-circuit voltage of 4.4 volts and current densities as high as 50 milliamperes/cm^2 were observed. Efficiencies of 91% charge/discharge and theoretical energy densities of 320 watt-hr/kg were measured. These last two parameters were greater than could be achieved with either lithium-doped polyacetylene/perchlorate-doped polyacetylene or potassium-doped poly(*p*-phenylene)/arsenic pentafluoride-doped poly(*p*-phenylene) electrode combinations.

Two further developments are worthy of mention which manipulate one or more of the parameters which determine the performance of a battery: cell configuration, dopant ions, electrodes, electrolytes, and their concentration, and solvents. One of these concerns the electrolyte itself. *Fast ion* conducting electrolytes are composed of a polymer [polyethylene oxide (PEO)] and either an alkali metal perchlorate (e.g., $LiClO_4$ or $NaClO_4$) or an alkali metal thiocyanate (e.g., NaSCN or KSCN) complex. When these are used as solid-state electrolytes in batteries, thin stable sheets can be readily constructed which have electrical conductivity of 10^{-4} ohm^{-1} cm^{-1} and power densities of about 100 watt/l. For these systems the mechanism of both anion and cation conduction appears to be interstice-interstice hopping versus main chain hopping. Work by Fontanella and coworkers (Refs. 189–191) on the solid-state electrolyte $PEO_8-LiClO_4$ shows that electrical conductivity decreases with increasing pressure as vacant holes are "squeezed out" and ion mobility is reduced. It was further reasoned that, if main chain hopping was the mechanism, the conductivity would increase since the chains would be closer together. In addition to high-energy-density batteries, fast ion conductors may find applications as fuel cells and electrolytic cells.

In another approach a polymer battery has been sought which uses water as the solvent. In the literal sense, research by MacDiarmid et al. (Ref. 100) has indicated that a century-old ink, polyaniline, may be converted into a stable, crystalline, conductive polymer. This has been accomplished by treating a solution of aqueous hydrochloric acid and aniline with ammonium peroxydisulfate. Further acid washing of the powder completed the protonation. Chemical analysis of the thoroughly dried dark blue-black powder indicated that each monomer unit now contained one acid molecule. In practice, polyaniline and neutral water would substitute for the lead and sulfuric acid which are used in the conventional battery. Both the convenience of using water as the electrolyte and the economy of the polymer conversion make this approach attractive.

In the future more structure-property data will be required if these so-called synmetals are to be better understood. This includes such fundamental properties as the molecular weight distribution, degree of cross-linking, extent of branching, effect of end groups, and influence of residual catalysts. Practical issues which might lead to loss in performance or battery failure, such as the extent of postpolymerization reactions, electrical hysteresis, electrical fatigue, and polymer aging, also require investigation.

SUMMARY

This chapter has surveyed the applications of various metal- or metalloid-filled polymers. A brief history of composites was presented along with the typical properties of particulate- and fiber-reinforced composites. While particulate-filled composites increase the modulus of elasticity, only fiber-reinforced composites also increase the tensile and impact strengths. Applications of carbon-fiber and glass-filled composites as well as other material combinations were discussed. The characteristics and properties of metalloid-filled polymers, metal-filled polymers, and conductive polymers were then reviewed. In the case of carbon fillers, the importance of particle size, surface area, and high structure were indicated. With regard to metal-loaded composites the influence of particle size ratios, shape, oxide coatings, processing parameters, etc. were discussed. Using laboratory-generated composites, some of the more important properties were highlighted: electrical resistivity, thermal conductivity, elastic modulus, tensile strength, and elongation. The state of the art of conductive polymers was summarized, indicating the general families which have been discovered, the three major processing techniques which have been used, and the properties which some of these materials possess. In the last regard, the influences of dopant level, doping agent, complexed-metal ion, anisotropy, reaction time, and polymer chain length were all considered. The remaining portion of the chapter considered the general properties of, various forms of, and specific applications of metal-filled polymers. In one table the properties of 10 polymer/metal composite products were surveyed; in one figure the mechanical, physical, thermal, and electrical properties of four commercial products were illustrated as a function of filler loading. Metal-filled polymers were then subdivided into 10 common forms: bulk materials; laminations; foams; caulking compounds; sealants; adhesives and solders; tapes; paints, coatings, and lacquers; greases; and powders. Typical properties and applications were cited. Finally, 10 specific applications were presented in detail, 5 of which were more traditional or established (heating element, radiation shielding, aesthetic, magnetic, and electrostatic discharge applications) and 5 of which were current or novel applications (electromagnetic/radio-frequency interference shielding, switches, transducers, transparent composites, and batteries). Having completed this chapter, the reader is prepared to understand in more detail the nature of the fillers, the electrical conduction theories, the mechanical reinforcement theories, the nature of the mesophase, and nonmetallic composites. These topics will be presented in the remaining five chapters.

NOMENCLATURE

dB	decibels, i.e., the unit of measure of electric or magnetic energy attentuation which indicates shielding effectiveness (SE)
GFRP	graphite fiber-reinforced polymers
I_x/I_0	intensity ratio, i.e., the attentuation which results when a shielding material is placed between an electromagnetic energy source and an intensity-measuring device.
K_C	thermal conductivity (K) of composite (C) in W/(m-K)
K_P	thermal conductivity (K) of polymer (P) in W/(m-K)
L/D	aspect ratio, i.e., the quotient of length (L) divided by the diameter (D) of a fiber
M_n	number average molecular weight
n	polymer chain length
NTC	negative temperature coefficient
PTC	positive temperature coefficient

R_m	effective radius (R) of a metal (or metalloid) particle (m)
R_p	effective radius (R) of a polymer particle (P)
SE	shielding effectiveness in decibels (dB)
V	volume percent loading of metal (or metalloid) particles in a composite
V_c	critical volume percent loading of metal (or metalloid) particles which is required for continuity
V/L	voltage gradient, i.e., the potential difference (V) across a given length (L)
x	thickness of a radiation shield
Z	atomic number
ρ	electrical volume resistivity (ohm-cm)
σ	electrical volume conductivity (ohm^{-1} cm^{-1})
$-\mu$	absorption coefficient of a material

REFERENCES

1. R. Buchanan, Am. Lab. April: 56 (1983).
2. M. R. Urist, *Fundamental and Clinical Bone Physiology*, Lippincott, Philadelphia (1980).
3. W. L. Jongebloed, I. Molenaar, and J. Arends, JEOL News, *13e*(2): 14 (1976).
4. L. H. Van Vlack, in *Elements of Materials Science*, Addison-Wesley, Reading, Mass. (1964), Chap. 11.
5. W. V. Titow and B. J. Lanham, *Reinforced Thermoplastics*, Wiley, New York (1975).
6. J. E. Theberge, Recent Product Advances in Thermplastic Composites, LNP Corporation Report.
7. D. B. V. Parker, in *Fillers for Plastics* (W. C. Wake, ed.), Iliffe, London (1971), p. 129.
8. H. L. Thomas, Ind. Res. Dev., *23*(5): 141 (1981).
9. R. G. Quynn, High Technol., *1*(1): 50 (1981).
10. Met. Prog., *113*(5): 58 (1978).
11. J. Aveston, R. A. Mercer, and J. M. Sillwood, Fibre-reinforced Cements—Scientific Foundations for Specifications: Composites—Standards Testing and Design, National Physical Laboratory Conf. Proc. (April 1974), p. 93.

12. *Designing with Fibre Reinforced Materials* (I. Mech. E. Conference Publications, 1977–1979), Mechanical Engineering Publications, London (1977).

13. W. J. Fletcher and P. R. Tully, Proc. 23d ANTEC SPE, p. 537.

14. S. Sahu and L. J. Broutman, Polym. Eng. Sci., *12*(2): 91 (1972).

15. P. K. Mallick and L. J. Broutman, Mater. Sci. Eng., *18*: 63 (1975).

16. R. Heimel, L. Zichner, and K. Müller, in *Transplantalager und Implantatlager bei verschiedenen Operationsverfahren*, Springer-Verlag, Berlin (1980).

17. R. D. Tetik, J. O. Galante, and W. Rostoker, J. Biomed. Mater. Res., *8*: 231 (1974).

18. R. M. Pilliar, R. Blackwell, I. Macnab, and H. U. Cameron, J. Biomed. Mater. Res., *10*: 893 (1976).

19. Chem. Eng. News, *59*(24): 28 (1981).

20. *Fibre Cements and Fibre Concretes* (D. J. Hannant, ed.), Wiley, New York (1978).

21. Met. Prog., *117*(1): 38 (1980).

22. Met. Prog., *107*(6): 66 (1975).

23. Met. Prog., *127*(1): 46 (1985).

24. F. M. Lewis, Rubber Chem. Technol., *35*: 1253 (1962).

25. E. M. Dannenberg, in *Kirk-Othmer Encyclopedia of Chemical Technology*, 3d ed., Vol. 4, Wiley-Interscience, New York (1978), pp. 631–666.

26. B. B. S. T. Boonstra and E. M. Dannenberg, Ind. Eng. Chem. *46*(1): 218 (1954).

27. F. A. Heckman, unpublished work.

28. A. I. Medalia and F. A. Heckman, J. Colloid Interface Sci., *36*: 173 (1971).

29. A. I. Medalia and F. A. Heckman, Carbon, *7*: 567 (1969).

30. A. S. Blanchard, in *Applied Science of Rubber* (W. J. S. Naunton, ed.), Edward Arnold, London (1961).

31. T. Emma, GTE Laboratories, Waltham, Mass.

32. L. K. H. van Beek and B. I. C. F. van Pul, Carbon, *2*: 121 (1964).

33. E. O. Forster, IEEE Trans. Power Appar. Syst., *PAS-90*: 913 (1971).

34. J. A. Manson and L. H. Sperling, in *Polymer Blends and Composites*, Chap. 10, Plenum, New York (1976).

35. W. F. Verhelst, K. G. Wothuis, A. Voet, P. Erhburger, and J. B. Donnet, Rubber Chem. Technol., *50*: 735 (1977).

36. E. K. Sichel, J. I. Gittleman, and P. Sheng, Phys. Rev. B., *18*: 5712 (1978).

37. E. K. Sichel, J. I. Gittleman, and P. Sheng, in *Carbon Black-Polymer Composites: The Physics of Electrically Conducting Composites* (Plastics Engineering, Vol. 3) (E. K. Sichel, ed.), Dekker, New York (1982), p. 53.

38. J. Kost, M. Narkis, and A. Foux, Polym. Eng. Sci., *23*(10): 567 (1983).

39. M. Narkis, A. Ram, and F. Flashner, Polym. Eng. Sci., *18*(8): 649 (1978).

40. D. M. Bigg, private communication.

41. R. M. Simon, in *Conductive Polymers* (Polymer Science and Technology, vol. 15) (R. B. Seymour, ed.), Plenum, New York (1981), p. 54.

42. J.-P. Reboul, in *Carbon Black-Polymer Composites: The Physics of Electrically Conducting Composites* (Plastics Engineering, vol. 3) (E. K. Sichel, ed.), Dekker, New York (1982), p. 81.

43. P. Datta and R. N. Friel, Polym. Prepr., Am. Chem. Soc. Div. Polym. Chem., *24*(1): 193 (1983).

44. B. Jachym, Z. Dreger, K. Kozlowski, and T. Stupkowski, Acta Polym., *30*: 10, 616 (1979).

45. H. Kawamoto, in *Carbon-Black Polymer Composites: The Physics of Electrically Conducting Composites* (E. K. Sichel, ed.), Dekker, New York (1982), p. 135.

46. Hochberg, Method of Making Electrically Conductive Polystyrene Articles, U.S. Patent 2,721,357 (1955).

47. Coler, Manufacture of Conductive Plastics, U.S. Patent 2,761,854 (1956).

48. R. P. Kusy, J. Appl. Phys., *48*(12): 5301 (1977).

49. J. Gurland, in *Plansee Proc.* (F. Benesovsky, ed.), Springer, Vienna (1962), p. 507.

50. J. Gurland, Trans. Metall. Soc., AIME, *236*: 642 (1966).

51. S. Kirkpartrick, Rev. Mod. Phys., *45*(4): 574 (1973).

52. S. K. Bhattacharya and A. C. D. Chaklader, Polym.-Plast. Technol. Eng., *19*(1): 21 (1982).

53. R. P. Kusy, unpublished work.

54. R. P. Kusy and D. T. Turner, Nature *229*(2): 58 (1971).

55. A. Malliaris and D. T. Turner, J. Appl. Phys., *42*(2): 614 (1971).

56. J. W. Cahn, Acta Metall., *14*: 477 (1966).

57. F. Bueche, J. Appl. Phys., *43*: 4837 (1972).

58. S. K. Bhattacharya, S. Basu, and S. K. De, Composites, *9*(3): 177 (1978).

59. J.-I. Yamaki, O. Maeda, and Y. Katayama, Rev. Electr. Commun. Lab., *26*(3–4): 616 (1978).

60. A. Kusy, E. Listkiewicz, and D. Plis, Arch. Electrotech., *31*: 327 (1982).

61. Conductive Composites for EMI Shielding: An Emerging Technology. A Proposed Research Program, Proposal No. 542-H-8348, Battelle-Columbus Laboratories (1978).

62. S. K. Bhattacharya, Polymer, *20*: 1166 (1979).

63. *Handbook of Engineering Fundamentals* (O. W. Esbach and M. Souders, eds.), Wiley, New York (1975), p. 604.

64. V. Ye Gul', and M. G. Golumbera, Kolloidn, Zh., *29*(1): 62 (1967).

65. R. P. Kusy, Ph.D. dissertation, Dept of Metallurgical Engineering, Drexel University, Philadelphia (1973).

66. Data of Aluminum Company Research Laboratories, in J. Delmonte's *Metal-Filled Plastics*, Reinhold, New York (1961), p. 68.

67. J. E. Scheer, B. S. thesis, Dept. of Metallurgical Engineering, Drexel Univeristy, Philadelphia, 1969.

68. S. K. Bhattacharya and A. C. D. Chaklader, Polym.-Plast. Technol. Eng., *20*(1): 35 (1983).

69. R. P. Kusy and R. D. Corneliussen, Polym. Eng. Sci., *15*(2): 107 (1975).

70. F. F. T. de Araujo, K. W. Garrett, and H. M. Rosenberg, International Conference on Composite Materials, Geneva (April 1975).

71. R. E. Meredith and C. W. Tobias, J. Appl. Phys., *31*: 1270 (1960).

72. D. M. Bigg and E. J. Bradbury, in *Conductive Polymers* (Polymer Science and Technology, vol. 15) (R. B. Seymour, ed.) Plenum Press, New York (1981), p. 23.

73. R. J. Gordon, M. S. dissertation, Dept. of Materials Engineering, Drexel University, Philadelphia, 1982.

74. R. P. Kusy and D. T. Turner, SPE J., *29*(7): 56 (1973).

75. S. K. Bhattacharya, S. Basu, and S. K. De, J. Mater. Sci., *13*: 2109 (1978).

76. D. M. Bigg, Composites, April: 95 (1979).

77. H. R. Simonds and J. M. Church, *A Concise Guide to Plastics*, Reinhold, New York (1964).

78. L. A. Igonin, Yu V. Orchinnikov, and S. A. Arzhakov, Dokl. Akad. Nauk SSSR. *120*: 1062 (1958).

79. J. Delmonte, in *Handbook of Fillers and Reinforcements for Plastics* (H. S. Katz and J. V. Milewski, eds.), Van Nostrand Reinhold, New York (1978), p. 185.

80. V. Ye Gul', *Conducting Polymer Materials*, English Translation JPRS 48359, Joint Publications Research Service, Department of Commerce (July 1969).

81. A. G. MacDiarmid and A. J. Heeger, Synth. Metals, *1*: 101 (1979).

82. R. H. Baughman, J. L. Bredas, R. R. Chance, R. L. Elsenbaumer, and L. W. Shacklette, Chem. Rev., *82*(2): 209 (1982).

83. B. Wasserman, M. S. Dresselhaus, and G. E. Wnek, Polym. Prepr., Am. Chem. Soc. Div. Polym. Chem., *25*(2): 286 (1984).

84. *Chemistry and Physics of One-Dimensional Metals* (H. J. Keller, ed.), Plenum, New York (1977).

85. T. J. Marks, Ann. N.Y. Acad. Sci., *313*: 594 (1978).

86. Synthesis and Properties of Low-Dimensional Materials (J. S. Miller and A. J. Epstein, eds.), Ann. N.Y. Acad. Sci., *313* (1978).

87. *Highly Conducting One-Dimensional Solids* (J. T. Devreese, V. E. Evrard, and V. E. Van Doren, eds.), Plenum, New York (1979).

88. *Molecular Metals* (W. E. Hatfield, ed.), Plenum, New York (1979).

89. J. B. Torrance, Acc. Chem. Res., *12*: 79 (1979).

90. H. Shirakawa and S. Ikeda, Polym. J., *2*: 231 (1971).

91. H. Shirakawa, T. Ito, and S. Ikeda, Polym. J., *4*: 460 (1973).

92. T. Ito, H. Shirakawa, and S. Ikeda, J. Polym. Sci. Chem., *12*: 11 (1974).

93. T. Ito, H. Shirakawa, and S. Ikeda, J. Polym. Sci. Chem., *13*: 1943 (1975).

94. H. Shirakawa, E. J. Louis, A. G. MacDiarmid, C. K. Chiang, and A. J. Heeger, J. Chem. Soc. Chem. Commun., 578 (1977).

95. G. E. Wnek, J. C. W. Chien, F. E. Karasz, M. A. Druy, Y. W. Park, A. G. MacDiarmid, and A. J. Heeger, J. Polym. Sci. Polym. Lett., *17*: 779 (1979).

96. W. Deits, P. Cukor, and M. Rubner, in *Conductive Polymers* (Polymer Science and Technology, vol. 15) (R. B. Seymour, ed.), Plenum, New York (1981), p. 171.

97. J. C. W. Chien, Polym. Prepr., Am. Chem. Soc. Div. Polym. Chem., *25*(2): 262 (1984).

98. C. E. Carraher, Jr., R. J. Linville, T. A. Manek, H. S. Blaxall, J. R. Taylor, and L. P. Torre, in *Conductive Polymers* (Polymer Science and Technology, vol. 15) (R. B. Seymour, ed.), Plenum, New York (1981), p. 82.

99. G. E. Wnek, J. C. W. Chien, F. E. Karasz, and C. P. Lillya, Polymer, *20*: 1441 (1979).

100. Ind. Chem. News, *5*(12): 5 (1984).

101. D. Bhaumik and J. E. Mark, Polym. Prepr., Am. Chem. Soc. Div. Polym. Chem., *25*(2): 266 (1984).

102. M. J. Rice, Phys. Lett., *71A*: 152 (1979).

103. W. P. Su, J. R. Schrieffer, and A. J. Heeger, Phys. Rev. B., *22*: 2099 (1980).

104. Y. Tomkiewicz, T. D. Schultz, H. B. Broom, T. C. Clarke, and G. B. Street, Phys. Rev. Lett., *43*: 1532 (1979).

105. J. A. Pople and S. H. Walmsley, Mol. Phys., *5*: 15 (1962).

106. K. Mortensen, M. L. W. Thewalt, Y. Tomkiewicz, T. C. Clarke, G. B. Street, Phys. Rev. Lett., *45*(6): 490 (1980).

107. M. Rolland, M. Aldissi, P. Bernier, M. Cadene, and F. Schue, Nature, *294*: 60 (1981).

108. J. K. Jeszka, J. Ulanski, and M. Kryszewski, Nature, *289*: 390 (1981).

109. Conductive Polymers and Plastic Compositions: Prospectus for a Multiple-Client Study, Skeist Laboratories, Inc. (1983).

110. K. K. Kanazawa, A. F. Diaz, R. H. Geiss, W. D. Gill, J. F. Kwak, J. A. Logan, J. F. Rabolt, and G. B. Street, J. Chem. Soc. Chem. Comm., 854 (1979).

111. T. Yamamoto, K. Sanechika, A. Yamamoto, J. Polym. Sci., Polym. Lett., *18*: 9 (1980).

112. H. W. Gibson, F. C. Bailey, J. M. Pochan, A. J. Epstein, and H. Rommelmann, Org. Coatings Plastics Chem., *42*: 603 (1980).

113. D. M. Ivory, G. G. Miller, J. M. Sowa, L. W. Shacklette, R. R. Chance, and R. H. Baughman, J. Chem. Phys., *71*: 1506 (1979).

114. R. H. Baughman, D. M. Ivory, G. G. Miller, L. W. Shacklette, and R. R. Chance, Org. Coatings Plastics Chem., *41*: 139 (1979).

115. C. K. Chiang, Y. W. Park, A. J. Heeger, H. Shirakawa, E. J. Louis, and A. G. MacDiarmid, J. Chem. Phys., *69*: 5098 (1978).

116. R. H. Baughman, J. L. Bredas, R. R. Chance, H. Eckhardt, R. L. Elsenbaumer, D. M. Ivory, G. G. Miller, A. F. Preiziosi, L. W. Shacklette, in *Conductive Polymers* (Polymer Science and Technology, vol. 15) (R. B. Seymour, ed.), Plenum, New York (1981), p. 137.

117. G. W. Wnek, J. W. Capistran, J. C. W. Chien, L. C. Dickinson, R. Gable, R. Gooding, K. Gourley, F. E. Karasz, C. P. Lillya, and K.-D. Yao, in *Conductive Polymers* (Polymer Science and Technology, vol. 15) (R. B. Seymour, ed.), Plenum, New York (1981), p. 183.

118. H. Yau and S. I. Stupp, Polym. Prepr., Am. Chem. Soc. Div. Polym. Chem., *25*(2): 276 (1984).

119. S. I. Yaniger, M. J. Kletter, and A. G. MacDiarmid, Polym. Prepr., Am. Chem. Soc. Div. Polym. Chem., *25*(2): 264 (1984).

120. R. R. Chance, L. W. Shacklette, G. G. Miller, D. M. Ivory, J. M. Sowa, R. L. Elsenbaumer, and R. H. Baughman, J. Chem. Soc. Chem. Commun., 348 (1980).

121. J. F. Rabolt, T. C. Clarke, K. K. Kanazawa, J. R. Reynolds, and G. B. Street, J. Chem. Soc. Chem. Commun., 347 (1980).

122. C. W. Dirk, K. F. Schoch, and T. J. Marks, in *Conductive Polymers* (Polymer Science and Technology, vol. 15) (R. B. Seymour, ed.), Plenum, New York (1981), p. 209.

123. Chem. Eng. News, *60*(16): 29 (1982).

124. T. Ogawa, R. Cedeno, T. E. Herrera, B. Almaraz, and M. Inoue, in *Conductive Polymers* (Polymer Science and Technology, vol. 15) (R. B. Seymour, ed.), Plenum, New York (1981), p. 85.

125. R. R. Chance, L. W. Shacklette, H. Eckhardt, J. M. Sowa, R. L. Elsenbaumer, D. M. Ivory, G. G. Miller, and R. H. Baughman, in *Conductive Polymers* (Polymer Science and Technology, vol. 15) (R. B. Seymour, ed.), Plenum, New York (1981), p. 125.

126. L. W. Shacklette, H. Eckhardt, R. R. Chance, G. G. Miller, D. M. Ivory, and R. H. Baughman, in *Conductive Polymers* (Polymer Science and Technology, vol. 15) (R. B. Seymour, ed.), Plenum, New York (1981), p. 115.

127. C. W. Dirk, J. W. Lyding, K. F. Schoch, Jr., C. R. Kannewurf, and T. J. Marks, Org. Coatings Plastics Chem., *43*: 646 (1980).

128. D. R. Gagnon, J. D. Capistran, F. E. Karasz, R. W. Lenz, Polym. Prepr., Am. Chem. Soc. Div. Polym. Chem., *25*(2): 284 (1984).

129. A. J. Epstein, H. W. Gibson, P. M. Chaikin, W. G. Clark, and G. Gruner, Phys. Rev. Lett., *45*: 1730 (1980).

130. A. J. Epstein, H. W. Gibson, P. M. Chaikin, W. G. Clark, and G. Gruner, Chem. Scr., *17*: 135 (1981).

131. D. A. Upson and G. A. Campbell, in *Conductive Polymers* (Polymer Science and Technology, vol. 15) (R. B. Seymour, ed.), Plenum, New York (1981), p. 71.

132. Product Data, Wilson-Fiberfil International.

133. Product Data, RTP Co.

134. D. M. Bigg and D. E. Stutz, Ind. Res. Dev., *21*(7): 103 (1979).

135. J. Regan, Polym.-Plast. Technol. Eng., *18*(1): 47 (1982).

136. I. L. Kamel, in *Metals Handbook*, 9th ed., vol. 7, *Powder Metallurgy*, American Society for Metals (1984), p. 606.

137. M. Narkis, A. Marmur, J. Yacubowicz, and A. Vaxman, 40th Annual Conf., Reinforced Plastics/Composites Institute, Society of the Plastics Industry, Session 8-H: 1 (1985).

138. R. R. Singh and R. W. Roberts, Polym. Composites, *6*(1): 58 (1985).

139. R. M. Simon, in *Conductive Polymers* (Polymer Science and Technology, vol. 15) (R. B. Seymour, ed.), Plenum, New York (1981), p. 49.

140. Technical Bulletin: Cho-Therm Thermal Interface Products, Chomerics, Inc. (1983).

141. M. Narkis, A. Ram, and Z. Stein, Polym. Eng. Sci., *21*(16): 1049 (1981).

142. Technical Support Package, NASA Tech. Briefs, *5*(2): GSC-12553 (1980).

143. N. R. Lazear and W. J. Phillips, Polycarbonate/ABS Blends for EMI Shielding, Mobay Chemical Corporation.

144. Technical Bulletin: Eccoshield SV-F, Emerson & Cuming.

145. Product Data, Emerson & Cuming.

146. H. E. Chandler, Metal Prog., *126*(3): 19 (1984).

147. Technical Bulletin 3-2-5E: Eccobond Solder 70C, Emerson & Cuming, 1974.

148. A. Sternfield, Mod. Plast. Int., *12*: 48 (1982).

149. N. Peach, Plast. Eng., *40*:(4): 21 (1984).

150. Product Data, Chromerics, Inc.

151. P. C. Julien, *Carbon Black-Polymer Composites: The Physics of Electrically Conducting Composites* (Plastics Engineering, vol. 3) (E. K. Sichel, ed.), Dekker, New York (1982), p. 189.

152. D. T. Turner, A. Malliaris, and R. P. Kusy, Radiation Cross-linking of Polymers with Segregated Metallic Particles, Annual Report, June 1, 1969–May 31, 1970, U.S. Atomic Energy Commission, (January 1972), NYO-4110.

153. M. Narkis, Mod. Plast., May: 96 (1983).

154. M. Steinberg, private communication.

155. E. W. Collings, *The Physical Metallurgy of Titanium Alloys*, American Society for Metals, Metals Park, Ohio (1984), Chap. 1.

156. G. R. Henning, Colloquium on the Effects of Radiation on Materials, John Hopkins University, Baltimore (March 1957).

157. *Radiation Detection and Measurement* (G. F. Knoll, ed.), Wiley, New York (1979).

158. H. Henry, *Radiation Protection*, Wiley-Interscience, New York (1969), p. 485.

159. H. Blatz, *Introduction to Radiological Health*, McGraw-Hill, New York (1964).

160. Stanford Research Institute Rept. 361, The Industrial Uses of Radioactive Fission Products.

161. S. C. Gujrathi and J. M. D'Auria, Nucl. Instrum. Methods, *100*: 445 (1972).

162. R. P. Kusy and W. W. Ambrose, unpublished work.

163. Colloidal Dispersion of Metals in Plastics, U.S. Patent 2,947,646 (1960).

164. A. M. Litman and N. E. Fowler, Polym. Plast. Technol. Eng., *18*(1): 93 (1982).

165. Product Literature, Conductive Thermoplastics, RTP Co.

166. Video Disk, U.K. Patent 1,574,595 (1982).

167. R. M. Simon, Polym.-Plast. Technol. Eng., *17*(1): 1 (1981).

168. Federal Register, FCC Docket 20780, Part 15, Subpart J, 1979.

169. Promotional Literature, Radiation Testing Service, Chomerics, Inc. (1982).

170. S. Gerbig, Wilson-Fiberfil International, private communication.

171. A. Genova, Plastics Design & Processing, October/November (1982).

172. Staff Report, Plast. Technol., February: 42 (1985).

173. Electrical Shielding and Sealing Gasket, U.S. Patent 3,140,342 (1963).

174. Technical Bulletin No. 14, Chomerics, Inc. (1981).

175. D. T. Turner, A. Malliaris, and R. P. Kusy, Radiation Crosslinking of Polymers with Segregated Metallic Particles, Annual Report, June 1, 1970–May 31, 1971, U.S. Atomic Energy Commission (January 1973), COO-3035-1.

176. F. Bueche, J. Appl. Phys., *44*: 532 (1973).

177. F. Bueche, J. Polym. Sci. (Phys.), *11*:1319 (1973).

178. J. Meyer, Polym. Eng. Sci., *13*:462 (1973).

179. J. Meyer, Polym. Eng. Sci., *14*: 706 (1974).

180. M. Narkis, A. Ram, and Z. Stein, J. Appl. Polym. Sci., *25*: 1515 (1980).

181. Conductive Plastic Product, U.S. Patent 2,761,849 (1950).

182. R. D. Corneliussen and R. P. Kusy, 6th St. Louis Symposium on Composite Materials and Engineering Design (May 1972).

183. J. Y. Josefowicz and C. C. Yang, in *Conductive Polymers* (Polymer Science and Technology, vol. 15) (R. B. Seymour, ed.), Plenum, New York (1981), p. 95.

184. J. R. Ellis and R. S. Schotland, Polym. Prepr., Am. Chem. Soc. Div. Polym. Chem., *23*(1): 134 (1982).

185. Chem. Eng. News, *59*(41): 34 (1981).

186. P. J. Nigrey, A. G. MacDiarmid, and A. J. Heeger, J. Chem. Soc. Chem. Comm., 594 (1979).

187. P. J. Nigrey, D. MacInnes, Jr., D. P. Nairns, A. G. MacDiarmid, and A. J. Heeger, in *Conductive Polymers* (Polymer Science and Technology, vol. 15) (R. B. Seymour, ed.), Plenum New York (1981), p. 227.

188. R. L. Elsenbaumer, L. W. Shacklette, J. M. Sowa, R. R. Chance, D. M. Ivory, G. G. Miller, and R. H. Baughman, Polym. Prepr., Am. Chem. Soc. Div. Polym. Chem., *23*(1): 132 (1982).

189. J. J. Fontanella, M. C. Wintersgill, J. P. Calame, F. P. Pursel, D. R. Figueroa, and C. G. Andeen, Solid State Ionics, *9&10*: 1139 (1983).

190. J. J. Fontanella and M. C. Wintersgill, Technical Rept. 1, Office of Naval Research, Contract N00014-82-AF-00001 (1982).

191. J. J. Fontanella and M. C. Wintersgill, Technical Rept. 8, Office of Naval Research, Contract N00014-83-AF-00001 (1983).

2

Production Technology for Metallic Fillers

JOHN DELMONTE

Delsen Testing Laboratories, Inc.
Glendale, California

To understand the full significance of powder metal technology and the development of composites of metal powders and polymers, there are characteristics of metal powders which should be examined before combination with organic polymers. In general, one refers to metal powder as those particles possessing a size of 0.1 μm (100 nm) or larger, and these are the subject of this chapter.

Because particle sizes and shapes of metal powders, their surface chemistry, and their physical behavior influence their processing, it is appropriate to review some of the manufacturing methods for powdered metals (P/M) and their products. Attention will then be focused on commercially available P/M materials in combination with organic polymers. Considerable attention has been given to iron and its alloys because of their greater availability and lower cost.

Much of the physical data on the preparation and characteristics of finely powdered metals has been reported by members of the Metal Powder Industries Federation (MPIF) and American Powder Metallurgy Institute (Ref. 1). To assist the design and materials engineer, materials standards for structural parts have recently been issued (Ref. 2). To establish a frame of reference for the polymer/metal powder composites, some of the physical properties of P/M structural parts are presented in this chapter.

Available data on the quantities of metal powders produced in North America are reproduced below (Refs. 1, 2):

Iron and steel powder	1983 data on other metal powders
1965—60,000 tons/year	Copper and copper base—19,400 tons
1970—110,000 tons/year	Aluminum—28,000 tons
1975—150,000 tons/year	Molybdenum—1,900 tons
1984—216,000 tons/year	Tungsten—1,800 tons
	Nickel—12,700 tons
	Tin—800 tons

Applications of metal powders include:

High strength structural parts for automotive and aerospace components
High-wear-resistant surfaces for machine tools
High-temperature-resistant and wear-resistant parts for engine components
Load-transmitting components such as cams, gears, sprockets, worm gears, key ways, ratchets
Farm and garden equipment, where economies are realized through minimizing machining and the attainment of close tolerance parts
Controlled porosity metal parts, for application as bearings and in the manufacture of filtration devices

Applications of metal powder-filled polymers (which comprise powdered resins or liquid polymer systems) include:

Electrical conductive materials for electronic industries, including adhesive pastes and semiconductors.
Electromagnetic shielding for many electrical and electronic devices. New Federal Communications Commission (FCC) standards will be applicable to consumer devices and will entail conversion of moldable insulating plastics to semi-electrically conductive products through the addition of metallic fillers.
Radiation shielding associated with nuclear power devices.
Magnetic tapes for computer data storage and replication.
Magnetizable powders in liquid resin fluids which "freeze up" under applied magnetic fields.
Photocopier ink for office copying equipment.
High-performance parts that are lighter in weight than conventional metal, in the aerospace industry.

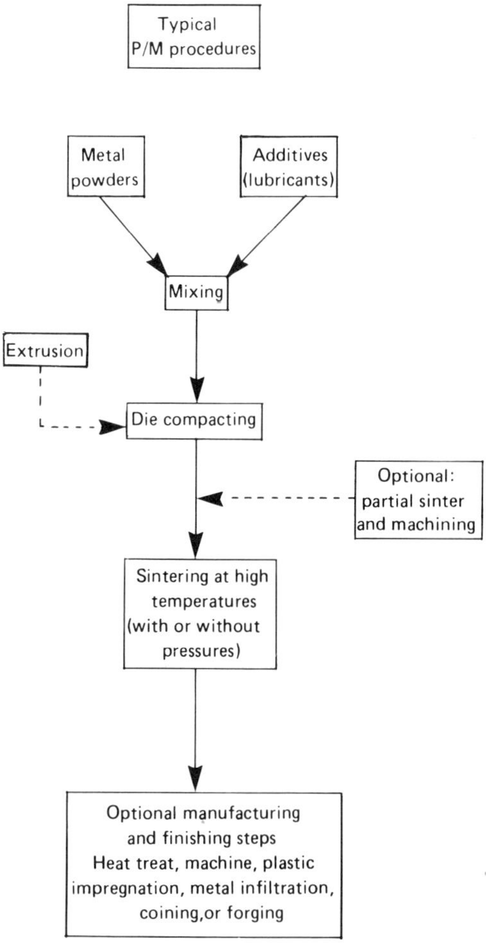

FIGURE 1 Procedures for powder metallurgy compared with metal-filled plastics.

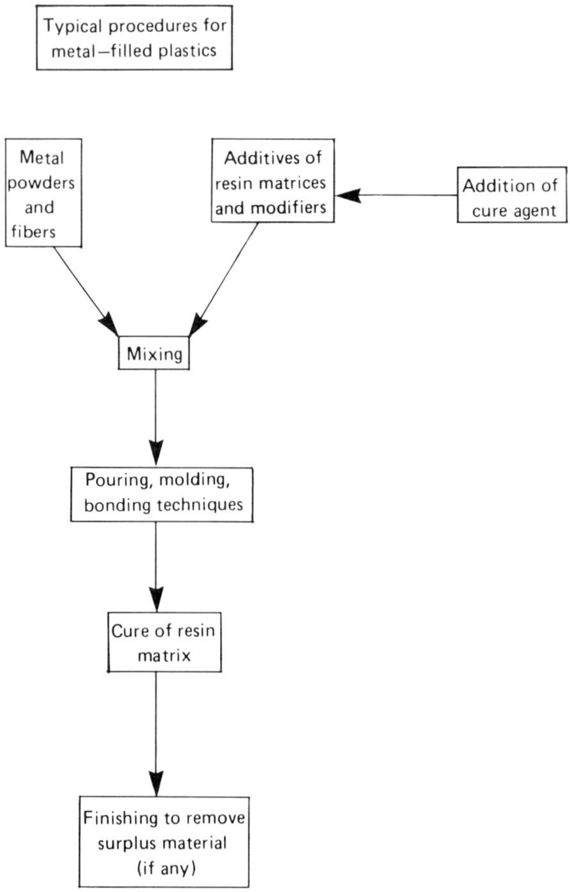

FIGURE 1 (continued)

See Figure 1 for a comparison of processes for P/M and plastic/metal composites.

METHODS FOR THE MANUFACTURE OF METAL POWDERS

Most metals and their alloys have been produced in powdered form. The characteristics of typical examples are noted in following sections on ASTM Standards and from bulletins of manufacturers of metal powders. The processes for preparing these powders are summarized below.

Mechanical attrition: Grinding and breaking up of the more brittle metals take place in combinations of ball mills, impact mills, gyratory mills, and other attaining fine particles. Screening and air classification will separate the particle sizes. A minute amount of contaminants from the grinding machines must be expected.

Atomization: Streams of molten metal are broken up into fine particles by hot gases or hot liquids impinging on the molten material. Argon gas atomization accompanied by gas recirculation and purification yields powder particles with an oxygen level less than 100 ppm.

Condensation of metal vapor: For some rare metals, condensation of vaporized metals, under vacuum and closely controlled temperatures, permits the collection of finely divided metal on a cooler surface.

Electrolytic deposition from solution: Electrodeposited metals or displacement due to difference in electropotential will form adherent masses of metal. More brittle dendritic growths may be broken up mechanically. The use of pure chemicals in solution permits attainment of metal powders of high purity.

Reduction of metal compounds: By reducing mostly oxide powders in solid or gaseous media, metal powders can also be prepared.

Centrifugal atomization: Molten metal is tangentially ejected as droplets from a rapidly spinning crucible, disk, etc., solidifying into particles while flying away in a gas or vacuum. In addition to the rotating electrode process, plasma rotating electrodes and electron beam rotating processes are used for superalloys (Ref. 5).

Decompositions of metal hydrides and metal carbonyls afford further processes for obtaining fine metal powders of high purity and uniform sphericity.

In the past few years electrically conductive fibers from both metals and graphite have become available. These fibers are relevant to the combinations of polymers and metal fillers because they not only bridge gaps between the particulate matter and enhance conductivity, but also contribute significantly to improvement in impact strength and to shear strength. Hence, consideration of metallic fillers includes the use of metal fibers (or graphite fibers and metallized fibers) as additives to particles in developing metal powder-filled polymers.

TYPICAL ASTM STANDARDS

There has been considerable effort on the part of many individuals and organizations to establish standard methods for evaluating and testing metal powders. Under the auspices of Committee B-9 of the American Society for Testing and Materials (ASTM), basic materials and testing specifications have been established (Ref. 3). These specifications provide useful data on the characterization of metal powders. Because there is growing interest in metal powders and their applications, particularly in combination with organic polymers, there are continuing efforts to develop data and guidelines which will assist industrial users to evaluate their development efforts. While in the earlier years of P/M products emphasis was placed on structural products and oil-impregnated bearings, the growing requirements of electronics industries have led to an emphasis on electrical conductivity and effective radiation shielding. In fact, the exigency of federal requirements has led to the promulgation of emergency standards for radiation shielding (ASTM-0-ES7-83).

For reference purposes, the following typical standards for metal powders are cited because they contribute to an understanding of the diverse roles of these materials (Ref. 3). A brief summation of the pertinent subject matter is tabulated after the specifications below. The materials standards for P/M structural parts (MPIF standard 35) have been issued by the Metal Powder Industries Federation for 1984–1985. They contain details on P/M material properties (Ref. 2).

ASTM-B-243-76: Definition of terms used in powder metallurgy (selected items pertinent to this chapter):

1. Atomized metal powder: Usually produced by dispersion of a molten metal by a rapidly moving gas or liquid stream or by mechanical means.
2. Carbonyl powder: A metal powder prepared by thermal decomposition of a metal carbonyl.
3. Chemical precipitated metal powder: Produced by the reduction of a metal from a solution of its salts either by the addition of another metal higher in the electomotive series or by other reducing agents.
4. Compact: An object produced by the compression of metal powder, generally while confined in a die, with or without non-metallic constituents.
5. Dendritic powder particles, usually of electrolytic origin, having a typical pine tree structure.
6. Electrolytic powder: Produced by electrolytic deposition or by pulverizing an electrodeposit.
7. Hydrogen loss: Loss in weight of metal powder or of a compact caused by heating a representative sample for a specified time and temperature in a purified hydrogen atmosphere—broadly, a measure of oxygen content of the sample (not applicable to hydride-forming elements).
8. Metal powder: Discrete particles of elemental metals or alloys, normally in the range of 0.1 to 1000 μm.
9. Porosity: The amount of pores (voids) expressed as a percentage of the total volume of the powder metallurgy part.

10. Prealloyed powder: Powder consisting of two or more elements which are alloyed in the powder manufacturing process and in which particles are of the same nominal composition throughout.
11. Sieve classification: The separation of powder into particle size ranges by the use of graded sieves.
12. Sinter: The bonding of adjacent surfaces of particles in a mass of powder, or a compact, by heating at temperatures lower than the melting point of the metal particles.

ASTM-B-282-70 (1976): Sintered brass structural parts prepared from brass alloys principally of 77.0 to 80.0% copper, 1.0 to 2.0% lead, and the balance zinc.

ASTM-B310 (1983): Sintered carbon steel structural parts prepared from three grades, containing uniformly distributed carbon (not case carbonized).

Grade	Minimum iron	Max. acid insolubles	Carbon content	Total other elements by differences (max.)
A	97.75	0.5	0.25 max.	1.50
B	97.40	0.5	0.26–0.60%	1.50
C	97.10	0.5	0.61–0.90%	1.50

Representative data, though not part of the specifications, are shown in Table 1 for a few selected examples.

ASTM-B426 (1983): Sintered copper steel structural parts in which there are controlled amounts of copper and carbon. (The presence of copper permits shrinkage control of the sintered compacts.) Chemical requirements are as given in Table 2.

ASTM-B-484-70 (1976): Sintered nickel steel structural parts are comprised of iron powders to which controlled amounts of nickel and carbon have been added.

Typical properties are included for reference, although they do not constitute part of the specification. A short part of the table is included indicating the importance of the density of the structural part and heat treatment (see Table 3). The heat treatment consists of oil quenching from 1500 to 1600°F (816 to 871°C) and tempering at 400–600°F (204–316°C).

The data in Table 3 are abstracted from the published data in ASTM-B-484 specifications and MPIF Standard 35 and illustrate the strengths that are realized. The influence of the density of the compact and the alloying effects of the nickel powders should be noted for these particular products. Nickel steel P/M products

TABLE 1

	Tensile strength				
	As sintered		Quenched and tempered		Percent elongation as sintered
	psi	MPa	psi	MPa	
	Type II, density 6.10–6.50 g/cm^3				
Grade A	20,000	138	—	—	3.0
Grade B	26,000	179	40,000	276	2.0
Grade C	34,000	234	50,000	345	0.5
	Type IV, density 6.9–7.3 g/cm^3				
Grade A	30,000	207	—	—	8.0
Grade B	50,000	345	70,000	482	4.0
Grade C	60,000	414	80,000	552	2.0

TABLE 2

	Composition (%)			
Element	Grade 1	Grade 2	Grade 3	Grade 4
Copper	1.0–2.5	2.5–6.0	6.0–11.0	18.0–22.0
Combined carbons	0.6–0.9	0.6–0.9	0.6–0.9	0.6–0.9
Iron, min.	94.6	91.1	86.1	75.1
Acid insolubles, max.	0.5	0.5	0.5	0.5
Total other elements by differences, max.	1.5	1.5	1.5	1.5

TABLE 3

	Typical properties		
	Tensile strength		
	psi	MPa	Elongation (%) in 25 mm
Grade 2			
B, density 6.4–6.8 g/cm^3	45,000	310	3.0
B, density 6.8–7.2 g/cm^3	62,000	427	4.5
B, density 7.2 g/cm^3 (min.)	74,000	510	6.0
Grade 2 (heat treated)			
B, density 6.4–6.8 g/cm^3	112,000	772	0.5
B, density 6.8–7.2 g/cm^3	154,000	1,062	0.5[a]
B, density 7.2 g/cm^3 (min.)	180,000	1,241	0.5[a]

[a]From MPIF Standard 35 (1985).

TABLE 4

	Composition (%)		
Element	Grade 1	Grade 2	Grade 3
Copper	0.5 max.	1.5–2.5	3.5–5.0
Magnesium	0.4–1.2	0.2–1.2	0.2–0.8
Silicon	0.2–0.8	0.5 max.	1.2 max.
Aluminum, min.	96.0	94.3	91.5
Total of other elements (max.)	1.5	1.5	1.5

represent strong structural units. The additions of organic polymers are made for other purposes. Furthermore, sintering and heat treatment of metals are beyond the temperature capability of organic polymers.

ASTM-B-595-73 (1979): Sintered aluminum structural parts. This covers parts made primarily from aluminum powders to which control amounts of elemental copper, magnesium, and silicon have been added by blending. Chemical requirements are given in Table 4.

Three types are established by their densities: type I, 2.30 to 2.4 g/cm^3; type II, 2.45 to 2.60 g/cm^3; and type III, 2.60 g/cm^3 min. Optimum tensile strengths after heat treatment of sintered aluminum structural parts vary between 30,000 and 40,000 psi and elongations of 2.0 to 4.0%.

TYPICAL COMMERCIAL POWDERS SELECTED FOR COMPARISON

Iron Powders (Hoeganaes Corporation)

High-strength P/M parts from iron powders are developed typically from Ancormet 101. When mixed with lubricant (zinc stearate is typical), good elongation and high strength are obtained with a small amount of shrinkage after sintering for 30 min at 1121°C in a disassociated ammonia atmosphere. For compacts prepared at different densities by pressing, the following data are reported by Hoeganaes.

Density g/cm^2	6.0	6.4	6.8
Ultimate tensile (R.T.) strength			
ksi	20	28	36
MPa	140	196	252
Elongation, %	5.6	7.5	10.0
Unnotched Charpy, ft. lb.	4.5	8.0	15.0
Size change, %, from die dimensions	−0.22	−0.20	−0.12

Small amounts of copper powder (typically 4.0 to 5.0%) and 0.90 to 1.0% graphite in combination with copper offer a good balance of properties. Copper powder along will decrease the shrinkage of P/M compacts during sintering, while in combination with graphite powder better strengths are achieved. Nickel powder will increase hardenability and impact strength. The addition of organic polymers prior to sintering is not recommended, as their thermal decomposition will lead to more porosity. The use of graphite fibers in lieu of graphite powder is an area unreported in the literature on P/M powders at this date.

Ancorsteel 1000B, comprising 99.3% iron powder, has the following screen analysis:

Sieve analysis	%
Deposited on −40 screen	0
−40/+60	Trace
−60/+100	12.9
−100/+200	37.0
−200/+325	25.4
−325/	24.7

Nickel Powders (INCO)

High-purity nickel powders are produced from nickel carbonyl. Three different types of nickel powders are produced:

1. Individual spherical particles, about 5 μm in diameter, with spiking surfaces—general-purpose grade.
2. Strings of spherical particles, 2 to 3 μm in diameter, also characterized by spiking surfaces. Filamentary particles are produced in varying densities.
3. Larger coarse particles of assorted shapes in the range of 10–20 μm diameter. This grade is identified as high-density nickel powder (HDNP).

The selection of the appropriate grade of powder is influenced by the nature of the applications. Some examples follow.

General-purpose grade: From 1.8 to 2.7 g/cm^3 is used in nickel alloy steel P/M parts, isostatic pressed or direct rolled into nickel strips. Also used for welding electrode coatings and filters. The chemical grade suitable for manufacture of nickel chemicals is processed through a 325 mesh sieve.

Low-density grades: Apparent densities in four types from 0.5 to 1.0 g/cm^3 permit the production of sintered nickel products of controlled porosity. End use applications for these products include nickel alkaline battery plates, fuel cell electrodes, and metal filters.

High-density grade: Apparent density of 3.6 to 4.4 g/cm^3; distinguished by its usefulness in milling with polymer resins to produce conductive plastics. It is also employed in manufacture of nickel chemicals.

Prealloyed Metal Powders (Stellite Division of Cabot Corporation)

Prealloyed powders, rather than powder blends, are atomized to ensure uniformity of particle sizes. The powders are produced in cobalt-, nickel-, and iron-base compositions to provide high-temperature service wear resistance and resistance to corrosion. A typical alloy, stellite No. 31, is compared in Table 5 with respect to casting and preparation of P/M parts. Data on the mean coefficient of thermal expansion are important in blends of powdered metals.

Prealloyed Metal Powders (Nuclear Metals, Inc.)

Clean, spherical, prealloyed metal powders are produced by a rotating electrode process (REP) in a ceramic-free environment. This processing detail ensures absence of contaminants in certain critical applications. Titanium and superalloy powders prepared in this

TABLE 5

	Stellite No. 31	
	P/M	Cast
Chemical composition (major components)	Nickel, 10.5%; chromium, 25.5%; manganese, 1%; iron, 2%; carbon, 0.5%; silicon, 1.0%; balance cobalt	
Ultimate tensile strength, ksi (MPa)		
At room temperature	120 (828)	108 (745)
At 1200°C	89 (614)	76 (521)
Elongation (%) in 25 mm		
At room temperature	4	9
At 1200°C	13	12
Mean coefficient of thermal expansion, 21–871°C (μm/m-°C)	16.6	
Density at 22°C (g/cm^3)	8.45	8.61

manner are used for high-quality aircraft engine parts. Particle sizes may be adjusted in the range of 500 to 44 μm.

The REP powders of low-carbon steel and aluminum are either coated with a polymer or oxidized and mixed with carbon black toners for dry developer systems for printed matter. The closely sized spherical characteristics provide a uniform dispersion of toner onto the electrostatically charged copy plate.

Teledyne (Los Angeles)

Teledyne produces a large number of specialty metals. Zirconium, a highly corrosion-resistant metal, is transparent to neutrons (low neutron capture cross section). It is used for fuel tubes and structural parts in nuclear power reactors and in the form of foil in photo flash tubes.

Hafnium, derived as a by-product of zirconium, is used for control rods in nuclear reactors as it absorbs neutrons.

Tungsten powder is available and used in electric light bulb filaments and inert-gas welding electrodes.

Molybdenum in powder form is shaped into compacts by P/M.

Tantalum, a most corrosion-resistant metal, is used for medical implants shaped by P/M.

Columbium (also niobium) is a valuable alloy agent for special steels.

Niobium alloyed with titanium develops superconducting characteristics for high-strength magnetics, as reported by Teledyne.

Nonferrous Powders (Alcan Aluminum Corp.)

In recent years, Alcan has placed emphasis on the development of bronze filter particles. Spherical copper particles made by atomization were coated with tin that would diffuse into the copper during heating, forming a low-melting alloy at the surface which would facilitate bonding of particles. The HP, A, and P grades are described below. Average densities are in the range 4.5 to 5.5 g/cm^3.

Grades	% of tin	Description
HP	2.5–8.0	Coarsest filter particles with maximum pore diameters of 100–700 μm
A	9.5–10.5	Fine powders which yield pore diameters of 30–130 μm
P	9.5–10.5	Atomized prealloyed bronze powders with pore diameters varying between 10 and 180 μm

Testing methods for air permeability and pressure drops were developed by the Metal Powder Industries Federation. Alcan also offers a wide variety of metal powders of silver, tin, antimony, bismuth, silicon, and manganese.

Powdered Metallic and Intermetallic Compounds (Consolidated Astronautics, Inc.)

More than 250 metallic and intermetallic powders and compositions are offered and priced by this organization. Mesh sizes most frequently listed are 100 and 300 mesh. Emphasis appears to be on chemical-, thermal-, and wear-resistant requirements. Indicated are applications in the aerospace industry, additives to solid propulsion fuels, superalloys and ultrahigh-temperature refractories, fuel refining, etc. The 1985 prices of a few selected metal powders are: molybdenum, $35/lb; columbium, $105/lb; chromium, $36/lb; hafnium, $235/lb; tantalum, $310/lb; and cobalt, $34/lb.

Superalloys (RST Process)

Considerable development effort has been accorded to the preparation of composites of metal powders by rapid solidification technology (RST). Combinations of metals and glasses, which in many instances appear as highly improbable mixtures, are quenched very rapidly at cooling rates up to 1 million °C/sec. Brittle ribbons of amorphous metals are pulverized into powder and hot isotactically pressed at high temperatures to form blanks which demonstrate remarkable durability (Ref. 4). One of these new alloys (53.2% nickel, 36.1% molybdenum, 9% iron, and 1.7% boron) has demonstrated a service life two to three times that of a cobalt-base die in extruding copper billets at 1500°F (833°C). It is quite probable that some of the new metallic glassy alloy powders prepared by the RST process will find functional roles with organic polymers.

In the past decade, Ni-base superalloys have played a leading role for gas turbine engines because of their enhanced high-temperature strength. Metallurgical improvements are realized through solid solution strengthening based on the alloying of Cr, Co, Mo, etc. and through precipitation hardening due to formation of a second phase Ni_3Al with dissolved titanium. Atomization processes, such as argon gas atomization, accompanied by a gas recirculation/purification system, attain an oxygen level less than 100 ppm, important for good-quality superalloy powders. It is estimated that production of P/M powders of superalloys is about 500 tons/year [5].

BLENDS OF METALLIC FILLERS AND POLYMERS

Higher heat-resistant thermoplastic resins are usually preferred in the polymer blends with metallic fillers. Hence, commercial blends include polyphenylene oxides, Nylon 6/6, elastomeric silicones, polycarbonates, and polyphenylene sulfides. Cost considerations may influence the selection of lower priced thermoplastics such as ABS or polypropylene. As polymers with higher glass transition temperatures become available, there will be an increasing interest in using them in blends with metallic fillers. The wide disparity between organic polymers and metallic materials is evident not only in electrical insulating qualities but also in thermal expansion and thermal conductivity characteristics. In finely powdered mixtures of the two classes of materials these differences are ameliorated to some extent.

Thermosetting resin systems favor the one-component epoxies (with latent curing agents) for blending with metal fillers. The cure shrinkage is low compared with most thermoplastics, and heat resistance is high. Other thermosets such as phenolic resins, polyesters, and curable elastomeric rubberlike compounds are also used. Extruding-mixing equipment is effective on dry blends of materials; critical timing at higher processing temperatures is a major consideration for the thermosets, though less critical for the thermoplstic blends. Always a factor to consider in blending is the difference in density between organic polymers and metallic fillers. As the viscosity of the mixture is lowered during processing, the heavier metallic particles will tend to separate. Iron, copper, aluminum, and silver powders or flakes most frequently appear in commercial blends with the organic polymers. Specific volume loadings of metal powders or flakes are blended into liquid polymers, aqueous dispersions of polymer systems, or under heat and pressure of compounding, extruding, or milling equipment. Commercial compounds of metal polymer composites are available from several sources.

Combinations of metal particles, carbon or graphite fibers, metal fibers, metal-coated fibers or films which have been chopped to specific sizes, metal-coated foams, etc. provide a limitless variety of techniques for introducing metals into polymer mixes. Recent developments in electromagnetic interference (EMI) shielding have shown advantages in material utilization through combinations of metal particles and electrical conductive fibers, establishing networks with reduced volume resistivity (Ref. 6). Stainless steel fibers 7 μm in diameter are most commonly used for EMI shielding applications.

Several organizations have become active in preparing blends of metal powders and polymers for molding materials, adhesives, tapes, coatings, gaskets, and pastes. The expertise that has developed reflects not only the marketing needs but also the experiences and skills aquired in producing these blends. Unless large volumes of metal/powder blends are required for in-house consumption, the commercially available blends fulfill special manufacturing needs of plastics molders and processors. In most instances, specific end uses are designated for the polymer/metal blends. An example noted here is from the LNP Corporation of Malvern, Pa.*

1. EMI-attenuating thermoplastic composites (EMI-X540) comprise 40% aluminum flake-filled polycarbonate. They are used as EMI and RFI shielding for electrical and electronic application. The inherent surface resistivity (10^1 ohm/sq.) and volume resistivity (10^1 ohm-cm) (ASTM-D257) provide 35 to 40 dB of attenuation over a frequency range of 0.5 to 1000 MHz. Basic properties at 25°C using ASTM test methods:

Specific gravity	1.54
Tensile strength	6,400 psi
Tensile elongation	2%
Flexural strength	12,500 psi
Heat distortion temperature	140°C

2. Statically conductive composites suitable for the electrostatic discharge of polymer and paper films are the Stat-kon molding compounds of LNP Corporation. In general, the surface resistivities of these compounds are several orders of magnitude higher than those of the EMI shielding materials.

Properties at 25°C	Polypropylene	Nylon 6/6	Mod. polyphenylene oxide
Specific gravity	1.0	1.15	1.12
Tensile strength (psi)	3,000	9,500	6,000
Tensile elongation (%)	11	4	3
Flexural strength (psi)	3,600	14,500	9,000

*RTP (Winona, Minn.), Wilson Fiberfill (Evansville, Ind.), and Emerson and Cuming (Canton, Mass.) are other typical suppliers.

Polymer-Dominant Composites

Finely divided metallic particles have contributed to numerous industrial applications when dispersed in organic polymer matrices. When polymer volume predominates in the composite and the metal particles are insulated from one another, flow and cure of the polymer under heat and pressure will influence production methods such as compression molding, injection molding, and extrusion processes. Mechanical properties such as tensile strength and elongation will follow those of the polymers whether they are rigid or flexible. Electrical properties will demonstrate the presence of the metal in higher dielectric constants and higher loss factors at various frequencies. Thermal conductivity increases with increasing volume of metal particles, which also function as a heat sink to dissipate electrical heat losses (Ref. 7).

Metal-Dominant Composites

With the metal particles predominant in the volume of the polymer/metal composites, the character and properties of the product change dramatically. Improvements in electrical conductivity will be much more apparent, as well as effectiveness in electromagnetic shielding and in magnetic devices.

Improved tensile strength of the composite will not be apparent until there is some welding together of the metal particles as by sintering, but the temperature of this process is usually beyond the thermal endurance of the polymer. However, the physical presence of metal particles in the polymer/metal composite may be manifested as follows:

1. The physical shape of the metal particles, particularly mixed granular sizes and those of a fibrous nature, which would respond to compressive stresses, could lead to more effective packing and contribute to higher compressive strength, higher transverse strength, and higher moduli of elasticity.

2. Metal composites use polymers as aids to enhance their own specific functions. For example, control of porosity in an M/P compact may be obtained by using metal particles of different sizes and through their distribution and the pressures used during thermal sintering. The prior introduction of polymer particles will also predetermine voids that will form during sintering, as the plastic material decomposes and disappears. The M/P processes can develop porous castings without the use of temporary polymer inclusions. These porous compacts are widely used as self-lubricating bearings, after they have been filled with oil lubricants. The availability of silicone polymer lubricants and tetrafluoroethylene for this purpose will extend even further the use of self-lubricating bearings. Prior introduction of polymer particles and fibers into metal powder blends will facilitate the planning and geometry of filter designs.

3. There are interesting designs in which the material base may contain specific areas which reflect polymer dominance, while on the same base other areas may rely on basic metal characteristics. These hybrid composites suggest new trends in metal/polymer composites. This activity will be further enhanced as ion implantation techniques are introduced.

To indicate industry awareness of the versatility of metal/polymer composites, a few examples of recent patents and papers are noted here. It will be observed that Japanese contributions appeared most prominantly during the time interval covered. There are many more references to automotive products such as mufflers and sound-deadening structures, though these appeared to be primarily M/P products because of their high-temperature requirements.

1. Metal powders either deposited on a tough flexible polymer film or mixed with tough, high-elongation polymer bases will yield flexible, electroconductive members. These have potential application for electromagnetic shielding, electroconductivity, or static charge modifiers using lightweight thermoplastic bases (Ref. 8).

2. Metallized graphite fibers and metallized glass fibers are available for use in composites requiring enhanced electroconductivity. For example, copper-ion plating on graphite fibers has been used for Cu-Sn composite reinforcement (Ref. 9). These have an advantage over metallized organic polymer fibers because of superior heat resistance and an advantage over fibers of metal because of lower weight. Silver-coated spheres of plastics and of glass have been used in electroconductivity plastics for adhesive coatings, though mixing processes have been known to fragment the fragile silver coating, reducing metal-to-metal contacts. Mica particles electrodes coated with copper plus silane coupling agents have been melt-kneaded and injection-molded with polypropylene to form sheets with electromagnetic shielding (Ref. 10).

3. The metals which may be shaped at the lower temperatures of plastics molding have been brought together by hot isotatic pressing where acrylic polymers, lead, and tin have been combined (Ref. 11). Furthermore, injection molding of a low-melting alloy of Bi(48), Pb(28.5), Sb(9.0), and Sn(14.5) with an organic polymer has been used to produce an electromagnetic wave insulating cabinet (Ref. 12) (Toshiba).

4. Solutions of certain organometallic compounds were deposited on the surfaces of insulating substrates. Laser (argon ion)-induced deposition of metal spots on the substrate by localized pyrolysis. Molybdenum and chromium spots as small as 10 μm were cited (Ref. 13).

5. Magnetic coatings for magnetic disks and tapes are described by Fuji Photo Film Company (Ref. 14). These were comprised of reduced iron oxide to fine powdered iron (300 Å), which in conjunction with aluminum oxide were applied with a polyvinyl chloride-acetate copolymer.

6. For the large-scale production of fine electronic components, processing frequently takes place during the manufacturing operation. For example, Matsushita Electric Company describes the vapor deposition of a polyester film on a 1000 Å cobalt-nickel film at 5×10^{-5} torr vacuum. To achieve a large improvement in corrosion resistance, further coating with *p*-nitrosophenol took place with an electron beam for units of a magnetic recording medium (Ref. 15).

7. Composites of brass screens and Nylon 6 (containing 28% glass beads and reinforcing glass fabric) were press-molded into a box which offered electromagnetic shielding with an attenuation of 50 decibels for radiation between 100 and 600 MHz (Ref. 16).

ELECTRICAL CONDUCTIVITY IN METAL/POLYMER COMPOSITES

Electrical conductivity in metal/polymer composites influences applications for adhesives as well as electromagnetic shielding. With the processing advantages of polymers, such as molding and shaping under heat and pressure, many new uses are being developed for P/M. Polymer systems are used in industrial and consumer products where light weight, appearance, feel, and lower cost will enhance their appeal. Methods for achieving controlled electroconductivity in metal/polymer composites continue to proliferate, now that research has opened prospects for attainment of good conductivity at a molecular level in the polymer. Current methods for attaining good electrical conductivity may be categorized as follows:

1. High-volume loading of metal particles. Finely divided gold, silver, and copper appear frequently in conductive adhesives for bonding to electronic components. Epoxy resin systems are generally used in adhesive formulations. Recent patents have given empirical data for establishing specific blends of two or more particle sizes for optimum results—good adhesion and high electrical conductivity. For a random distribution, 35–40 vol % metal will effect a large drop in electrical resistivity.

2. Addition of specific grades of carbon black and/or graphite fibers has been used in some formulations to reduce the quantities of expensive silver and gold powders used. As the highly oriented, high-modulus graphite fibers prepared from pitch become available, with electrical conductivities surpassing that of copper, the use of such fibers in metal/polymer composites will increase.

3. Lesser amounts of metal particles have been found necessary when large polymer particles are used, which assist formation of a reticulated network in which the conductive metal particles concentrate outside the insulative spheres. This presupposes that such spheres will retain their physical identity.

4. Metallic fibers represent valuable additions to conductive formulations because they not only multiply contacts with the discrete

metal particles but also promote improved bond strength. The efficacy of this technique has already been demonstrated in EMI shielding. Advantages of high aspect ratios of metallic fibers are recognized.

5. Electrocoating or vacuum coating of foils and fibers of polymers (or carbon and glass fibers) with a number of metal coatings (aluminum, nickel, copper, for example) is relevant to this discussion, when they are chopped into short lengths for inclusion in polymer/metal blends.

6. A metal powder-rich polymer composite may have its electrical conductivity increased at specific surface areas through the use of a laser beam to eliminate portions of the insulative polymer matrix by evaporation or thermal decomposition.

7. Metal flakes of aluminum and copper, chopped metal ribbons, and chopped metallized polymer foils find applications where electrical conductivity is important. Aluminum flakes, for example, were originally developed for radar chaff.

8. Electroconductive coatings prepared from metal powders and solutions of polymers are numerous. They will deposit films with a high volume of metal particles. Mixing in aqueous dispersions or solutions of polymers will permit the initial inclusion of more solids. Nickel-acrylic solvent coatings and nickel-urethane coatings are offered for more EMI shielding and metal/acrylic latex coatings are offered for thermoplastics such as polycarbonates and polyphenylene oxides. Some polymers such as polyethylene and polypropylene are not readily soluble in solvents and are blended with powdered metals as polymer particles.

For processing of metal/polymer composites and the problems which arise, recent publications may be reviewed. Kame (Ref. 17) calls attention to metal particle segregation along flow lines which appear during injection molding of thermoplastics.

For specific analyses of electrical conductivity, radiation shielding, and applications, attention is invited to the accompanying chapters.

SUMMARY

Details on the production technology for metallic fillers are available for the many materials and products of the Metal Powder Industries Federation. Iron and steel powder, which comprised the bulk of metallic powders produced, amounted to 216,200 tons in 1984 in the United States. Powdered metal parts and products developed sales of over $2 billion.

For comparison, representative ASTM standards are presented which not only report chemical control features of specific metallic powders, but also list physical properties in some instances. Both ferrous and nonferrous fillers are reviewed. The metal powders prepared by rapid solidification technology are also cited not only for their potential as M/P products but also for their innate qualities in polymer blends. The expanding role of metallic fibers concomitant with metal particles is also noted.

Commercial products of blends of powdered metals and polymers are examined, together with their physical and electrical properties. This is a new and growing segment of the plastics industry, wherein plastics material suppliers are enlarging their markets by offering blends of both thermosetting and thermoplastic polymers with metallic fillers.

Examples of the growing technology of polymer/metal composites are drawn from recent chemical patent and chemical literature.

REFERENCES

1. Metal Powder Industries Federation, Princeton, N.J.
2. MPIF Standard 35 for P/M Structural Parts (1984–1985).
3. American Society for Testing and Materials, *Annual Book of ASTM Standards*, Part 9 (Philadelphia) (1981).
4. H. Chandler, *Metal. Prog.*, *121*:38 (January 1982).
5. Kusaka, Kato, and Horata, Production Methods of Super-Alloy Powders, Japan-U.S. Seminar on Super Alloys (December 1984).
6. J. Noble and R. P. Tolareau, Stainless Steel Fibers, SPE RETEC (June 1984).
7. J. Delmonte, *Metal Filled Plastics*, Van Nostrand Reinhold, New York (1961).
8. C.A. 100-18050J. (1984); Japanese Patent 59,15.441 (January 26, 1984).
9. C.A. 100-178,9951Z (1984); Showa Al. Corp., Japanese Patent 58,204,140, 28 (November 1983).
10. Kuraray Co., Japanese Patent 59,86,638 (May 18, 1984).
11. C.A. 100-11078R (1984); Acta Metall. *31*(11):1820–1983 (1984).
12. C.A. 100-157901A (1984); Toshiba Chemical Products, Japanese Patent 58,220,714 (December 22, 1983).
13. C.A. 100-178795J. (1984); Appl. Phys. Lett., *44*:755 (1984).
14. C.A. 100-184,640W (1984); Fuji Photo Film Corp., Japanese Patent 58,200,423 (November 22, 1983).

15. C.A. 100-160,721X (1984); Matsushita Electric Co., Japanese Patent 58,189,832 (November 5, 1983).

16. C.A. 101-132,108C (1984); Allied Corp., Japanese Patent 59,98,841 (June 7, 1984).

17. I. Kamel, Metal Powders Used in Polymers, in *Metals Handbook*, 9th ed., Vol. 7, American Society of Metals, Toledo, Ohio (1984).

3

Electrical Properties of Metal-Filled Polymer Composites

DONALD M. BIGG

Battelle Columbus Laboratories
Columbus, Ohio

INTRODUCTION

The primary reason for adding metal particles to a polymer is to enhance the electrical and/or thermal conductance characteristics of the polymer. In general, such important polymer properties as low density, strength, and impact resistance are reduced by the addition of metallic particles. The thermal conductivity of such a composite may increase to 10 times that of the matrix polymer by the addition of metal particles (Ref. 1). Normally high insulating materials, polymers filled with metals can become excellent conductors of electricity. Of these two property modifications, the conversion of a polymer from an electrical insulator to a current carrier is the most important.

The rapid growth of electronic devices has created a need to isolate the electromagnetic radiation emissions of such devices. These emissions can be at any frequency above 10^4 Hz. Those between 30 MHz and 1 GHz are of the most concern, however, because they can interfere with broadcast communications. In response to this, the Federal Communications Commission (FCC) has issued regulations limiting the field strength of devices emitting radiation between 30 MHz and 1 GHz (Ref. 2).

The FCC has designated two classes of emitting devices, class A and class B, which are subject to slightly different regulations because of their different potential for affecting broadcast signals. Class A devices are those intended for use in a commercial, industrial, or business environment, and not for use by the general public. Such devices should be less likely to interfere with public broadcasting than class B devices. Class B devices are those intended for use by the general public, and include personal computers and electronic games. If there is a doubt concerning a device's status it is considered a class B device. Table 1 summarizes the permissible field strengths at specified distances for the two classes of devices over the frequency range 30 MHz to 1 GHz.

One method of isolating electromagnetic radiation in this region of the spectrum is to attenuate the signal at the device's structural housing. This requires the housing to be electrically conductive. Because of such desirable features as low cost, low density, corrosion resistance, attractive aesthetics, and ease of forming complex housing shapes, polymers are ideal materials for enclosing electronic devices. They are also transparent to electromagnetic radiation in the region of interest. Fortunately, there are a number of ways to modify a polymer to add sufficient electrical conductivity that it can effectively attenuate radiation between 30 MHz and 1 GHz.

A widely practiced approach is to coat the molded polymer housing with a conductive film, either with a conductive paint or by direct metal deposition (Ref. 3). This involves a secondary processing operation, which may add substantially to the cost of the housing. Another approach is to fill the polymer with conductive carbon particles (Ref. 4). Carbon black-filled polymers can produce a material which exhibits 35 dB of signal attenuation (Ref. 5). Such

TABLE 1 FCC Regulations Pertaining to Allowable EMI Emissions

Class	Frequency (MHz)	Distance (m)	Field strength (μV/m)
A	30–88	30	30
A	88–216	30	50
A	216–1000	30	70
B	30–88	3	100
B	88–216	3	150
B	216–1000	3	200

materials are quite brittle and must be reinforced with glass fibers. They are also subject to problems associated with part-to-part reproducibility. Carbon fiber-filled composites exhibit excellent shielding attenuation and mechanical properties. The cost of carbon fibers is still quite high, however, and these composites cannot always economically compete with conductive painted systems.

Metal particle-filled polymers offer another means of obtaining highly conductive composites. Because of the high density of metals relative to polymers and the lack of reinforcing ability of metal fibers, it is desirable to minimize the concentration of metal used to form an electrically conductive polymer composite. The remainder of this chapter will be devoted to describing efforts to achieve this goal, particularly with regard to producing composites which have sufficient electrical conductivity to satisfactorily shield electromagnetic radiation.

THEORY

The conductivity of any material is related to its atomic or molecular structure (Ref. 6). The electrons of an atom exist in different energy states corresponding to different orbits around the nucleus. Quantum mechanical considerations permit only a certain number of electrons in a given radial orbit. The electrons in a given radial orbit comprise a shell. If an orbit is filled, additional electrons cannot enter that shell or have an energy associated with that level. The outermost shell contains the valence electrons, whose orbits are usually not filled and are responsible for chemical bonding. As two atoms bond to each other, their electrons interpenetrate each other, forming molecular orbitals. The energy levels of these interpenetrating electrons may split into valence bands and conduction bands. These bands are separated by an energy difference or gap. The magnitude of the energy gap determines the conductivity of the material. In highly conductive compounds (i.e., metals) the valence band is not filled and overlaps the conduction band. Therefore, there are numerous energy levels into which an electron can be excited. Under these conditions, the addition of a small external energy raises some of the valence electrons into the conduction band, where they are free to flow. This type of conduction is referred to as band conduction (Refs. 7–9). For less conductive materials (i.e., graphite) the valence band is filled, but the energy gap is so small that it can be easily jumped (Refs. 9, 10). Materials conducting by these mechanisms are referred to as semiconductors. When the valence band is filled and the energy gap is too large to be easily jumped, the materials are insulators.

The addition of a conductor or semiconductor to an insulator affects the electrical properties of the composite according to the degree of filling and proximity of conductive particles to other conductive particles. Three situations are possible; no contact between conductive particles, close proximity, and physical contact (Ref. 11). When the conductive particles are isolated, the conductivity of the composite is changed only slightly. The composite remains an insulator, although its dielectric properties may change significantly (Ref. 12).

When the conductive particles are in close proximity, electrons can jump the gap between the particles, creating current flow. The ability of an electron to jump a gap, under a given voltage field, increases exponentially with decreasing gap size (Ref. 13). Gaps as large as 10 nm can be jumped (Ref. 14). The process of electron transport across an insulator gap is referred to as hopping or tunneling (Ref. 8). Tunneling is a special case of hopping in which the electrons can jump (or tunnel) from the valence band of ions or molecules on one side of the gap to the conduction band on the other side without an energy exchange. Hopping consists of the same type of electron flow between conductors across an insulating gap. However, the electron must have its energy level increased to that of an appropriate band from which it can jump across the gap. Hopping, therefore, involves an activation energy. The theoretical development of the hopping model is due to Mott (Ref. 15). The equation he derived for the probability, P, that an electron will hop from one energy state to another through the insulator is of the form

$$P = a\tau\, e^{-\Delta E/kT} + e^{\Delta E_p/kT} \qquad (1)$$

where a is a constant, E the activation energy needed to hop the energy gap, T the sample temperature, E_p the local polarization energy across the barrier, and τ the tunneling factor. The factor τ is of the form

$$\tau = e^{-\alpha r} \qquad (2)$$

where α is a constant and r the spatial distance through which the electron must tunnel. The exponential form of the tunneling factor shows the importance of maintaining close proximity between adjacent conductive particles. The presence of an activation energy in Mott's equation also shows that temperature effects are to be expected.

The third condition which is possible when a conductive filler is introduced into an insulating matrix is that the conductive particles will physically contact each other to form a continuous network throughout the composite. Under this condition the composite conducts through the particle network by the conduction mechanism of the particles. Composites using metal particles as the conductor would then exhibit band-type conduction. Band conduction and hopping conduction can be differentiated by the ac and dc behavior of the composite (Ref. 7). A composite which conducts by a hopping mechanism will exhibit a higher ac conductivity than dc conductivity. It will also show an increase in conductivity with increasing frequency.

Scarisbrick claims that in highly loaded composites actual touching does occur (Ref. 11). The basis for his conclusion is that the current-voltage characteristics of such composites are linear (ohmic), whereas the nonohmic behavior observed in other composites is due to the electron hopping mechanism. Reboul provides data which refute this claim (Refs. 16, 17). He shows that even composites with resistivities as high as 10^7 ohm-cm exhibit linear voltage-current characteristics except at the electrodes. The total nonohmic behavior of composites is claimed to be caused by extended space charge distributions near the electrodes. Reboul asserts that these space charge distributions are generated by local polarization of the matrix material.

It is probable that metal particles and fibrous conductive fillers conduct by hopping, although actual physical contact between these particles is also possible. Their much larger size makes them less likely to be isolated from other particles and more likely to penetrate a thin film. On the other hand, the 10-nm coating through which electrons can pierce is not very thick, and good wetting between polymer and filler should provide a coating at least that thick. Gul and co-workers present evidence that, depending on filler loading, processing conditions, polymer-metal wettability, and temperature, both conditions may occur (Ref. 13).

Prediction of Network Formation

The behavior of composites made with an insulating matrix and conductive filler is quite interesting and extremely important to understand. The reason for these comments is that the composite changes from an insulator to a conductor over a very narrow range of filler concentration. A typical curve of resistivity versus volume concentration of filler is shown in Figure 1 (Ref. 18). At low filler concentrations the composite remains an effective insulator. At a critical volume fraction, the composite's resistivity falls sharply to a level at which the composite can readily conduct electricity. Increases in filler content above the critical loading do not appreciably reduce the resistivity of the composite. Gurland has shown that the sharp change from an insulator to a conductor is due to the formation of a network among the conducting filler particles (Ref. 18). As explained previously, this network does not necessarily imply physical contact between adjacent particles. Network formation has most frequently been treated as a percolation process. The percolation

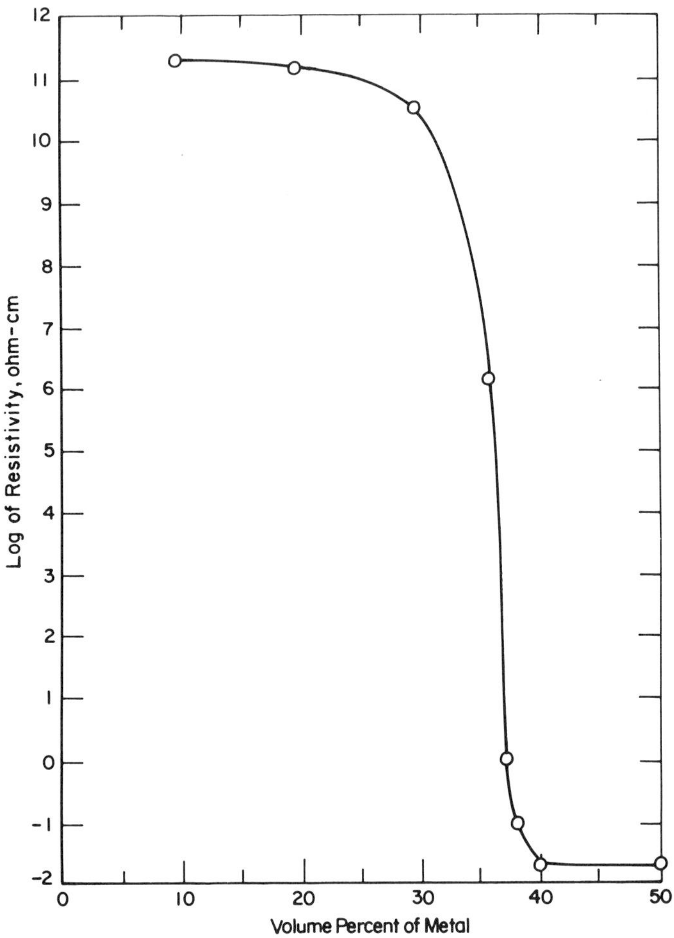

FIGURE 1 Resistivity versus volume fraction of silver spheres in Bakelite. (Reprinted from Reference 18 by courtesy of the American Society for Metals and the Matallurgical Society of AIME.)

model simply refers to a means of continuous network formation through a lattice, taking into account the relative concentrations of the two materials comprising the network. It is a statistical representation most frequently analyzed by Monte Carlo techniques (Refs. 18–20). Although this approach has not been completely successful in predicting the performance of real composites, it is worthwhile to examine it, some of its implications, and the reasons for its shortcomings.

The probability of a continuous network being formed by filler particles in a matrix is related to the statistical average number of contacts each particle makes with neighboring particles and the maximum number of contacts per particle that are sterically possible (Ref. 20). At the critical point of network formation

$$C_p = P_c Z \tag{3}$$

where C_p is the critical number of contacts per particle, P_c is the critical probability of network formation, and Z is the maximum number of possible contacts, or coordination number. Values of Z for several packing arrangements of spherical particles are listed in Table 2. For the ordered packing arrangements in this table, Monte Carlo techniques have been used to determine the critical probability, P_c, of network formation (Refs. 18–20). These are also shown in Table 2. From these values of P_c and Z, the critical average number of contacts per particle for network formation, C_p, has been calculated as 1.5 for each case investigated. One of the shortcomings of the percolation model is that it requires a specified lattice from which the statistical analysis of P_c is made. In a truly random situation a lattice cannot be specified. Fortunately, Gurland showed experimentally that the value of C_p for a random dispersion of

TABLE 2 Critical Contacts per Particle Determined from Bond Percolation Theory

Type of lattice	Z	ϕ_m	P_c	C_p
Face-centered cubic	12	0.74	0.125	1.50
Body-centered cubic	8	0.68	0.183	1.46
Simple cubic	6	0.52	0.254	1.52
Diamond	4	0.34	0.389	1.56
Random	6	0.637	—	—

spheres in a matrix is 1.5 (Ref. 18). Based on his experimental work and the constancy of C_p calculated for various lattice arrangements by the percolation model, C_p is apparently a constant, at least for spherical particles or close approximations to spherical particles.

It is obvious that the average number of contacts per filler particle is related to the volume loading of filler in the matrix. If this relationship were known, the critical concentration at which a continuous network is formed could be determined. In general terms

$$C = f(\phi) \tag{4}$$

and

$$C_p = f(\phi_c) \tag{5}$$

where C is the average number of contacts per particle, ϕ the volume fraction of filler, and ϕ_c the critical volume fraction for network formation (Ref. 19). Bueche assumed the linear relationship

$$f(\phi) = \frac{\phi Z}{\phi_m} \tag{6}$$

where ϕ_m is the maximum packing fraction of the filler in the matrix (Refs. 21,22). Combining Equations (5) and (6), he arrived at

$$\phi_c = \frac{C_p \phi_m}{Z} \tag{7}$$

For a random mixture of spheres C_p equals 1.5, Z equals 6, and ϕ_m equals 0.637. Therefore, ϕ_c is 0.159. This value is much lower than the experimental value of 0.38 obtained by Gurland. Jantzen proposed the following model for $f(\phi)$ (Ref. 19):

$$f(\phi) = \left(\frac{Z\phi}{1 - \phi}\right)\left(\frac{1 - \phi_m}{\phi_m}\right) \tag{8}$$

At the critical loading for network formation

$$\phi_c = \left[1 + \frac{Z}{C_p}\left(\frac{1 - \phi_m}{\phi_m}\right)\right]^{-1} \tag{9}$$

For the random sphere case (in which C_p equals 1.5, Z equals 6, and ϕ_m equals 0.637), ϕ_c is 0.305. Jantzen claimed that this represents the point at which network formation begins and the sharp drop in resistivity starts. However, the value of C_p was determined at the point where the network has been formed. For practical considerations, it is better to predict the value of ϕ_c where network formation is complete and the lower resistivity has been achieved. Another model for $f(\phi)$ which attempts to do this is

$$f(\phi) = \frac{Z\phi}{\phi_m/(1 - \phi_m) - \phi} \tag{10}$$

From this

$$\phi_c = \left[\left(1 + \frac{Z}{C_p}\right)\left(\frac{1 - \phi_m}{\phi_m}\right)\right]^{-1} \tag{11}$$

When C_p equals 1.5, Z equals 6, and ϕ_m equals 0.637, ϕ_c equals 0.35, which is close to the value of 0.38 found by Gurland. These equations do not adequately account for preparation technique, particle size distribution, or particle-particle interaction effects.

Predictions for flakes, fibers, and other irregular shapes are even less well developed. For irregular fillers with a completely random distribution, the critical value of ϕ_c is the same as for spheres (Ref. 23). This is seen in Figure 2. This curve represents the percentage of salt removed from an NaCl-polyethylene composite at various volume loadings. The salt was roll milled into the polyethylene, and then extruded into a sheet. The salt in this sheet was then extracted from the sheet by submerging the composite in 80°C water for 48 hr. Fresh water replaced the salt water solution every 2 hr, except for the overnight periods. The results, shown in Figure 2, indicate that a continuous network is not formed until the volume loading of salt is approximately 38%, the same as for electrical network formation with conductive spheres.

Yamaki et al. (Ref. 24) developed a model for the conductivity of fibers in a three-dimensional composite, using the following set of assumptions:

1. Fibers with the same length and radius are dispersed uniformly in the polymer.
2. The fiber center is restricted to lattice points.
3. The conductivity of the polymer matrix is negligibly small compared to the conductivity of the fiber.
4. The electrical conduction is made by the contact of one fiber with another located in any spatial direction.
5. The created paths are multiply connected to opposite surfaces of composite cube.

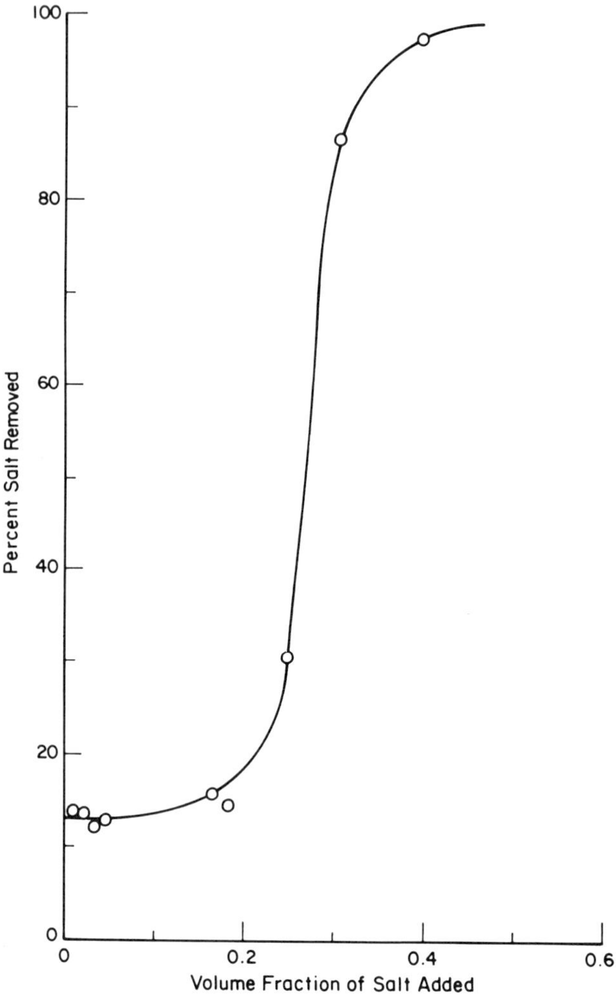

FIGURE 2 Percent salt leached from a polyethylene-salt composite as a function of salt concentration (Ref. 23).

Their theoretical development is based on a percolation model assuming a regular, rather than random, lattice structure. This development was based on a complex statistical approach which was claimed to provide good agreement between actual and predicted conductivity values. Actual comparisons were not presented, however.

A simpler approach is to investigate the effect of particle geometry on the maximum packing fraction of the filler. At the maximum packing fraction, a network has already been formed. This value then represents an upper bound for the concentration of a particular filler required to form a constructive network. Milewski investigated the effect of fiber aspect ratio on the maximum packing fraction in a completely random distribution (Ref. 25). A plot of his results is shown in Figure 3. From this graph, it is clear that the concentration of fibers required to produce a continuous network decreases very sharply with increasing aspect ratio. For

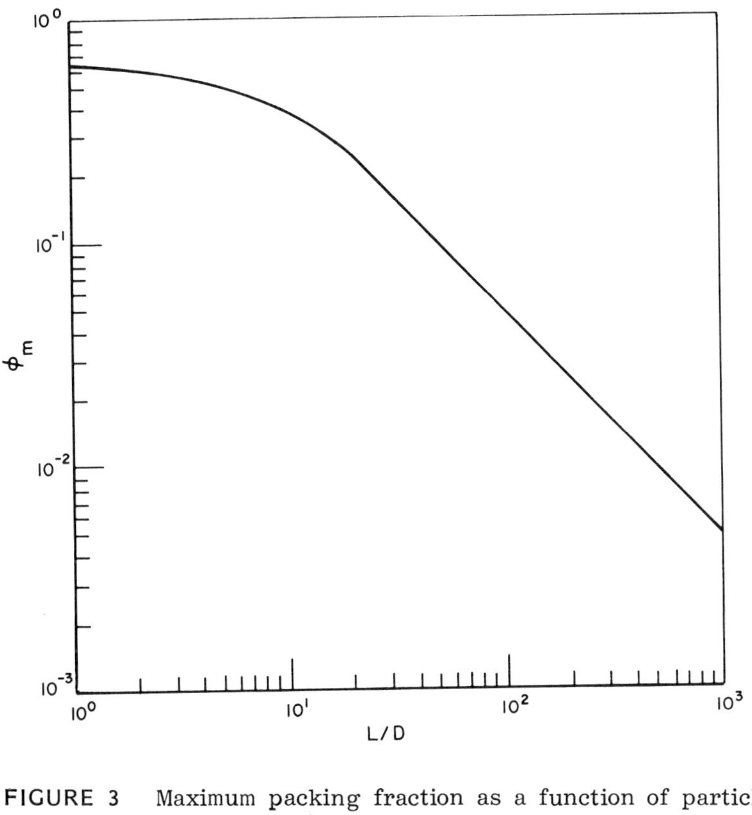

FIGURE 3 Maximum packing fraction as a function of particle aspect ratio for three-dimensionally random dispersions.

spherical particles, the critical concentration for network formation of 0.38 is much less than the maximum packing fraction of 0.637. At lower values of ϕ_m, therefore, the critical network concentration should be correspondingly lower. Because the relationship between aspect ratio and maximum packing fraction is an exponential one, long fibers require very low concentrations to achieve a continuous network. Fibers having an aspect ratio of 1000:1 require a maximum packing fraction value of only 0.005 to reach their three-dimensional packing limit.

EMI Shielding

Electromagnetic radiation can be divided into near-field and far-field regions. In the near field, the electromagnetic signal can be predominantly an electric vector, or predominantly a magnetic vector depending on the source antenna. In the far field, plane waves exist in which the electric and magnetic vectors have a ratio equal to 377 ohms, are in phase, and are orthogonal to each other. Figure 4 shows the relationship between frequency, distance from the source, and impedance. Near-field signals are generally lower-frequency signals at closer distances to the source than far-field signals.

As was shown in Reference 26, plane wave (far-field) radiation is of most concern in measuring shielding effectiveness. Shielding effectiveness is defined as SE = 10 log P_T/P_R in decibels, where P_T is transmitted power and P_R is received power. Figure 5 shows the relationship between shielding effectiveness and percent P_R. Plane

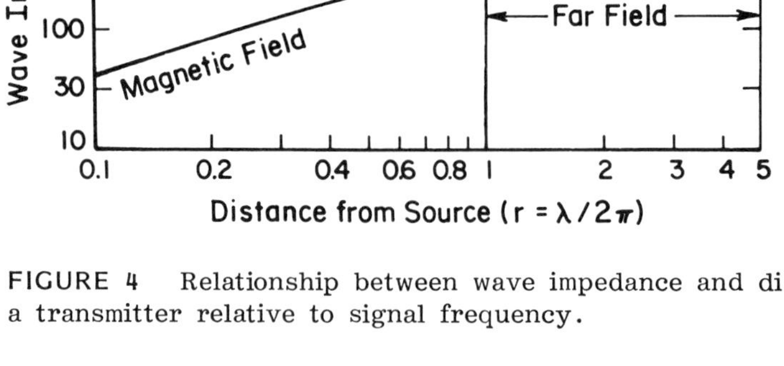

FIGURE 4 Relationship between wave impedance and distance from a transmitter relative to signal frequency.

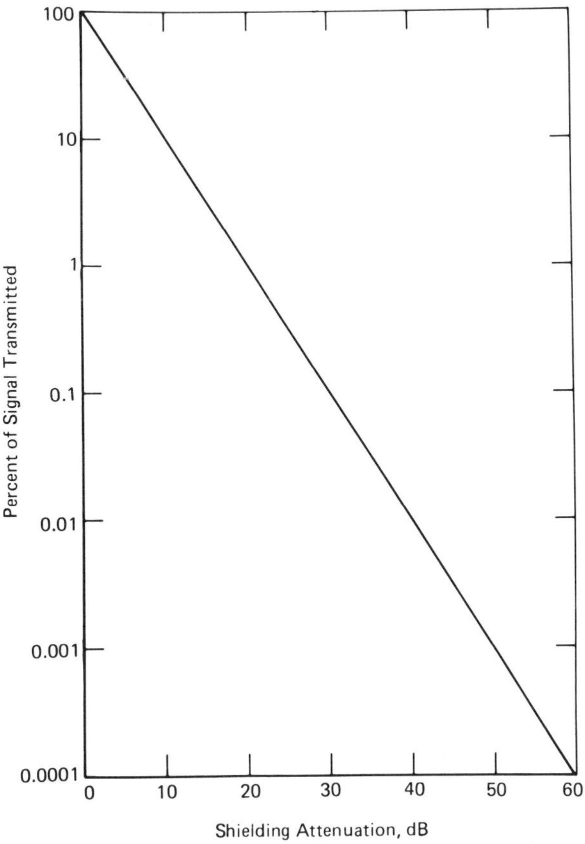

FIGURE 5 Relationship between shielding effectiveness in decibels and percent of transmitted signal that is attenuated.

wave shielding attenuation in the sum of three additive contributions; R_1, direct reflection; R_2, internal rereflection; and A, absorption. Contribution R_2 is negligible when A is over 10 dB. For practical applications the shielding effectiveness must exceed 20 dB. For plane-wave radiation and A greater than 10 dB, R_1 and A are given by the following equations (Ref. 27):

$$R_1 = 108.2 + 10 \log_{10} \frac{10^6}{\rho \mu f} \qquad (12)$$

$$A = 1.33t \sqrt{\mu f/\rho} \qquad (13)$$

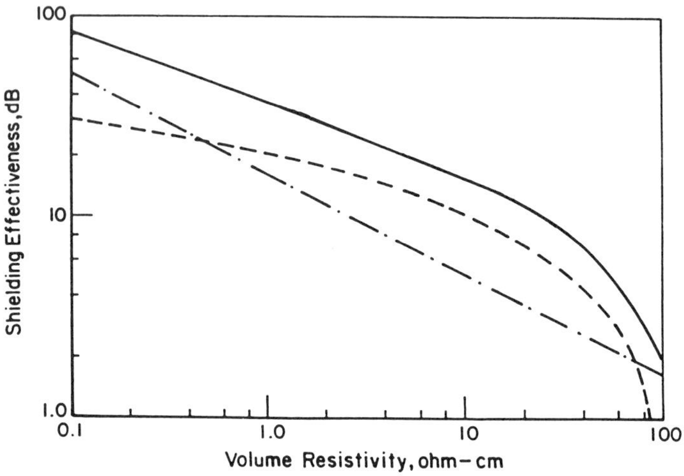

FIGURE 6 Theoretical prediction of plane wave shielding attenuation versus volume resistivity for a 3-mm-thick material at 1 GHz. (——) Total attenuation, (---) attenuation due to reflection, (-··-) attenuation due to absorption.

where ρ is the sample's resistivity relative to copper, μ the permeability of the sample relative to space (μ is usually 1.0), f the frequency of the radiation in hertz, and t the thickness of the sample in centimeters. The individual and total contributions of these shielding mechanisms are shown in Figure 6 for impinging radiation at 1 GHz.

TESTING

Resistivity

The resistivity of a material is defined as the electrical resistance between two parallel surfaces of a 1-cm cube and is expressed in ohm-cm. Three classes of materials can be specified according to their resistivities. Insulators are materials having resistivities greater than 10^{10} ohm-centimeters. Most polymers meet this criterion. Semiconductors have resistivities between 10^{10} and 10^2 ohm-cm. Materials with resistivities below 10^2 ohm-cm are conductors. A unique feature of resistivity as a material property is that its value can vary over 40 orders of magnitude between superconductors and highly insulating materials. Very few material properties have this range. One result of this wide range in possible values is that one instrument, or measuring system, cannot be used to measure the resistivity of all materials.

In principle, the resistivity of a material is determined by measuring the resistance across a length of the sample and dividing by the ratio of the sample's cross-sectional area to the length over which the voltage is applied. This is shown schematically in Figure 7. Two problems make this relatively simple procedure difficult for composite systems. The first problem is that of contact resistance at the electrode-sample interface. If this interface does not consist of the electrode in intimate contact with the sample, the interface resistance can exceed that of the sample, and the measured resistivity will be higher than that of the material (Ref. 10). To overcome this problem, the contact edges of the sample must be carefully cleaned and polished. In addition, a coating of a highly conductive silver paint is often used to fill in microvoids and provide an intimate electrical interface between the sample and the electrode. Pressure is often applied to improve interfacial contact, and in some cases conductive liquid electrodes are used. Molded-in metal electrodes have also been used.

The second potential problem in measuring the resistivity of composites is one normally associated with two-phase systems, namely a poor dispersion of conductive particles in the matrix. This can be manifested in several ways. First, most molded polymers that are filled have a resin-rich layer at the surface. This acts as an insulator to the interior of the molded part. This layer is usually removed prior to measuring resistivity. Molded-in metal electrodes are also used to overcome this problem. Second, when low concentrations

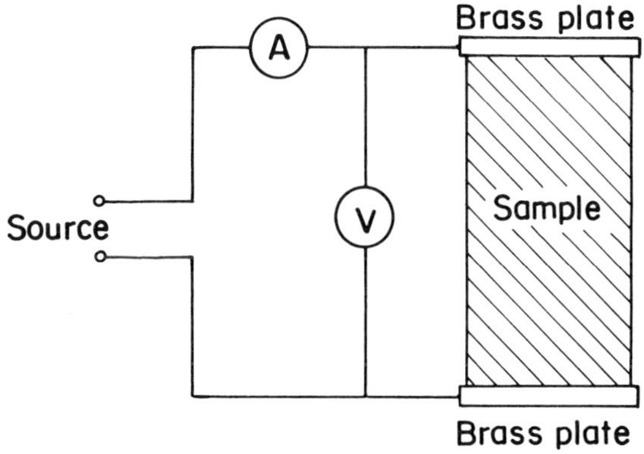

FIGURE 7 Schematic of simple volume resistivity measurement technique.

of conductive fibers are used to form the conductive network, they may not provide adequate electrical contact at the measuring interface. This is particularly noticeable in systems using metal-coated glass fibers. In these composites, the metal coating provides adequate contact surface area within the composite, but very little at the part's edges, where the coating is only exposed as a thin ring. Again, molded-in inserts or the application of a penetrating conductive paint is helpful in minimizing contact resistance in these composites. Third, in two-phase conductive systems, the conductor may be poorly dispersed within the matrix. An example of the type of poor dispersion which can occur is seen in Figure 8. Here all the conductive fibers are concentrated around the outer periphery. The measured resistivity will then be much lower than the resistivity that would have been obtained in a uniformly dispersed composite. Poor dispersions are sometimes difficult to judge by visual inspection or resistivity measurements without sacrificing the sample. Experience in knowing the level of resistivity that should be expected in a given composite is very useful in evaluating measured resistivities.

A commonly used system for measuring volume resistivity is the guarded electrode method shown schematically in Figure 9. This system is well suited for surface resistivity measurements as well as volume resistivity determinations. A metallic coating is applied to the indicated areas of the disk-shaped sample to provide effective electrodes. The volume resistivity is determined by the following equation:

$$\rho_v = \frac{A_e R_v}{t} \tag{14}$$

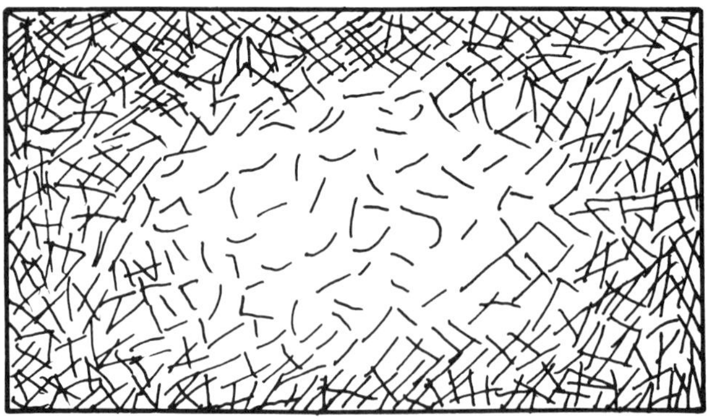

FIGURE 8 Dispersion of conductive fibers in a composite that produces conductivity but not shielding effectiveness.

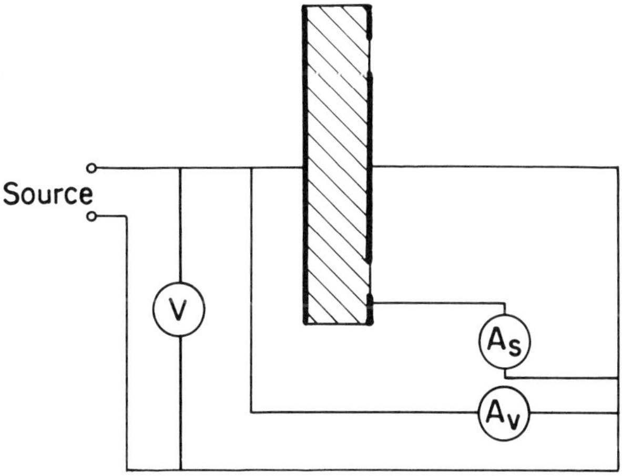

FIGURE 9 Schematic of the guarded electrode measurement technique for measuring volume and surface resistivities: R_V, voltage measured by voltmeter V divided by current measured by ammeter A_V; R_S, voltage divided by current measured by ammeter A_S.

where R_V is the measured resistance of the sample, A_e is the cross-sectional area of the interior electrode, and t is the sample thickness. If the surface resistivity is desired the circuit indicated by the dashed line in Figure 9 is used. The appropriate equation for surface resistivity is

$$\rho_s = \frac{2\pi R_s}{\ln(D_0/D_1)} \tag{15}$$

where R_S is the measured resistance, D_0 is the interior diameter of the guard electrode, and D_1 is the outside diameter of the interior electrode.

EMI Shielding: Transmission Line

Two basic techniques can be used to measure the shielding effectiveness of materials. Figure 10 shows a schematic of a coaxial transmission line system (Ref. 28). In this device a washer-shaped sample is electrically connected between the inner and outer conductors of a specially designed coaxial fixture, which is placed between the signal source and the detector. In this system, a signal is transmitted through the coaxial side of the sample fixture, which is

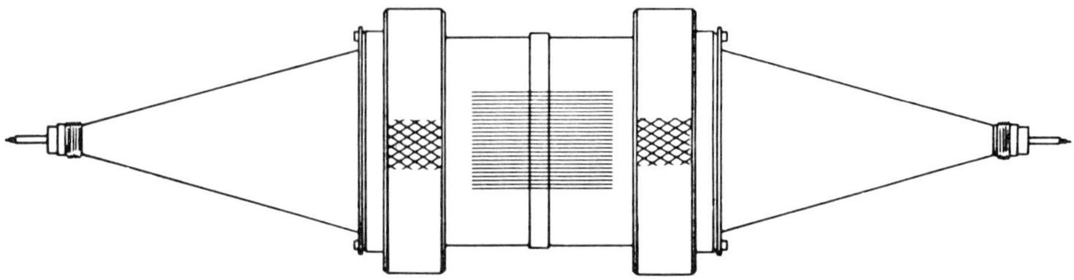

FIGURE 10 Schematic of coaxial transmission line specimen holder.

placed between the signal source and the detector. In this system, a signal is transmitted through the coaxial side of the sample fixture. The ratio of the power of the signal received with the sample present to the power received with no sample is the attenuation or shielding effectiveness of the sample. A good conductor will shunt most of the signal to the outer conductor of the cable. This system has a distinct advantage in that the signal is always a plane wave. Therefore, the specimen is always tested in the far field, regardless of the transmitting frequencey.

This system has no resonant frequencies below 5 GHz for a 10-cm-diameter sample holder, a frequent problem in other systems. The measurement system is shown in the block diagram of Figure 11. Measurements can be made at single frequencies or by sweeping the range of frequencies of interest. In the single-frequency mode, a signal of the desired frequency is generated by a signal generator. This signal bypasses the sample holder to a variable attenuator, which is set for maximum signal attenuation. The signal is then transmitted through the sample and the attenuator level reduced

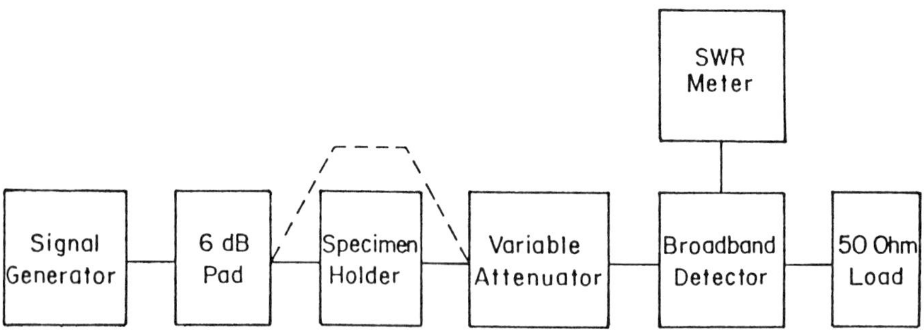

FIGURE 11 Block diagram of coaxial transmission line device used to measure plane wave EMI shielding effectiveness.

until it is the same as it was when the signal bypassed the sample. The attenuation change on the attenuator represents the shielding effectiveness of the sample.

The swept-frequency method presents a curve of shielding behavior over the entire frequency range on an oscilloscope or similar plotting device. The signal generator is replaced by a tracking generator, which is driven by a spectrum analyzer. A variable attenuator is not required in this mode as the spectrum analyzer can also measure the signal attenuation. The spectrum analyzer supplies the attenuation response of the system with and without the sample in place over the entire frequency range. The difference between the two curves is the shielding attenuation of the sample at any frequency.

As simple as the measurement technique is, the mechanics of the sample holder are largely responsible for the accuracy of shielding effectiveness measurements. The annular section between the inner and outer conductors must be large enough to contain a sample representative of the material being tested. The sample must also make good electrical contact at both the inner and outer surfaces, or the measured shielding effectiveness will be lower than the sample is capable of providing. The design feature which ensures adequate electrical contact is the use of flexible walls in both the inner and outer conducting portions of the sample holder (Ref. 28). The flexible walls are obtained by machining a slot in the sample holder sections as shown in Figure 10. A carefully machined sample is then always under compression while it is in the sample holder. Conductive silver paint is also put on the sample edges to reduce sample/holder resistance.

The dynamic range of this system is determined primarily by the power level of the signal generator and the sensitivity of the receiver. It is also important that all coaxial electrical connections be properly shielded. With standard laboratory care a 1-watt signal will have a dynamic range of approximately 80 dB.

EMI Shielding: Dual Chamber

A schematic of a dual-chamber facility for measuring the EMI shielding behavior of plastic composites is shown in Figure 12. The primary structure is a 1.8-m steel cube. The cube is welded light tight along all 12 seams and 4 corners. A 0.7-m-square opening is located on one face of the cube. Over this opening a steel plate is bolted which contains the various apertures for signal impingement on test samples. The large opening is designed to permit access to the interior of the box. The interior of the box is lined with microwave absorber. This material absorbs 99% of all impinging radiation above 300 MHz and is partially effective at lower frequencies.

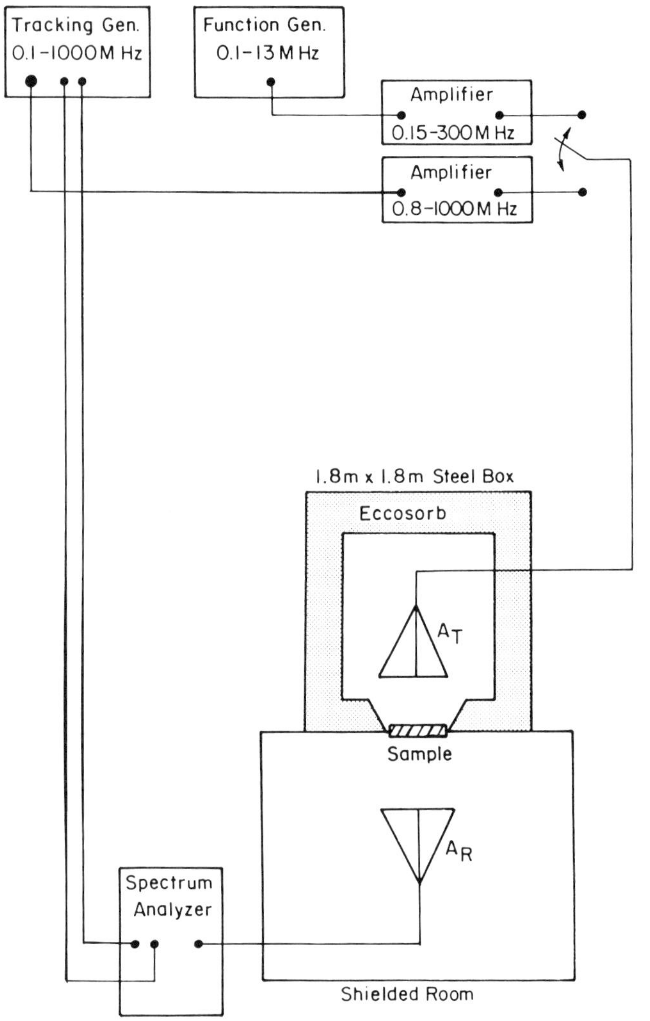

FIGURE 12 Schematic of dual-chamber facility for measuring EMI shielding effectiveness.

Also inside the box are the transmitting antennas. Most measurements are made in the electric field mode, although magnetic field measurements can also be made. The low-frequency electric field antenna has a nominal transmission range between 10 kHz and 32 MHz. The high-frequency antenna is a log spiral antenna, which has an operating range between 200 MHz and 1 GHz.

The transmitted signals are generated by a spectrum analyzer driving a tracking generator at frequencies above 100 MHz. This device is capable of generating and analyzing signals between 10 kHz and 1.8 GHz. The low-frequency signal, 32 MHz, is generated by a function generator prior to final amplification.

The preamplifiers are isolated in a screened copper cage. Each of the electrical connections is made through a shielded conduit. Signal entry into the box is through a blind-passage shielded junction. Isolation of the receiving antenna is important since outside radiation sources can interfere with the signal received by the antenna. When the box is against the screened chamber electrical contact is achieved through flexible silver finger stock, which seals the junction between the box and the screened chamber. The receiving antennas are the same as the transmitting antennas. The signal received by these antennas is fed directly to the spectrum analyzer.

Shielding attenuation measurements are made by transmitting a signal of fixed frequency and measuring the strength of the signal received with no sample in the aperture. The transmitted signal strength is usually between 1 and 3 watts. The strength of the signal received with no sample present is set as the baseline on the spectrum analyzer. A sample is then inserted and its attenuation recorded as the change in decibels of the signal received relative to the baseline signal. The spectrum analyzer makes the measurement

$$SE = 10 \log \frac{P_T}{P_R} \tag{16}$$

directly, with P_T being the power of the transmitted signal and P_R the power of the received signal. The operating range of the system is determined by inserting a copper plate as a test panel and measuring the shielding attenuation. Since copper has a shielding effectiveness much greater than 100 dB, a lower value indicates the degree of leakage present in the sample-holding fixture. Typically, the described system has a dynamic range (SE of the copper plate) of 80 dB. Since most of the samples investigated have shielding effectiveness values in the range between 20 and 60 dB the system is sufficiently well sealed to permit accurate measurement of shielding effectiveness.

Sample Handling

In this system the main aperature opening is 9 × 14 cm to accommodate standard 10 × 15 cm molded test plaques. This aperture is part of the 0.7 × 0.7 m plate which covers the main opening on the face of the box. Other plates with larger openings have been built to accommodate larger test samples. Smaller test samples are not recommended because the strength of the signal that passes through the opening drops sharply if the minimum sample dimension drops below 8 cm.

Many test samples must be pretreated to accurately measure the shielding effectiveness of the bulk sample. The principal source of potential measurement error is signal leakage around the edge at the sample-test facility interface. For that reason many samples are lightly sanded and cleaned along the edges and contact perimeter to remove the resin-rich layer which is frequently obtained in molded composites. The edges and adjacent contact periphery are then painted with a highly conductive silver paint to produce an electrode which makes good electrical contact with the test chamber. This procedure can increase the indicated shielding effectiveness by as much as 10 dB. It is most critical for composites which provide 20 to 30 dB of attenuation. Samples which are more conductive usually, although not always, have less surface insulating layer. The measured shielding effectiveness of samples which attenuate 40–60 dB usually does not change significantly when the samples are pretreated in the manner described.

To demonstrate the importance of making good electrical contact between the sample and the test box, several layered plaques were tested. These plaques, 3 mm thick, consisted of a conductive layer and an insulating layer, each approximately 1.5 mm thick. When the conductive layer was in contact with the box the recorded shielding attenuation was 38 dB. The recorded shielding effectiveness dropped to 18 dB when the insulating layer was held against the test chamber. In this instance leakage through the 1.5-mm gap created by insulating layer was appreciable. Experience has shown that long gaps, even if thin, permit more leakage than a random hole or short gap.

Measurement Calibration

A coaxial, or transmission line, test device measures the shielding effectiveness of test samples under far-field conditions at all transmission frequencies (Ref. 28). The major shortcoming of the coaxial device is that it cannot measure the shielding effectiveness of molded parts or evaluate the effect of vent holes on shielding effectiveness. It does make an excellent facility for calibrating dual-chamber facilities such as the one described in this chapter. Such a calibration was made by vapor-coating a polyethylene terephthalate film with a thin coating of gold. The coating was such that the surface resistivity of the film was 10 ohm/sq. The shielding effectiveness of this film was then measured in the dual-chamber facility and a coaxial cable device over the frequency range between 0.1 and 1000 MHz. The results, shown in Figure 13, indicate that the box system operates in the far field at 1000 MHz. These data also show that in the near field the measured shielding effectiveness in the electric field mode is greater than the value measured for plane-wave radiation. This makes plane-wave measurements the preferred technique for calibration. Near-field measurements cannot adequately describe the shielding ability of a material subjected to far-field radiation since test facilities of different size will give different results. If the plane-wave shielding effectiveness is known and satisfactory, near-field radiation will give greater shielding effectiveness than the plane-wave radiation in the electric field mode.

One other calibration evaluation was made. A test plaque was injection-molded and tested in the dual-chamber facility. The same molding compound was used to injection-mold a small 14 × 9 × 6 cm box to judge the effectiveness of testing a three-dimensional structure. The wall thickness of the molded box was the same as that of the molded plaque, 3 mm. As shown in Figure 14, this box was flanged for good contact to the test facility. The shielding effectiveness of this box was then measured in the test facility. Finally, a transmitter was built to fit inside the molded box and it was tested in an open field site similar to that specified by the FCC. A photograph of the transmitter is also shown in Figure 14. Filler dispersion

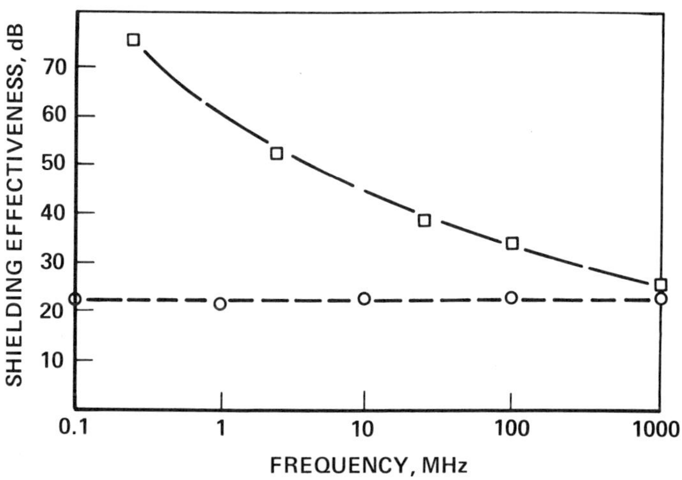

FIGURE 13 Shielding effectiveness of a 10-ohm/sq, gold-coated polymer film measured by transmission line (○) and dual-chamber (□) systems.

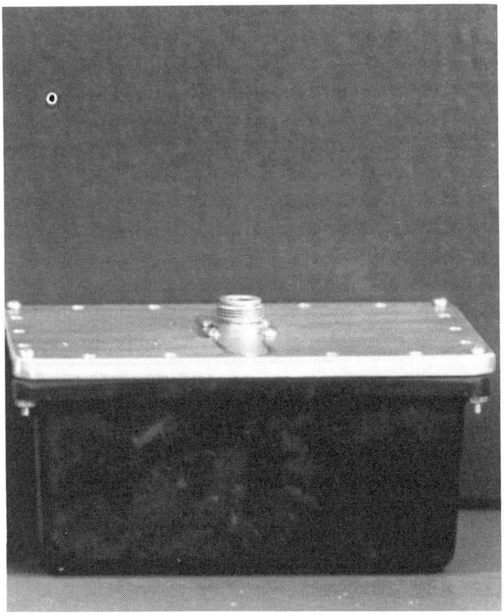

(a)

FIGURE 14 Injection-molded composite structure with interior transmission antenna: (a) inside structure showing coaxial cable junction; (b) outside structure.

TABLE 3 Correlation of Various Shielding Effectiveness Tests Under Plane Wave Conditions

Sample	Test facility	Frequency (MHz)	SE (dB)
Molded plaque	Dual chamber	1000	32
Molded box	Dual chamber	1000	29
Molded box	Free space	1000	30

(b)

FIGURE 14 (continued)

problems were minimized by using Ketjenblack EC conductive carbon black in an acrylonitrile-butadiene (ABS) matrix. The concentration level of the black was 8.2 vol %. Table 3 shows the data obtained for the plaque, molded box, and molded box in the open field site. The very close agreement between the three values demonstrates that the electrical characteristics molded into a flat plaque can be incorporated into a three-dimensional structure, that a three-dimensional structure can be evaluated for shielding effectiveness in a dual-chamber facility, and that data obtained in such a facility are comparable to those obtained in an open field test site. This latter verification makes laboratory testing a useful venture.

In addition to agreeing with other measurement systems, the data generated by the system described in this report correlate very well with predictions based on plane wave theory. The predicted shielding effectiveness of a film having a surface resistivity of 10 ohm/sq subjected to plane wave radiation is 26 dB (Ref. 29). As Figure 13 shows, the dual-chamber facility measured a shielding attenuation of 25 dB for such a film.

MATERIALS

Metal Particles

Since metals are the quintessential conductors of electricity, it is not surprising that they have been used as fillers to create conductive plastic composites. In comparison to conductive carbon, the resistivities of metals are much lower, while their densities are higher. Only aluminum has a density comparable to carbon. Table 4 summarizes the resistivities and densities of the metals most commonly used as conductive fillers. Until recently, the most common form of metal fillers was either spherical or irregularly shaped isotropic particles. These particles were typically 100 μm or less in diameter. These small isotropic particles are needed in high concentrations to produce an electrically conductive composite. Mechanical properties and density advantages are usually lost because of the high filler loadings required. In the past few years, metal fibers, flakes, and ribbons have become available. In these anisotropic forms metals are more attractive as fillers because reduced concentration levels are possible.

Several factors influence the use of metals in plastics. Copper, for example, catalyzes the degradation of many polymers. Iron, copper, and aluminum oxidize rapidly at the particle's surface. These oxide layers are nonconductors and therefore dramatically change the electrical properties of the composite. This effect is most significant with the smallest particles, because the oxide layers comprise a greater fraction of particle than they do on larger particles. Prosvirin and Savitskii (Ref. 30) reported that the resistivity of oxidized aluminum particles may be as high as 10^5 ohm-cm. Sheer and Turner also reported that the oxide layer on aluminum particles will effectively insulate the metal core, making composites nonconductive up to very high filler concentrations (Ref. 31). In both cases, the particles used were under 100 μm in size. Since all metals except gold oxidize, the long-term electrical stability of metal-plastic composites must be investigated. Of the frequently used metals, silver is the most conductive. It is also the most expensive. Stainless steel is very resistant to oxidation, although it is less conductive than other metals (it is still more conductive than carbon-based conductors) and more dense. Aluminum is attractive because it is a low-cost, low-density, highly conductive metal. Surface oxidation is the primary electrical concern when using aluminum. Nickel is also of interest because of its low rate of oxidation.

Metal-Coated Glass

Metal-coated glass fillers have the advantage of the low density of glass combined with the high electrical conductivity of the metal coating. Metal-coated glass fibers were investigated as a reinforcement

TABLE 4 Summary of Properties of Metal Fillers Used to Produce Conductive Composites

Metal	Density (g/cm^3)	Resistivity (ohm-cm)	Size	Form	Source
Copper	8.82	1.7×10^{-6}	∿44 μm	Dendritic	U.S. Bronze Powders, Inc.
			∿100 μm	Dendritic	Reference 53
			∿58 μm	Flake	Belmont Smelting & Refining Co.
			<50 μm	Electrolytic	Assam Carbon Products, Reference 53
			<100 μm	Spherical	Reference 66
Nickel	8.90	6.84×10^{-6}	4–7 μm	Spiky equiaxed	Inter. Nickel Co.
			52–74 μm	Irregular	Glidden
			∿4 μm	Spherical	Reference 13
			10–30 μm	Irregular	Reference 13
			∿20 μm	Irregular	Reference 49
			0.001 mm × 2 × 2 μm	Flakes	Reference 44
Aluminum	2.70	2.8×10^{-6}	∿5 μm	Irregular	Alcan Metal Powders
			<15 μm	Smooth anisodiametric	Reference 30
			∿100 μm	Irregular	Reference 41
			<100 μm	Spherical	Reference 42
			0.001 mm × 0.1 mm × 3–300 μm	Ribbon	M.B.A.-Tracor
			0.1 mm × 1.25 μm	Fibers	Transmet Corporation
			0.025 mm × 1.0 mm × 1.25 μm	Flakes	Transmet Corporation
Iron	7.86	10×10^{-6}	25–100 μm	Irregular	Reference 41

TABLE 4 (continued)

Metal	Density (g/cm^c)	Resistivity (ohm-cm)	Size	Form	Source
Stainless steel	7.80	74×10^{-6}	4–50 μm diam.	Fibers	N. V. Bekaert, Brunswick
			4–79 μm diam.	Fibers	Brunswick
Silver	10.5	1.59×10^{-6}	∿100 μm	Spherical	Reference 42
			∿100 μm	Spherical	Reference 42
			75–100 μm	Electrolyzed polyhedral	Reference 66

for tires during the 1950s. Their commercial use never materialized, however, due to poor adhesion between the metal and glass. In the November 1975 issue of Plastics Design and Processing, IBM announced that metal-coated glass fibers had been used in sheet molding compounds and injection-molded articles to produce an electrically conductive plastic composite having essentially the same mechanical properties as the unfilled resin (Ref. 32). The apparent key to achieving good adhesion between the metal coating and the glass fibers was the prior removal of the glass sizing. In their announcement IBM indicated that nickel had been deposited on the glass fibers by high-rate electron beam vacuum deposition. An agreement was made between IBM and Lundy Electronics by which Lundy manufactures the coated glass fibers. Lundy's fibers are coated with aluminum instead of nickel. The fibers are supplied as a roving with less than 100% coated area.

M.B.A. Tracor also manufactures an aluminum-coated glass fiber in either continuous roving or chopped fiber. The coating thickness is approximately 2.5 μm on standard 12.5-μm-diameter E glass fibers. The process that M.B.A Tracor uses to coat their glass fibers is to pass the as-spun fibers directly into a bath of molten aluminum. A variety of other techniques are possible for coating metals on glass. Vacuum deposition, electroless plating, flame spraying, and electric arc discharge are some of the possible techniques. Economics and metal-glass adhesion are the primary factors which determine the choice of technique.

Potters Industries has investigated silver-coated glass particles to replace silver in conductive silver coatings. Such particles have a significant cost advantage over pure silver particles of the same size. The filler particles investigated included spheres, flakes, short chopped fibers, irregular particles, and hollow spheres. Only coated spheres are offered commercially. These include spheres of diameter less than 44 μm and those between 53 and 105 μm. The coating thicknesses are between 0.02 and 0.1 μm. The total silver concentration is between 4 and 12%.

POLYMERS

All polymers can be filled with metals to achieve a low-resistivity composite. Some polymers are more suited for the applications for which conductive composites are intended. For example, most business machines are molded with either ABS or polycarbonate because of the excellent impact properties of those polymers. Specific polymer characteristics must be considered when selecting a metal filler for a given polymer matrix. For example, copper catalyzes a degradation reaction in polycarbonate. Polyolefins and ABS do not readily wet most metals. The improvement in wetting behavior that is obtained by using a modified polypropylene with aluminum fiber and flake-filled composites is shown in Reference 33.

Another factor which influences the composite behavior of metal-filled polymers is the morphology of the polymer. Two-phase polymers, such as semicrystalline polymers and certain phase-separated block and graft copolymers, can segregate small fillers into amorphous regions of semicrystalline polymers and soft or uncross-linked regions of the block and graft copolymers. In such polymers the critical concentration required to reach network formation is much less than in a purely amorphous polymer where maximum dispersion is achieved. The larger the filler particle is, however, the lower is the segregation effect.

PROCESSING

Spherical particles and irregularly shaped particles can be mixed into the polymer by the same methods used to incorporate more common fillers. These techniques include internal batch mixing (as in a Banbury mixer) and extrusion compounding. Shaped fillers such as flakes or fibers must be handled more cautiously. Very small flakes can be handled as fillers, whereas large flakes and fibers must be added to the polymer in a manner designed to reduce damage to the particles. For example, internal mixing cannot be used without seriously damaging the particles. Desirable techniques include adding the filler to the melt in a compounding extruder (Ref. 34). This is justified since most of the damage to the particles occurs during melting of the polymer (Ref. 35) and by adding the filler to the melt this operation is avoided. Another technique is to use extrusion coating of continuous fiber roving, followed by chopping to the desired length (Refs. 36–38). This last process is particularly well suited to metal-coated glass fibers since these fibers have a strong tendency to aggregate when handled in the short chopped form (Ref. 39).

Conductive Composite Behavior

Metal-Filled Composites

Gurland was the first to present data showing that metal-filled polymers undergo a sharp transition from an insulator to a conductor at a critical volume concentration, ϕ_c, of metal particles (Ref. 18). Since that first report, many others have shown the sharp transition from insulator to conductor (Refs. 40,41,33,1,31). Different researchers have reported different critical volume concentrations, however. Gurland found that a composite of randomly dispersed silver spheres in a Bakelite matrix became conductive at a volume fraction of 0.38. Aharoni found that a composite of iron spheres in polyamide-imide became conductive at a volume loading of 0.20 (Ref. 40). Aharoni also claimed that the iron spheres were randomly dispersed in the matrix.

The Bakelite resin that Gurland used was a thermosetting phenol-formaldehyde resin. The technique used to disperse the silver spheres in the resin and the curing conditions were not discussed by Gurland in Reference 18, although the spheres were probably dry blended with the powdered resin and the mixture compression molded. The particle size and distribution of the silver particles was not published, but it can be guessed from the photomicrograph that was presented to show the random nature of the dispersion that the particles were between 25 and 200 μm in diameter. Aharoni used oxide-free iron spheres which were between 5 and 20 μm in diameter. His composites were prepared by stirring the particles into a dimethyl formamide solution of the polymer. The suspension was then cured for 45 min at 200°C. Aharoni states that microscopic examination revealed a random distribution of iron particles in the cured resin. De Araujo and Rosenberg found that at least 40 vol % of spherical metal particles was needed to induce conductivity in an epoxy matrix (Ref. 42). They investigated a wide range of metal powders including copper, silver, aluminum, tin, lead, bronze, and stainless steel. The particle size distribution of these powders was not given, although the particles were all reported to be less than 100 μm in diameter. The copper had been reduced in hydrogen at 600°C for 4 hr to remove any insulating oxide layer. Besides confirming Gurland's critical volume loading of approximately 0.40, de Araujo and Rosenberg found that the curing conditions significantly affected the composite's conductivity. Fast cures at higher temperatures caused thermal expansion of the resin, which separated adjacent metal particles, reducing conductivity. Slower cures at lower temperatures, with the same volume loading of filler, would produce composites having a higher conductivity. Since, even under their best curing conditions, they could not produce a conductive composite at volume loadings less than 40%, thermal movement of the metal particles cannot account for the difference in ϕ_c reported by Gurland and Aharoni.

One explanation of this puzzle was presented by Nicodemo et al., who made composites from aluminum and iron particles in a styrene-acrylonitrile copolymer, SAN (Ref. 41). Although the particles were irregular in shape, they had no characteristic dimension and could be treated as spheres. Both particles had approximately the same average particle size of 100 μm, although their size distributions were very different. The particle size distributions of these two powders are shown in Figure 15. The aluminum particles have a much broader particle size distribution than the iron particles. As Figure 16 shows, the differences in particle size distributions significantly affect the critical concentration at which the composite becomes conductive.

The broad particle size distribution of the aluminum powder induces conductivity in the composite site at a critical volume fraction of 0.40, while the narrow particle size distribution of the iron powder

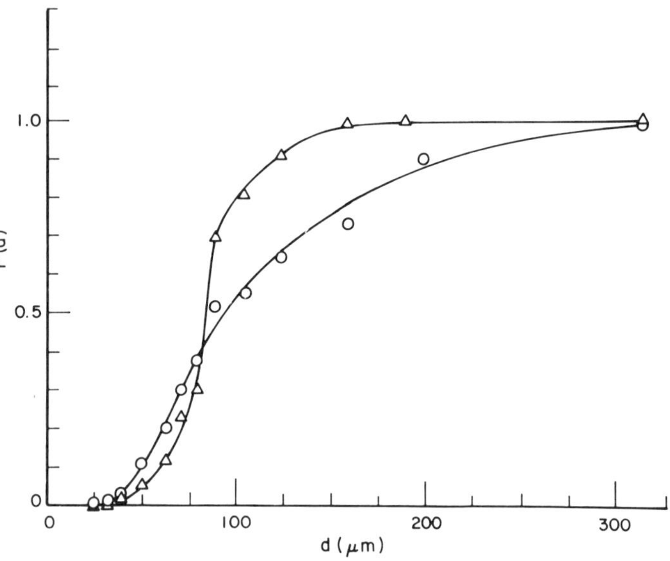

FIGURE 15 Cumulative size distribution versus particle diameter for iron (△) and aluminum (○) powders. (Reprinted from Reference 41 by courtesy of the Society of Plastics Engineers.)

causes the composite to become conductive at a volume fraction of 0.20. Reexamination of Gurland's and Aharoni's data shows similar differences in particle size distribution. It is known that a broad particle size distribution will increase the maximum loading of filler particles. At equal volume loadings, therefore, particles with a wide size distribution will have a lower probability of contacting a neighboring particle than those with a narrow size distribution. Another interesting aspect of the value of the critical concentration needed to induce electrical conductivity in a random dispersion of spherical or nondimensional particles in a matrix is related to the phenomenon of switching. A number of investigators have shown that some insulating composites will become highly conductive when subjected to a critical voltage. De Araujo reported that mixtures containing between 20 and 40 vol % filler that were insulating at low voltages became very conductive after being subjected to a critical higher voltage. The lower resistivity remained after the voltage was removed unless the sample was flooded by a high current flow, which then returned the sample to its insulating condition. While it is interesting that the switching became possible at a volume concentration of 20%, corresponding to Aharoni's measured value, the measurement technique Aharoni used did not involve a high-voltage switching mechanism. Others have also reported that switching will occur above a critical voltage. Kwan et al. reported that irregularly shaped particles switch at a lower voltage and volume loading than spherical particles because of the high electrical stresses built up at protrusions on the particle (Ref. 43). Such electrical stresses initially promote tunneling, but eventually lead to sputtering of the metal into the polymer to create an electrical path. This path is fragile and can be destroyed by high current flow.

The critical volume loading needed to induce electrical conductivity in a composite is also strongly dependent on the shape of the filler particles. Shorokhova and Kuzmin published volume resistivity data on nickel flakes dispersed in polystyrene (Ref. 44). The nickel flakes were square, 1 μm thick and 2 mm on each side. The effective aspect ratio for these flakes is 2000. A highly conductive sample was reported at volume loadings as low as 0.65%, although reproducibility of the resistance measured at different points on the sample was poor. At 10.6 vol % the point-to-point reproducibility of sample resistivity was 3%. Because the size of the flakes is of the same order of magnitude as the

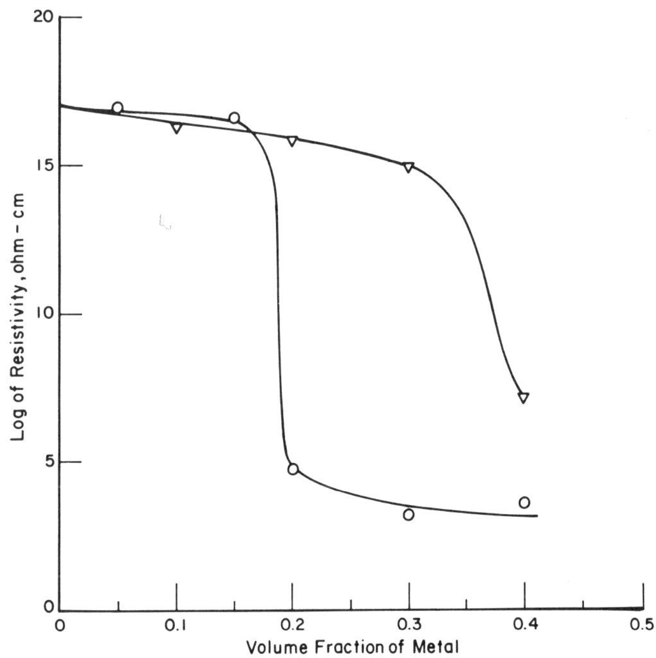

FIGURE 16 Volume resistivity versus volume fraction of iron (○) and aluminum (▽) powders in SAN polymer. (Reprinted from Reference 41 by courtesy of The Society of Plastics Engineers.)

thickness of the sample (5 mm), it is not surprising that there were conductive spots through the sample, even at the lowest filler loading. The sample specimens were compression molded at 150°C. Injection molding would have tended to align the flakes in the flow direction, thereby changing the sample's electrical response.

Flakes provide more stable conducting composites than nondimensional particles because their large surface area per unit volume imparts better particle-to-particle contact and less motion due to small-scale thermal expansion or contraction (Ref. 45). Marshall and Agarwal prepared conductive composites from copper flakes in epoxy resin (Ref. 46). Their flakes were specially prepared to remove and inhibit surface oxidation. The flakes were cleaned with a chlorinated solvent, such as trichloroethylene, followed by further cleaning in a mild acid for up to 24 hr. The flakes were then mixed into the epoxy and cured immediately. The copper flakes were approximately 58 μm in size. Their thickness was unspecified. The epoxy composites Marshall and Agarwal prepared were 1.6 mm thick. The ratio of sample thickness to flake length is then 27.1:1, compared to 2.50:1 for Shorokhova and Kuzmin. Marshall and Agarwal's composites, therefore, are more representative of a random dispersion of flakes in a polymer matrix than are those of Shorokhova and Kuzmin. Marshall and Agarwal indicated that their composites became conductive at a volume loading of 3.6%. Unfortunately, they did not report the thickness of the copper flakes, so that the aspect ratio of these flakes cannot be determined.

Researchers at Battelle also investigated the use of flakes in conductive composites (Ref. 33). Aluminum flakes 1.27 × 0.635 × 76.2 mm were roll milled into polypropylene at 200°C, then compression molded into 3-mm-thick test samples. These samples were found to become conductive at approximately 10 vol %. These experimental flakes have a high aspect ratio of 16.7:1. As part of this study, aluminum fibers were also investigated. Fibers having an aspect ratio of 12.5:1 induced conductivity in both polypropylene and thermosetting polyester at approximately 12 vol %. When the fiber aspect ratio was increased to 24:1 the critical volume loading dropped to 7%. Injection molding of aluminum flake-filled composites resulted in considerable damage to the shape of these flakes. Aluminum is very ductile and was easily deformed. As a result, the critical concentration needed to produce a conductive composite increases. This is shown in Figure 17, where the critical concentration of commercial aluminum flake that was injection molded is similar to that of the more carefully prepared experimental flake composite that was compression molded. Commercially available aluminum flakes have an initial aspect ratio three times that of the experimental flakes (50:1 compared to 16.7:1). These flakes were significantly damaged during melt compounding and injection molding. Chen showed that the flakes

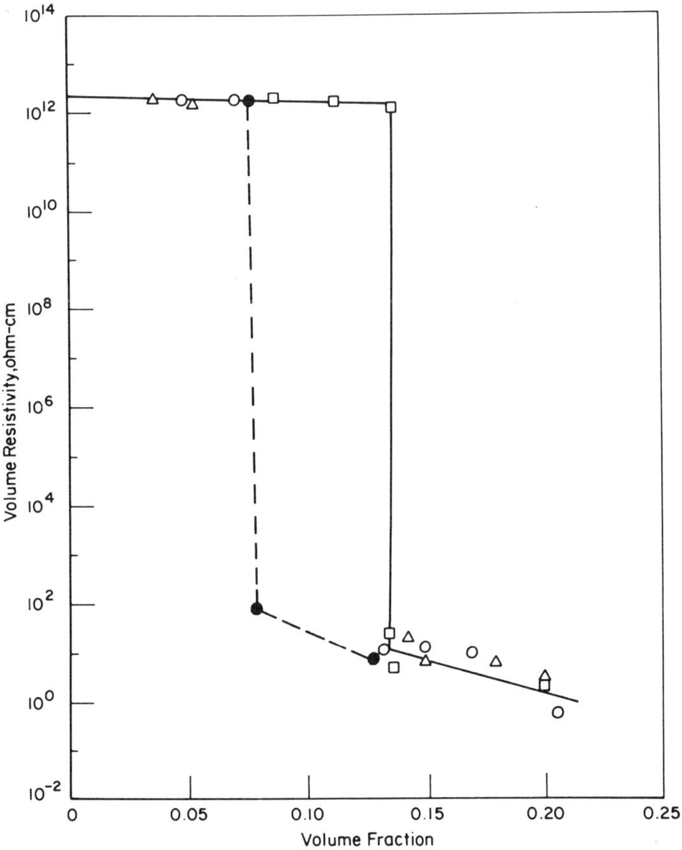

FIGURE 17 Volume resistivity versus volume fraction of aluminum flakes in various polymers. Open symbols represent injection-molded composites with 50:1 aspect ratio flakes. Filled symbols represent compression-molded composites 16:7:1 aspect ratio flakes. (□,●) polypropylene, (△) polycarbonate, (○) ABS.

"ball up" during processing (Ref. 47). He reported that a volume concentration of 23% was needed to produce a polycarbonate composite with a volume resistivity of 1.1 ohm-cm. Simon showed that bulk resistivity drops off rapidly with increasing mixing time (Ref. 48) (Figure 18).

Fibers with higher aspect ratios have also been investigated. Stainless steel fibers 8 μm in diameter and 6 mm long (aspect ratio, 750) are capable of producing composites which have a volume resistivity of 0.70 ohm-cm at 1 vol % (Ref. 39). These fibers are quite ductile, although they can also be broken during compounding and extrusion. Figure 19 shows the size distribution of 6-mm fibers removed from an injection-molded ABS composite. Both ductility and fiber breakage are evident in this photomicrograph. The overall effect of fiber aspect ratio on the critical concentration of conductive filler needed to induce conductivity is given in Figure 20. The general shape of this log-log curve is comparable to that shown in

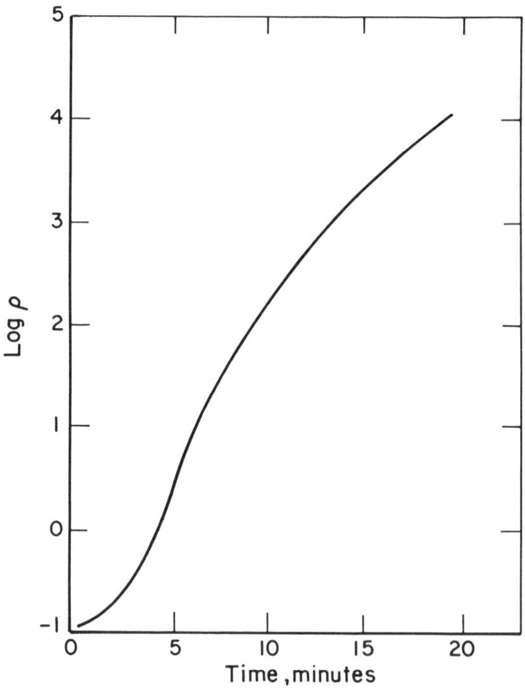

FIGURE 18 Volume resistivity of an aluminum flake-filled composite versus mixing time. (Reprinted from Reference 48 by courtesy of The Society of Plastics Engineers.)

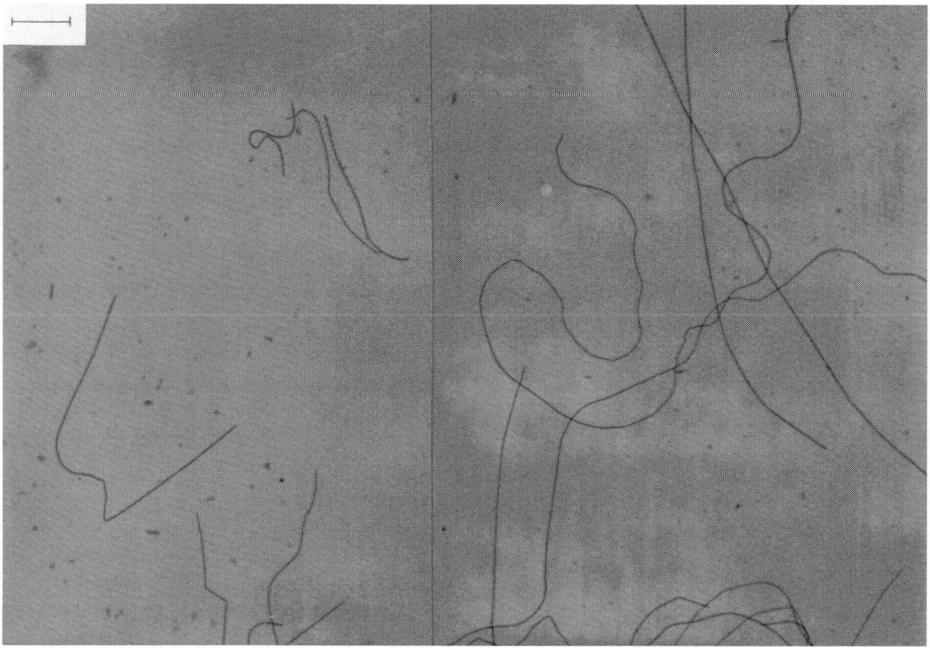

FIGURE 19 Photograph of stainless steel fibers removed from a compounded and injection-molded ABS composite. Initial fiber length was 6 mm. Compounded pellets were 6 mm long, 3 mm in diameter. Scale is 0.4 mm.

Figure 3 for the effect of aspect ratio on maximum packing fraction. These curves are not parallel, as might be expected, primarily because the molded samples are two-dimensional in geometry and the data presented in Figure 3 are for a three-dimensional random distribution of particles. In the two-dimensional case the maximum packing fraction is higher than in the three-dimensional case.

The principal conclusion from these studies is that dimensionally oriented filler particles (flakes and fibers) induce electrical conductivity in a randomly dispersed composite at much lower concentrations than spherical particles or irregular particles having an aspect ratio of 1. When nondimensional particles are used, a narrow particle size distribution initiates a conductive network at a lower filler loading than for particles with a broad size distribution. An alternative approach to using randomly dispersed fibers or flakes is to segregate the conductive filler from the matrix into a conducting network. This approach relies on special processing techniques to form the

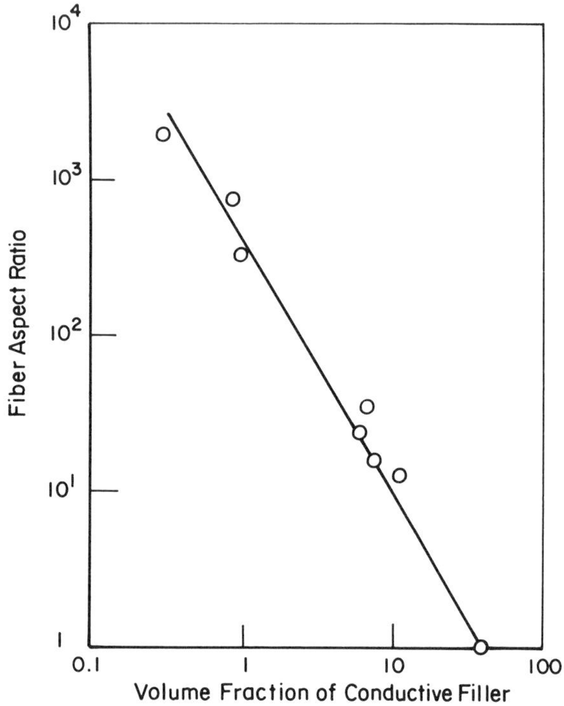

FIGURE 20 Relationship between fiber aspect ratio and minimum conductive filler concentration required to produce a composite having a volume resistivity below 100 ohm-cm.

conductive paths through the matrix, or the use of very small particles in semicrystalline polymers.

One processing scheme that has been successful is to align ferromagnetic particles in a magnetic field (Refs. 13,49,50). This requires a low-viscosity thermosetting matrix to permit particle alignment and connectivity. Nickel has been the preferred metal because of its low resistivity. Gul and co-workers found that a pulsating magnetic field aligned the particles more rapidly than a constant field. Increasing the field strength reduced the time required to reach the minimum composite resistivity. Irregularly shaped particles responded to the magnetic field better than spherical particles. Also, larger particles formed more connective networks than smaller particles at the same volume loading.

Another technique used to produce segregated networks in the polymer matrix is to dry blend resin particles with metal particles of a much smaller size (Refs. 31,51,52). During dry blending the metal particles will adhere to and coat the resin particles. The mixture can then be compression molded or compaction-sinter molded. The metal particles do not penetrate very far into the polymer, and the polymer does not flow enough to disrupt the metal particle network. Malliaris and Turner showed that when the diameter of the polymer particles was at least 10 times that of the metal particles, there was excellent coating of the polymer by the metal (Ref. 51). When the diameter of the polymer particles was two or less times that of the metal particles, the metal particles did not adhere to the polymer. Size ratios between these two limits gave mixed results, depending on the polymer and metal combination.

It was found that the greater the ratio of polymer particle size to metal particle size, the lower the critical volume fraction needed to induce conductivity. Kusy and Bhattacharya et al. independently developed theories predicting the critical volume loading of such compaction-sinter molded composites (Refs. 52,53). Kusy's equation for critical volume fraction is

$$\phi_c = \left[1 + \left(\frac{F}{2a_c}\right)\left(\frac{R_p}{R_m}\right)\right]^{-1} \tag{17}$$

in which F is the reciprocal of the planar packing factor. This represents the effective coverage that results when the surface of a primary particle is contacted by a specific packing arrangement of dispersed particles. Therefore, F depends on the packing arrangement assumed for the metal particles around the polymer particle. It has values of 1.11, 1.27, and 1.38 for particle arrangements having coordination numbers of 6, 4, and 3, respectively. Because F has such a narrow range, the particle arrangement has little effect on the predicted value of ϕ_c. The a_c in Eq. (17) represents the critical fraction of the surface of the polymer particle that must be covered to ensure network formation. Kusy relates a_c to the average number of contacts per particle, C, and the coordination number for the assumed packing arrangement as:

$$a_c = \frac{C}{Z} \tag{18}$$

R_p/R_m is the ratio of the radius of the polymer particle to that of the metal particle. Figure 21 shows the calculated values of R_p/R_m versus ϕ_c for two possible cases: (a) the average number of contacts per particle, C, is 1.5 and the packing geometry is simple cubic, and (b) C is 2.0 and the packing geometry is rectangular. Case (a) represents the minimum number of contacts per particle required to produce a network. Simple cubic packing was chosen

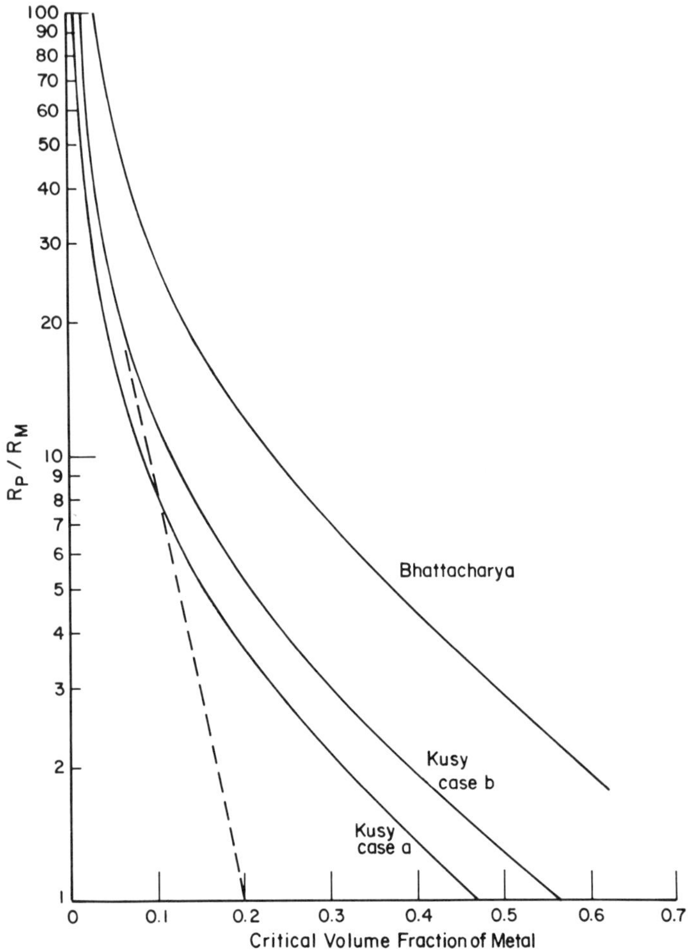

FIGURE 21 Theoretical predictions of critical volume fraction needed to induce bulk conductivity as a function of polymer-metal particle size ratios for compaction-sintering molded composites.

since it closely approximates random packing. Both random packing and simple cubic packing have a coordination number of 6. Case (b) represents a fully guaranteed network formation, C = 2.0, and a minimum square lattice coverage. The critical volume fraction is slightly lower at a given R_p/R_m ratio in case (a) than in case (b). The two curves are not very different, however. Kusy shows that experimental data verify his case (b).

Bhattacharya et al. (Ref. 53) developed a more general theory that can account for the three-dimensional geometry of filler particles. For the case of non-dimensional particles, similar to those assumed by Kusy, the resulting equation for critical volume fraction is

$$\phi_c = \frac{3R_m/R_p}{1 + 3R_m/R_p} \tag{19}$$

This equation is also plotted in Figure 21. It deviates significantly from that developed by Kusy, particularly at smaller R_p/R_m ratios. Neither theory has much validity at lower values of R_p/R_m. The case $R_p/R_m = 1$ was treated by Cahn, who found that ϕ_c should equal 0.20 (Ref. 54). In Figure 21 a dashed line has been drawn from $R_p/R_m = 1$ and $\phi_c = 0.20$ to tangentially connect wigh case (b) of Kusy's theory. Significant deviation between the dashed line and Kusy's case (b) begins at approximately $R_p/R_m = 10$. This is the ratio below which Malliaris and Turner found that the metal particles do not necessarily adhere to the polymer particles (Ref. 51). Segregation no longer persists when the metal does not adhere to the polymer. Thus is is not surprising that the theory begins to fail at the point where the network-forming mechanism begins to change.

It is important to realize that the theoretical considerations developed by Kusy, Bhattacharya, and Cahn are for the special case of dry-blended powders being compacted into a form by pressure and subsequently fused by sintering in place at an elevated temperature. This type of processing cannot be direclty compared to random mixing of particles into a fluid matrix followed by subsequent solidification. Cahn's analysis and Gurland's results do place limits on the critical volume fraction for randomly dispersed isotropic particles in a continuous matrix. Cahn's critical volume fraction of 0.20 corresponds to the value of Aharoni and Nicodemo for filler particles with a narrow size distribution. The essential assumption made by Cahn is that both polymer and filler particles have the same size and size distribution. This situation is only attainable by compaction-sintering molding.

Another way to form a segregated metal network in a plastic matrix is to deposit a continuous metal coating on plastic particles.

Hochberg describes a process in which polystyrene powder is coated with silver by chemical deposition in a silver nitrate solution (Ref. 55). The coated powders are then compression molded below the glass transition of the polymer at high pressures. Molding by this method breaks some of the metal sheath around the polymer particles, permitting polymer particle-to-particle fusion while maintaining the metallic network.

Crosby and Hornbeck patented essentially the same concept in England 22 years later (Ref. 56). The principal difference in their approach is that they vacuum coated pellets (as opposed to powder particles) and were able to injection mold their coated particles while retaining electrical conductivity. Their coatings were as thick as 0.2 μm and formed flakes during the molding process.

Techniques which rely on segregating the conductive medium into a continuous network achieve the goal of reducing the volume concentration of conductor needed. They do so without a substantial sacrifice in the mechanical properties of the composite. Unfortunately, the techniques used to achieve the segregated network involve special processing methods which do not lend themselves to high-volume production. Compaction-sintering molding is a slow process. It also leads to molded parts having substantial part-to-part variability in physical properties. Even compression molding of these samples, while providing greater part-to-part consistency, is slow in comparison to injection molding.

Natural segregation of submicrometer filler particles occurs in semicrystalline polymers and phase-separated polymers containing a cross-linked phase (Ref. 5). This natural segregation can increase the conductivity of systems filled with small conductive particles. For example, the bulk resistivity of a polyphenylene oxide/polystyrene blend filled with 20 vol % nickel flakes was found to be 29 ohm-cm. At a concentration of 11 vol % a Nylon 6/6 composite had a volume resistivity of 2.3 ohm-cm.

Metal-Coated Glass Fibers and Sphere Composites

The data for metal-coated glass fibers suggest that the electrical properties of composites made from these fibers are very sensitive to processing conditions. Bradish reported that bulk molding compound (BMC), a thermosetting polyester formulation, became electrically conductive when 7.5 vol % aluminum-coated E-glass was used with fibers that were 12.5 mm long (Ref. 57). The initial aspect ratio of these fibers was approximately 700:1. The extent of fiber breakage was not determined. When the fiber length was increased to 25.4 mm for an initial aspect ratio of approximately 1400:1, the critical volume loading was reduced to 5.0 vol %. While no specific resistivity data were given for these compounds, the article indicated that it was less than 10 ohm-cm.

Davenport also reported on the conductivity of aluminum-coated glass fiber composites (Ref. 58). Although he did not describe experimental procedures, or even specific compound formulations, a curve of resistivity versus filler weight concentration was presented. Davenport's curve was transposed into resistivity versus volume fraction of aluminum-coated glass concentration by making two assumptions. The first assumption is that the polymer was a BMC compound. The second assumption is that the total glass/metal-coated glass level remained constant at 30% by weight. In short, the aluminum-coated glass replaced some of the glass used in a standard BMC compound. Based on these assumptions, Figure 22 shows the effect of aspect ratio on conductivity and critical volume loading as well as the sensitivity of resistivity to volume loading at a very low resistivity level. The lower plateau resistivity is only beginning to be developed with the shorter fibers. It should also be pointed out that the volume loading axis for Figure 22 has been expanded relative to previous plots, and in comparing Figure 22 to other fillers this should be

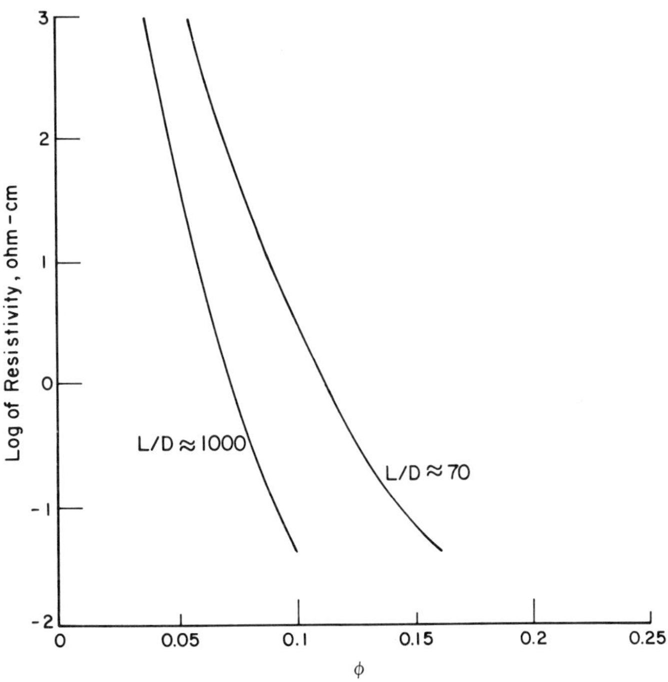

FIGURE 22 Relationship between volume resistivity and volume fraction of aluminized glass fibers for high and low aspect ratio fibers.

remembered. It is unfortunate that the molding technique, formulation, and resistivity measurement technique were not presented. Real data would also be useful, particularly in determining the reproducibility of composite resistivity. Davenport's and Bradish's observations are consistent. They both used the same aluminum-coated glass fibers, so this is not surprising.

Gerteisen et al. produced aluminum-coated glass-polycarbonate composites by injection molding extrusion-coated bundles of fiber tow (Ref. 59). They started with pellets 1.9 cm long and 0.3 cm in diameter having the fibers aligned along the length of the pellet. In this manner the fibers were fed to the injection-molding machine with an effective aspect ratio of 715:1. The average fiber aspect ratio after molding was not determined, however. The molded samples produced in this investigation did not exhibit surface conductivity until the volume concentration reached 23.4%. Examination of the samples revealed that they had resin-rich surfaces. The resistivity of the conductive sample was high at 2.0×10^4 ohm/sq. Samples with loadings of 13.3 vol % were not conductive at the surface unless they were foamed. Solid samples apparently had resin-rich surfaces, which insulated the conductive interior. Foaming increased the fiber content of the skin, thereby creating a conductive surface. While conductive, this surface still had a high resistivity, 2×10^4 ohm/sq. Chen also injection molded aluminum-coated glass fibers that had been extrusion coated with polycarbonate (Ref. 47). The initial aspect ratio of these fibers was alos 715:1. After injection molding, the fibers had an aspect ratio between 80 and 160:1. At a volume loading of 23% the resulting composite had a bulk resistivity of 15.7 ohm-cm. Moreover, the fibers still existed as bundles that had not been properly dispersed. A similar loading of coated fibers in Nylon 6/6 produced a better dispersion of fibers, but a much higher volume resistivity, 6.5×10^6 ohm-cm. The Nylon 6/6 wet the fibers better than the polycarbonate, resulting not only in a better dispersion but also more fiber damage. The fiber aspect ratio range was reduced to 90–120:1.

Using silver-coated milled glass fibers, Heinze and Ritter were unable to obtain a composite exhibiting good surface conductivity until the volume loading reached approximately 15% (Ref. 60). They hand-mixed the fibers into a thermosetting polyester to form a 1.27-cm-thick test sample. The fibers had an approximate aspect ratio of 150:1. They reported difficulty in mixing the fibers into the resin to achieve a void-free composite. The precured composite viscosity was apparently too high to easily disperse the fibers and devolatize the composite. This difficulty was overcome by using a mixture of the metallized fibers with metallized spheres. A composite conductivity of 0.8 ohm-cm was measured for a composite containing approximately 7.0 vol % fibers and 15 vol % spheres. The spheres had an average diameter less than 44 μm. By themselves, the coated spheres formed a composite having a resistivity of 25 ohm-cm at a volume loading of approximately 30%. This suggests that the particle size distribution of the spheres was relatively narrow.

Heinze and Ritter also investigated silver-coated ground glass particles and flakes (Ref. 60). The coated glass particles produced composites having a resistivity of 0.25 ohm-cm at a concentration of approximately 25 vol %. The particle size distribution of these particles and their geometric characteristics are unknown. A sieved fraction having a maximum size of 74 μm behaved exactly like a fraction having a maximum size of 210 μm. The silver-coated flakes also formed a composite with a resistivity of 0.6 ohm-cm at a volume loading of 25%. Concentrations below 25% were not reported for the coated particles or flakes. The aspect ratio of the flakes was not reported, although the large dimensions were between 105 and 250 μm.

The major difficulty in using metal-coated glass fibers appears to be processing them so that the composite's surface contains enough fibers to form a conductive surface. The formation of a resin-rich surface is not a detriment in most glass fiber-filled composites, where the appearance and chemical resistance of the resin are used to its best advantage. The lack of a conductive surface may not be a serious limitation if the interior is highly conductive and easily accessible by suitable grounding taps.

Shielding Behavior

The major problem associated with comparing EMI shielding attenuation data reported from different sources is that of variability due to the measurement facility. Earlier in this chapter detailed descriptions of two commonly used measurement systems were given. It was shown that data obtained from a dual-chamber facility can be correlated with those obtained in a transmission line device when both facilities are operated in the far field. This assumes that both systems are constructed and operated to minimize potential errors. Stutz showed that close correspondence among three different facilities could be obtained at 1 GHz (Ref. 28). Since a 1-GHz signal becomes a plane wave only 4 cm from its transmission point, data generated at this frequency should be appropriate for comparisons. In this chapter all reported shielding attenuation values are those generated at 1 GHz. An additional feature that is presented is the dynamic range of the test facility, if reported, since this represents the recorded sensitivity of the unit.

Even with these precautions, there are instances of significant variations in data reported from different measuring systems on presumably the same material. Table 5 summarizes data reported by Woodham and Gerteisen for three types of composites tested in three

TABLE 5 Comparison of Shielding Effectiveness of Injection-Molded Polycarbonate Composites Measured in Different Test Facilities

Filler	ϕ	Thickness (mm)	Dual-chamber room	Dual-chamber box	Transmission line
Zinc spray coat	—	—	60	53	46
Aluminized glass fibers	0.13	3	9	10	13
Aluminized glass fibers	0.13	6	33	11	—
Aluminum flakes	0.16	3	26	24	15
Aluminum flakes	0.16	6	40	18	—
Carbon fibers	0.18	3	42	32	31
Carbon fibers	0.18	6	40	28	—

different facilities (Ref. 61). A zinc-coated polycarbonate plaque was used to establish the dynamic range of these facilities. The reported dynamic ranges for the three experimental facilities were lower than the 70-dB level desired for satisfactory sensitivity. In particular, the dynamic range value of 46 dB for the transmission line device was lower than has been reported for that unit (Ref. 28). This may be due to the fact that the transmission line device used in these experiments was not designed to measure the shielding effectiveness of films and coatings. As Table 5 shows, there is a significant discrepancy in the recorded shielding attenuations in the dual-chamber room facility according to the thickness of the sample. The dual-chamber box facility shows similar readings for both the 3-mm-thick samples and 6-mm-thick samples. According to the plane wave theory, the less conductive samples should have increased in shielding effectiveness by approximately 3.5 dB, and the more conductive samples should have gained between 10 and 15 dB of attenuation by the increase in sample thickness. Transmission line values were obtained only for the 3-mm-thick samples. The data obtained with the transmission line device were more closely in line with those generated by the dual-chamber box system.

Woodham and Gerteisen concluded that the three test facilities did not closely correlate with each other. Several reasons for this example of poor correlation exist. The primary reason is the nature of conductive composites. It was pointed out earlier that polymers filled with conductive particles undergo a rapid transition from insulator to conductor. During this transition the volume resistivity of the composite changes by 12 to 16 orders of magnitude. It was also shown that the shielding effectiveness of composite samples tested in the far field correlates very closely with theoretical predictions based on sample resistivity (Refs. 39,5). This means that the shielding effectiveness of composite materials also changes very rapidly over a narrow range of filler concentration. Because it is difficult to generate narrowly reproducible resistivity data in the region of the transition from insulator to conductor, it is not surprising that difficulties will be experienced in obtaining narrowly reproducible shielding effectiveness data in the same region. To further complicate the situation, the levels of desirable shielding attenuations are encountered close to the transition concentration. In Woodham and Gerteisen's comparison of test facilities different samples were used in each facility (Ref. 62). The recorded variations in shielding effectiveness could represent variations among the different samples and not variations among the test facilities. This possibility is examined more closely in the following discussion.

Figures 23 through 26 show the relationship between shielding effectiveness and volume loading of conductive filler for four metallic fillers. The pertinent data describing the data points on these figures are presented in Tables 6 and 7. Resistivity values are

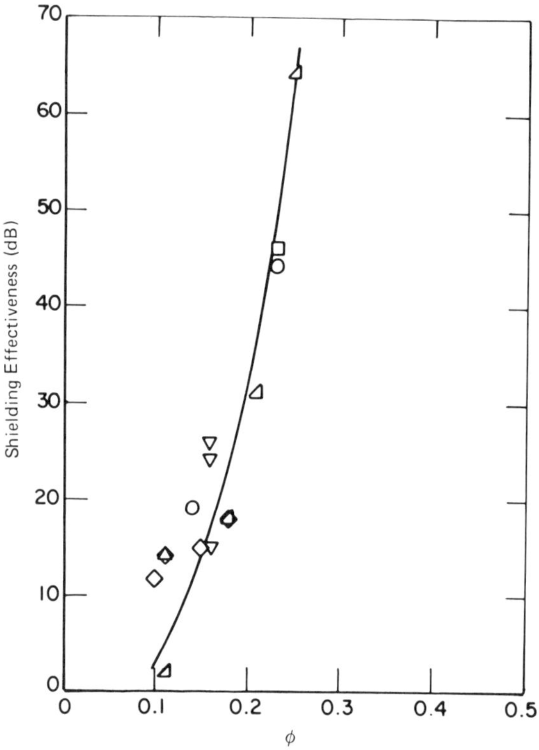

FIGURE 23 Shielding effectiveness versus volume fraction of aluminum flakes in various polymers. Composites are identified by symbols in Table 6.

presented only for data obtained by the author. These samples were carefully examined for uniformity of filler dispersion. Lack of filler uniformity is a major source of error in resistivity measurements. The data in Figures 23 through 26 show a sharp increase in shielding effectiveness with increasing filler content. The data of Woodham and Gerteisen are included in Figures 23 and 26. The variability in the data among the three test facilities appears as normal scatter when presented in this fashion.

These figures show the dramatic effect of fiber aspect ratio on the volume loading of filler required to achieve appreciable EMI shielding levels. This can be observed both within the same filler classes (Figures 24, 25, and 26) and between different filler types (compare Figure 24 to Figure 25).

De Bruyne and Van Loock made a comprehensive study of the effect of aspect ratio and concentration on shielding in the microwave region of the frequency spectrum (Refs. 64,65). They investigated steel fibers of various diameters, between 4 and 50 μm, and length combination in a silicone rubber matrix. Figure 27 shows the shielding behavior of composites containing fibers with aspect ratios of 125:1 and 250:1. From these curves the volume loading required to achieve a certain level of shielding using fibers of either size can be determined. The higher aspect ratio fibers exhibit significantly higher shielding characteristics than the lower aspect ratio fibers. For example, the fibers with an aspect ratio of 250:1 will provide shielding attenuation of 40 dB at approximately 0.4 vol %, while 1.5 vol % of the 125:1 L/D fibers is needed to provide 40 dB shielding attenuation.

Figure 28 presents the same information in a different manner. The shielding attenuation is plotted against aspect ratio at two volume loadings of fibers. Again, the effect of aspect ratio is demonstrated. De Bruyne and Van Loock reported that the dynamic range of their

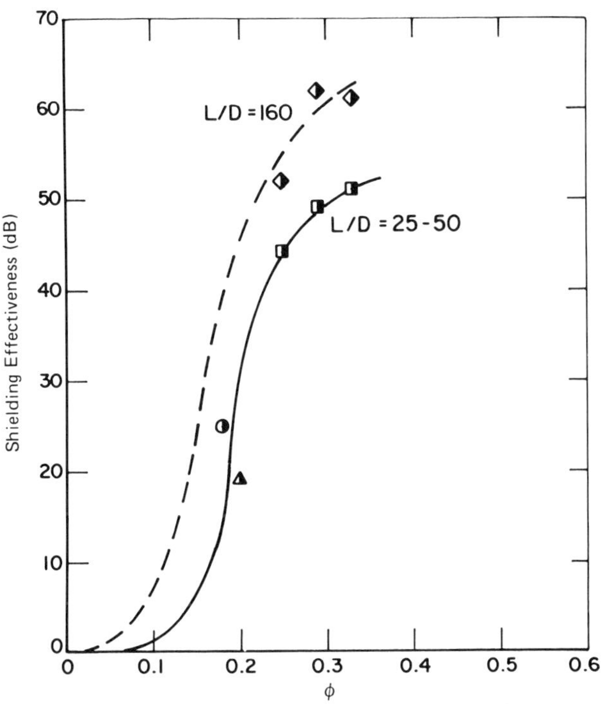

FIGURE 24 Shielding effectiveness versus volume fraction of aluminum fibers and ribbon in various polymers. Composites are identified by symbols in Table 6.

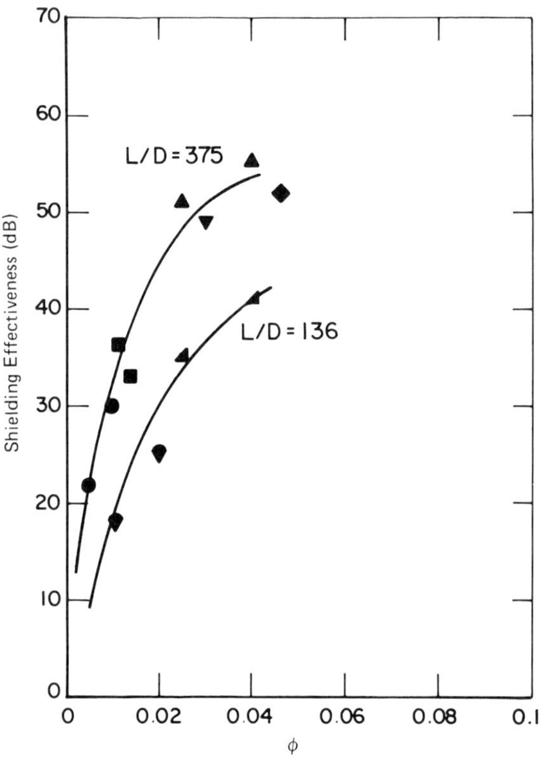

FIGURE 25 Shielding effectiveness versus volume fraction of stainless steel fibers in various polymers. Composites are identified by symbols in Table 6.

test facility was 60 dB. This accounts for the upper limit data at 60 dB, which do not correspond to the curves drawn on Figures 27 and 28. De Bruyne and Van Loock also noted the importance of making electrical contact between the sample and test facility. The samples reported on here were compressed 20% to fit the test fixture and ensure electrical contact. Because the silicone was compressed, these data cannot be directly compared to those for a rigid polymer. These composites were also not processed under shear. This explains why they were able to obtain over 40 dB of attenuation with 1250:1 aspect ratio fibers (similar in aspect ratio to fibers used by Davenport and Bradish) at only 0.25 vol % loading, while Davenport and Bradish required 6 vol %. Compressing the samples by 20% increases the effective volume loading by approximately 20%. Also, this mechanically connects fibers which may have been just beyond contact with an adjacent fiber.

The most prominent example of scatter is for the aluminized glass fiber composites. Two distinct regions can be observed in Figure 26. In processes which do not impart high shear to the fibers, where the high aspect ratios are maintained, the shielding effectiveness values rise sharply with increasing fiber content. Composites in which shear mixing was employed show a much lower increase in shielding effectiveness with increasing fiber content. These curves do not show the sharp transition from a nonshielding to a shielding composite that can be observed with the other composites. This is probably due to the difficulty in generating conductive surfaces in

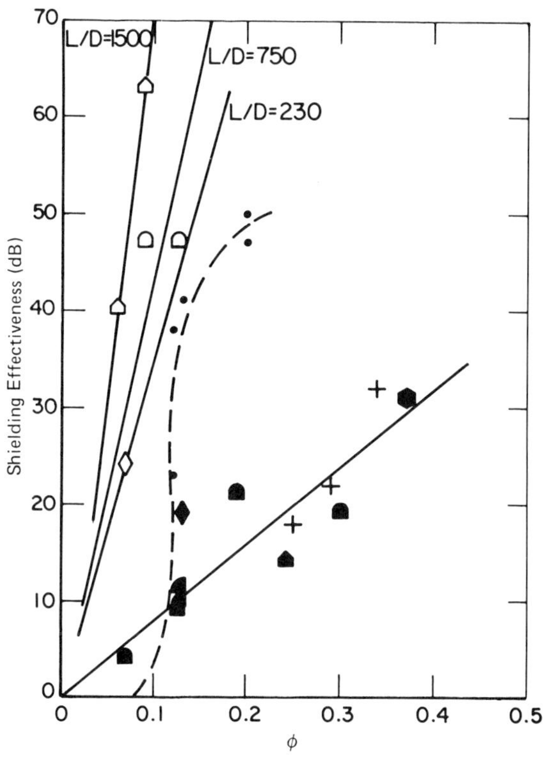

FIGURE 26 Shielding effectiveness versus volume fraction of aluminum coated glass fibers in various polymers. Composites are identified by symbols in Table 7. Dashed line and data points (•) are for injection-molded carbon fiber-filled composites.

TABLE 6 Shielding Effectiveness of Metal-Filled Polymer Composites

Filler	L/D	ϕ	Polymer	ρ (ohm-cm)	SE (dB)	DR (dB)	Symbol
Steel fibers	375	0.01	Polyester	2.2	30	70	●
Steel fibers	136	0.01	Polyester	7.2	18	70	♦
Steel fibers	375	0.005	Polyester	7.6	22	70	●
Steel fibers	136	0.02	Polyester	3.6	25	70	♦
Steel fibers	375	0.014	Epoxy	1.0	33	70	■
Steel fibers	375	0.011	Epoxy	2.6	36	70	■
Steel fibers	375	0.046	ABS	0.17	52	70	◆
Steel fibers	375	0.025	Polypropylene	—	51	70	▲
Steel fibers	136	0.025	Polypropylene	—	35	70	◢
Steel fibers	375	0.040	Polypropylene	0.21	55	70	▲
Steel fibers	136	0.040	Polypropylene	0.67	41	70	◢
Steel fibers	375	0.030	Polycarbonate	0.46	49	70	▼
Aluminum fibers	25	0.18	Polyester	—	25	70	◑
Aluminum fibers	50	0.25	PVC	—	44	70	◨ [63]
Aluminum fibers	50	0.29	PVC	—	49	70	◨ [63]
Aluminum fibers	50	0.33	PVC	—	51	70	◨ [63]

Aluminum fibers	40	0.20	Polycarbonate	—	19	70	▲	
Aluminum ribbon	160	0.25	PVC	—	52	70	◆	63
Aluminum ribbon	160	0.29	PVC	—	62	70	◆	63
Aluminum ribbon	160	0.33	PVC	—	61	70	◆	63
Aluminum flake	50	0.10	Polyester	25	12	70	◇	
Aluminum flake	50	0.18	Polyester	—	13	70	◇	
Aluminum flake	50	0.15	Polyester	8.1	15	70	◇	
Aluminum flake	50	0.14	ABS	6.6	19	70	○	
Aluminum flake	50	0.11	Epoxy	17.0	14	70	△	
Aluminum flake	50	0.23	ABS	0.31	44	70	○	
Aluminum flake	50	0.23	Polycarbonate	0.18	46	70	□	
Aluminum flake	50	0.16	Polycarbonate	—	26	60	▽	61
Aluminum flake	50	0.16	Polycarbonate	—	24	53	▽	61
Aluminum flake	50	0.16	Polycarbonate	—	15	46	▽	61
Aluminum flake	50	0.11	PVC	—	2	70	⊿	63
Aluminum flake	50	0.18	PVC	—	18	70	⊿	63
Aluminum flake	50	0.21	PVC	—	31	70	⊿	63
Aluminum flake	50	0.25	PVC	—	64	70	⊿	63

TABLE 7 Shielding Effectiveness of Aluminized Glass Fiber-Filled Composites

Polymer	ϕ	L/D	Molding technique	SE (dB)	DR (dB)	Symbol
PE	0.09	1500	Compression	63	70	⌂ 57
PE	0.06	1500	Compression	40	70	⌂ 57
PE	0.125	750	Compression	47	70	∩ 57
PE	0.09	750	Compression	47	70	∩ 57
PE	0.07	230	Compression	24	70	◇
ABS	0.24	230	Injection	14	70	
PP	0.07	230	Injection	4	70	■
PP	0.30	230	Injection	19	70	■
PP	0.19	750	Injection	21	70	■
PC	0.13	475	Injection	19	70	◆
PC	0.13	475	Injection	9	60	▼ 61
PC	0.13	475	Injection	10	53	▼ 61
PC	0.13	475	Injection	13	46	▼ 61
PC	0.13	475	Injection[a]	33	60	– 61
PC	0.13	475	Injection	11	53	– 61
PVC	0.25	70	Injection	18	70	+ 63
PVC	0.29	70	Injection	22	70	+ 63
PVC	0.34	70	Injection	32	70	+ 63

[a]Sample thickness is 6 mm. All other samples are 3 mm thick.

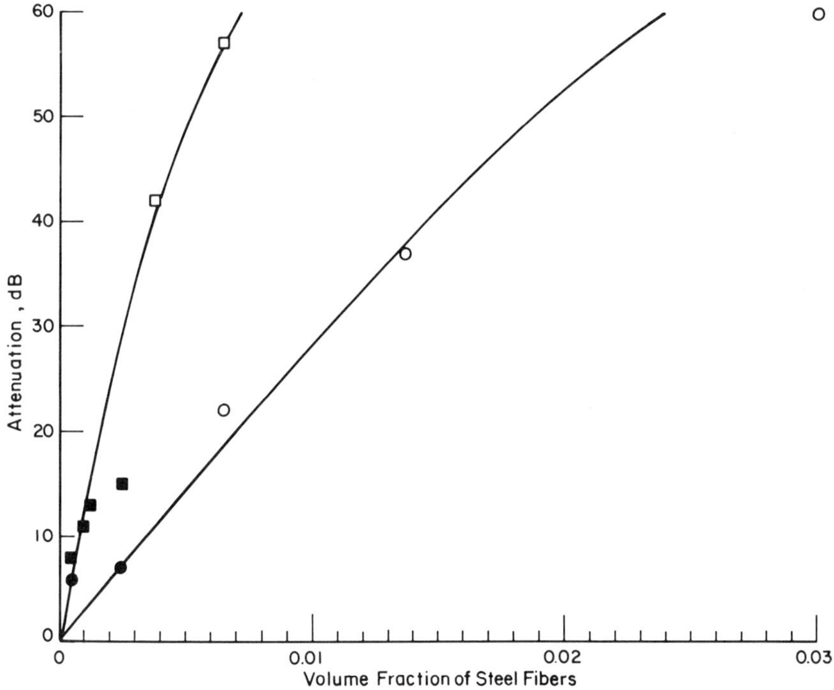

FIGURE 27 Effect of aspect ratio on shielding effectiveness of stainless steel fiber-silicone rubber composites as a function of volume fraction of fibers. (○,●) L/D = 125. (□,■) L/D = 250. Open symbols measured at 9 GHz; filled symbols measured at 2.45 GHz.

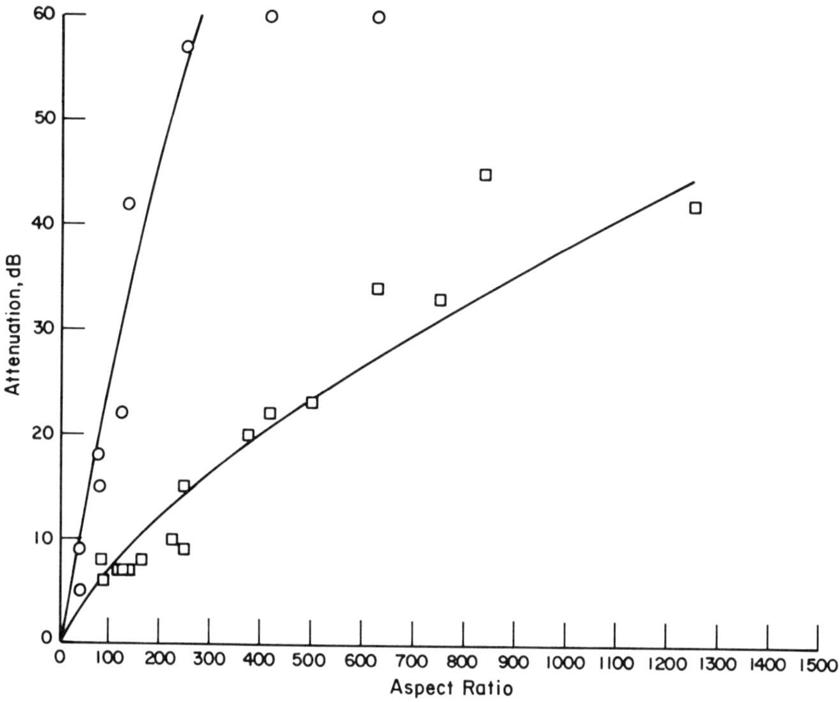

FIGURE 28 Shielding effectiveness of stainless steel fiber-silicone rubber composites as a function of fiber aspect ratio. (○) ϕ = 0.0065, measured at 9 GHz; (□) ϕ = 0.0025, measured at 2.45 GHz.

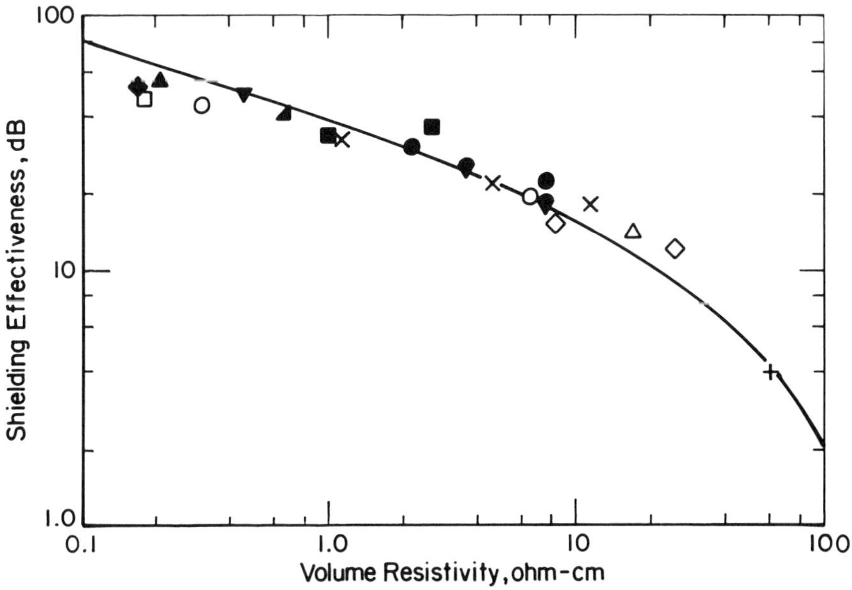

FIGURE 29 Shielding effectiveness of 3-mm-thick composites at 1 GHz versus volume resistivity. Data from Tables 6 and 7. Solid line is theoretical curve from Figure 6.

aluminized glass fiber-filled composites that are injection molded. Carbon fiber-filled composites having similar fiber sizes show a sharp increase in shielding effectiveness at approximately 10 vol %. Data for injection-molded carbon fiber-filled composites follow the dashed line in Figure 26. The wide range in scatter on this figure can also be attributed to difficulties in producing highly conductive surfaces which electrically connect to electrodes and shielding facilities. The same problem was observed when measuring resistivities of such samples.

Figure 29 shows the relationship between shielding effectiveness and volume resistivity for metal-filled composites. There is good agreement between the data and plane wave theory.

SUMMARY

Because they are excellent conductors of electricity, metal particles are suitable additives for inducing electrical conductivity in plastic materials. Metals are not effective reinforcements and are considerably more dense than polymers. Therefore, it is desirable to minimize their content in a conductive composite. It has been shown that highly conducting composites can be produced with low concentrations of metal fillers provided the aspect ratio of the filler particles can be kept high. An exponential relationship exists between the aspect ratio of a filler and the concentration of that filler required to create a conductive network. The ductility of metals and the relative fragility of particles of small diameter, or thickness, can lead to a substantial reduction in average particle aspect ratio during compounding and molding. Such considerations are leading to changes in traditional compounding and molding techniques as the use of metal-filled plastic composites grows.

NOMENCLATURE

A absorbed electromagnetic radiation (dB)

A_e area of contact electrode in Figure 9

C average number of contacts per particle

D_r diameter of guard electrode (cm)

D_o outside diameter of inner electrode (cm)

C_p critical number of contacts per particle for network formation

E activation energy for electron hopping

E_p local polarization energy across electron barrier

F reciprocal of planar packing factor

P_R power received by an antenna (watts)

P_T power transmitted by an antenna (watts)

P probability that an electron will hop from one energy state to another

P_c critical probability of network formation

R_1 direct reflection of electromagnetic radiation (dB)

R_2 internally rereflected electromagnetic radiation (dB)

R_s measured resistance in Eq. (15) (ohms)

R_v measured resistance in Eq. (14) (ohms)

SE shielding effectiveness (dB)

T temperature (K)

Z coordination number

a constant in Eq. (1)

a_c critical surface fraction of polymer particle that must be covered by metal

f frequency (Hz)

k Boltzmann constant

r distance through which an electron must jump from one energy band to another

t thickness (cm)

α constant in Eq. (2)

ϕ volume fraction of filler

ϕ_c critical volume fraction for network formation

ϕ_m maximum packing fraction of filler

τ tunneling factor defined in Eq. (2)

ρ resistivity of a composite relative to copper (ohm-cm)

ρ_v volume resistivity of a composite (ohm-cm)

ρ_s surface resistivity (ohm/sq)

μ magnetic permeability relative to free space

REFERENCES

1. R. P. Kusy and R. D. Corneliussen, Polym. Eng. Sci., *15*:107 (1975).
2. FCC Rules and Regulations, *11*, Part 15, Docket No. 20780 (July 1981).
3. J. J. Reilly, *EMI/RFI Shielding Plastics*, Society of Plastics Engineers, Brookfield, CT (1982), p. 57.
4. L. P. Fenocketti, *EMI/RFI Shielding Plastics*, Society of Plastics Engineers, Brookfield, CT (1982), p. 149.
5. D. M. Bigg, J. Rheol., *28*:501 (1984).
6. J. E. Hager, Mod. Plast., *47*(1):142 (1970).
7. R. E. C. Read and C. D. Stow, Br. J. Appl. Phys. (J. Phys. D.), Ser. 2, *2*:567 (1969).
8. R. E. Burgess, J. Polym. Sci. C, *17*:31 (1967).

9. E. O. Forster, IEEE Trans. Power Appar. Syst., *PAS-90*:913 (1971).

10. R. H. Norman, *Conductive Rubbers and Plastics*, Applied Science, Essex, England (1970).

11. R. M. Scarisbrick, J. Phys. D, *6*:2098 (1973).

12. M. Narkis, A. Ram, and F. Flashner, Polym. Eng. Sci., *18*: 649 (1978).

13. V. E. Gul, et al., Sov. Plast., (4):46 (1966).

14. M. H. Polley and B. B. S. T. Boonstra, Rubber Chem. Technol., *30*:170 (1957).

15. N. F. Mott, Adv. Phys. (Philos. Mag. Suppl.), *16*:49 (1967).

16. J. P. Reboul, J. Appl. Phys., *46*(7):2961 (1975).

17. J. P. Reboul and G. Moussalli, Int. J. Polym. Mater., 5:133 (1976).

18. J. Gurland, Trans. Metall. Soc. AIME, *236*:642 (1966).

19. J. Janzen, J. Appl. Phys., *46*:966 (1975).

20. H. Scher and R. Zallen, J. Chem. Phys., *53*:3759 (1970).

21. F. Bueche, J. Appl. Phys., *43*:4837 (1972).

22. F. Bueche, J. Polym. Sci. Phys. Ed., *11*:1319 (1973).

23. D. M. Bigg, Polym. Eng. Sci., *21*:76 (1981).

24. J. Yamaki, O. Maeda, and T. Katayama, Rev. Elec. Comm. Labor., *26*:616 (1978).

25. J. V. Milewski, Ind. Eng. Chem. Prod. Res. Dev., *17*:363 (1978).

26. D. M. Bigg, W. Mirick, and D. E. Stutz, Polym. Testing, *5*:169 (1984).

27. D. R. J. White, *EMI/EMC Handbook Series*, Don White Consultants, Germantown, MD (1971), vol. 4.

28. D. E. Stutz, Proc. 1981 Southern Prof. Prog., Session 21, p. 1 (1981).

29. F. W. Bradish, SPI Reinf. Plast./Comp. Inst. Tech. Pap., *31*:7-D (1976).

30. V. I. Prosvirin and A. Savitskii, Mech. Polym., (2):323 (1971).

31. J. E. Scheer and D. J. Turner, Adv. Chem., *99*:572 (1971).

32. Plast. Des. Prog., *15*(11):22 (1975).

33. D. M. Bigg, Composites, *10*:95 (1979).

34. M. J. Folkes, *Short Fibre Reinforced Thermoplastics*, Wiley, Letchworth, England (1982), p. 67.

35. J. M. Lunt and J. B. Shortall, Plast. Rubber Process., vol. 5 p. 37 (June 1980).

36. U. S. Patent 2,877,501 (1959). R. Bradt, patent holder.

37. U. S. Patent 3,042,570 (1962). R. Bradt, patent holder.

38. M. G. Bader and W. H. Bowyer, Composites, *4*:150 (1973).

39. D. M. Bigg and D. E. Stutz, Polym. Comp., *4*:40 (1983).

40. S. M. Aharoni, J. Appl. Phys., *43*:2463 (1972).

41. L. Nicodemo, et al., Polym. Eng. Sci., *18*:293 (1978).

42. F. F. T. De Araujo and H. M. Rostenberg, J. Phys. D: Appl. Phys., *9*:1025 (1976).

43. S. H. Kwan, F. G. Shir, and W. L. Tsui, J. Mater. Sci., *15*:2978 (1980).

44. V. I. Shorokhova and L. L. Kuzmin, Sov. Plast., (3):26 (1965).

45. J. W. Dally and G. A. Panizza, Exp. Mech., *12*:124 (1972).

46. D. W. Marshall and J. C. Agarwal, U.S. Patent 3,983,075 (1976).

47. K. H. Chen, *Advances in the Astronautical Sciences*, vol. 50, part 2 (H. M. Hsia, R. W. Longman, Y. L. Chou, eds.), American Astronomical Society (1983), p. 1487.

48. R. M. Simon, *EMI/RFI Shielding Plastics*, Society of Plastics Engineers, Brookfield, CT (1982), p. 77.

49. M. G. Golubeva, et al., Kolloid. Zh., *35*:745 (1973).

50. M. G. Golubeva, T. I. Shiskkina, and V. E. Gul, Kolloid. Zh., *35*:1145 (1973).

51. A. Malliaris and D. T. Turner, J. Appl. Phys., *42*:614 (1971).

52. R. P. Kusy, J. Appl. Phys., *48*:5301 (1977).

53. S. K. Bhattacharya, S. Basu, and S. K. De, Composites, *9*:177 (1978).

54. J. W. Cahn, Acta Metall., *14*:477 (1966).

55. F. Hochberg, U.S. Patent 2,721,357 (1955).

56. E. G. Crosby and F. C. Hornbeck, U.K. Patent 1,479,569 (1977).

57. F. W. Bradish, Annu. Tech. Conf. Reinf. Plast./Comp. Inst. SPI, *33*:4-A (1978).

58. D. E. Davenport, Plast. Eng., *34*(12):43 (1978).

59. S. R. Gerteisen, G. W. Woodham, and D. R. Wetzel, SPE ACTEC, *25*:766 (1979).

60. R. E. Heinze and J. R. Ritter, Annu. Tech. Conf. Reinf. Plast./Comp. Inst. SPI, *32*:8D (1977).

61. G. Woodham and S. Gerteisen, Annu. Tech. Conf. Reinf. Plast./Comp. Inst. SPI, *35*:3E (1980).

62. S. Gerteisen, personal communication.

63. L. Kleiner and A. S. Pazur, SPE ANTEC, *28*:788 (1982).

64. R. De Bruyne and W. Van Loock, J. Microwave Power, *12*:146 (1977).

65. R. De Bruyne and W. Van Loock, J. Microwave Power, *12*(4): 361 (1977).

66. R. Matsushita, M. Senna, and H. Kuno, J. Mater. Sci., *12*: 509 (1977).

4

Mechanical and Thermal Properties

T. S. CHOW AND RICHARD C. PENWELL

Xerox Corporation
Webster, New York

INTRODUCTION

The use of reinforced polymeric materials has been growing steadily and will continue to do so. More desirable mechanical and thermal properties are two major reasons for this growth. Intrinsic properties of polymers combined with the advantages of fillers have led to the creation of many new materials. These materials have one or a combination of excellent mechanical, chemical, thermal, or electrical properties at lighter weights than their predecessors. Recent advances in computer-aided engineering have added and will continue to add a greater level of sophistication to the design and use of filled polymeric materials. Particle size, orientation, and filler concentration will soon be assimilated into part design and melt flow software.

It is becoming more important to have a physical understanding of polymer-filler interactions so that the best models can be developed for use. Work to date on mechanical and thermal properties of polymers containing dispersed metallic or nonmetallic particles clearly indicates that not only are the material properties of the individual phases and percentage of filler important, but the size, shape, orientation, and state of adhesion between the filler and matrix are critical.

Polymer composites containing nonmetallic fillers, such as glass or graphite, have been extensively studied both theoretically and experimentally (Refs. 1–12). There are few range restrictions on most theoretical models and they should be directly applicable to composite systems with metallic fillers.

The major differences between metallic and nonmetallic fillers are the usually higher values of elastic modulus, thermal expansion coefficient, and thermal conductivity for metal fillers. These differences can affect the mechanical properties in two ways. First, the filler-to-matrix elastic modulus ratio E_f/E_m is approximately 20–100 for nonmetallic fillers and 200–1000 for metallic fillers. Second, the adhesion between the matrix and filler particles is more critically influenced in the case of metal-filled polymers. This is due to the greater internal stresses (Refs. 13,14) introduced during the fabrication process (Ref. 15).

Not a great deal of good experimental data is available for metal-filled composites. The picture can become very diffuse if one tries to model, understand, and control the adhesion between the matrix and filler, particle size and distribution, orientation and concentration. The focus in this chapter will be on providing a unified understanding and approach, with a strong theoretical emphasis, to the effect of fillers on mechanical and thermal properties of polymers. General properties and features will be illustrated primarily with nonmetallic fillers. A brief treatise on the effects of particle orientation is included. Detailed discussions of adhesion are not included. The first topic to be discussed is stress-strain behavior.

STRESS-STRAIN BEHAVIOR

When simple stress-strain behavior of polymeric composites is considered, the general effect of the filler is predictable. The ultimate stress and strain, though, do not lend themselves to the same precision as does the modulus. Ultimate stress and strain in composites depend a great deal not only on the matrix and filler properties, but on the fabrication process. Certainly, flaws, voids, and temperature profiles significantly affect fracture properties.

General trends can be seen in Figure 1. The tensile stresses for an epoxy resin and the same resin containing 14% aluminum oxide whiskers are given as a function of elongation. The properties of the materials in Figure 1 are: $E_f = 60 \times 10^6$ psi, $E_m = 3.5 \times 10^5$ psi ($E_f/E_m = 171$), ϕ (volume fraction) = 0.14, and ρ (aspect ratio) = 120 to 1000. For these materials, as ϕ increases, the composite modulus E increases, the ultimate strain e_y decreases, and the ultimate strength σ_y increases. Predictive models of these effects will be discussed in more detail in the following sections. The emphasis

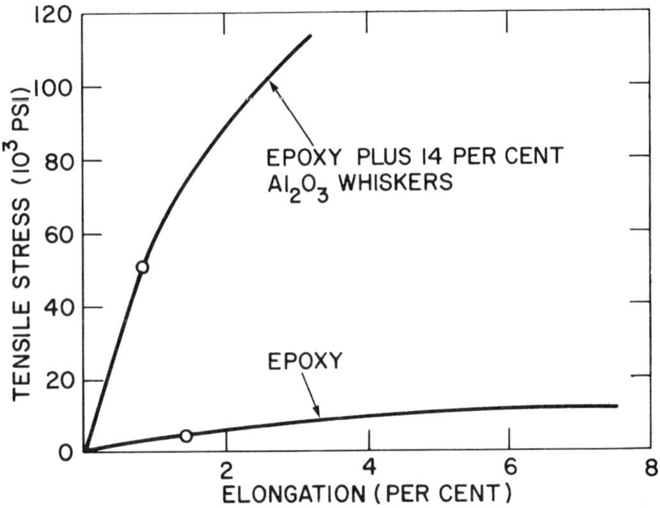

FIGURE 1 Stress-strain curves of unreinforced and reinforced epoxy resin. (From Kelly, Ref. 16.)

will be on the tensile properties: modulus and strength. Anisotropic moduli will also be discussed briefly.

TENSILE MODULUS

As seen in the preceding section, one effect of a filler in a polymeric matrix can be to increase the modulus. A brief survey of the important and recently developed relationships for composite Young's moduli is the basis of this section.

Begin by considering a two-phase composite that consists of randomly distributed particles aligned in a matrix. The adhesion between the filler and matrix is excellent and the filler is uniform in size. Both the filler and the matrix material are assumed to be homogeneous and isotropic. Young's moduli for the filler and matrix are E_f and E_m, μ_f and μ_m are the corresponding shear moduli, ϕ is the volume fraction of filler, and ρ is the aspect ratio. The initial consideration is determining the composite tensile modulus $E_{\|}$ in the longitudinal direction as a function of the aspect ratio and composition ϕ. The relationships for estimating the elastic moduli of two-phase materials are usually complex.

Bounds on the Modulus

The simplest cases for predicting the tensile modulus, which are the two bounds, have been discussed by Ashton et al. (Ref. 17) and Broutman and Krock (Ref. 18). The upper bound contains only the composition variable and is known as the mixture rule or series model. It assumes equal strains in the two phases under elastic deformation and is given by

$$E_{\parallel} = (1 - \phi)E_m + \phi E_f \tag{1}$$

If the stresses are assumed equal in the two phases, the parallel model describes the lower bound and is

$$E_{\parallel} = \left(\frac{1 - \phi}{E_m} + \frac{\phi}{E_f}\right)^{-1} \tag{2}$$

Equations (1) and (2) have been applied to various physical properties. The coefficient of thermal expansion as described by Holliday and Robinson (Ref. 19), the dielectric constants discussed by Von Hippel (Ref. 20), and the thermal conductivity discussed by Progelhof et al. (Ref. 6) are examples.

Takayanagi's Model

Takayanagi et al. (Ref. 21) have proposed a series-parallel model that combines Eqs. (1) and (2). Their relationship is

$$E_{\parallel} = \left[\frac{\psi}{\chi E_m + (1 - \chi)E_f} + \frac{1 - \psi}{E_f}\right]^{-1} \tag{3}$$

The matrix region is described by volume $\psi\chi$ and the filler region by $(1 - \psi\chi)$. Their relationship was developed to consider the modulus of crystalline polymers. The basic problem with the model is how to determine the values of ψ and χ.

The Cox Equation

The model by Cox (Ref. 22) is referred to as the shear-lag theory. Normal stresses in the matrix are not taken into account. The tensile stresses are carried by the filler material and the load is transferred from filler to filler particle through shear. The composite longitudinal Young's modulus $E_{\parallel}$ is given as

$$\frac{E_{\|}}{E_m} = 1 + \left[\frac{E_f}{E_m}\left(1 - \frac{\tanh z}{z}\right) - 1\right]\phi \tag{4}$$

where

$$z = \frac{\ell}{r_f}\left[\frac{\mu_m}{E_f}\frac{1}{\ln(R/r_f)^2}\right]^{1/2}$$

and ℓ is the filler length, r_f is the filler radius and R is the center-to-center distance of the filler particles. For hexagonal filler packing, Barham and Arridge (Ref. 23) give the following parametric relations

$$\phi = \frac{2\pi r_f^2}{\sqrt{3}R^2} \tag{5}$$

and

$$z = 2\rho\left[\frac{\mu_m}{E_f \ln(2\pi/\sqrt{3}\phi)}\right]^{1/2}$$

The filler shape is characterized by the aspect ratio $\rho = \ell/2r_f$, where ρ is much greater than one.

The Halpin-Tsai Equation

The Halpin-Tsai equation is a simple empirical expression reduced from Hermans' solution. It contains a geometric fitting parameter A, obtained through fits with numerical solutions from formal elasticity theory. More detail is given by Ashton et al. (Ref. 17). The composite moduli are expressed as

$$\frac{E_{\|}}{E_m} = \frac{1 + AB\phi}{1 - B\phi} \tag{6}$$

where

$$B = \frac{E_f/E_m - 1}{E_f/E_m + A}$$

For the tensile modulus, $A = 2(\ell/d)$, where ℓ/d is the aspect ratio. The self-consistent method, described by Hill (Ref. 24), Hermans (Ref. 25), and Law and McLaughlin (Ref. 26), which served as the basis for Eq. (6), has been more rigorously applied to short-fiber composites. However, their correlation has not yet been established.

The Chow Equation

Chow recently derived a general theory for elastic moduli (Refs. 27, 28) and thermal expansion coefficients (Ref. 29) of a two-phase anisotropic heterogeneous material. The anisotropic particle shape effect is characterized by the ratio of major to minor axes $\rho = c/a$ of a spheroid. For a uniaxially oriented structure, the tensile modulus is given by

$$\frac{E_{\parallel}}{E_m} = 1 + \frac{(k_f/k_m - 1)G_1 + 2(\mu_f/\mu_m - 1)K_1}{2K_1G_3 + G_1K_3}\,\phi \tag{7}$$

where k is the bulk modulus, μ is as defined earlier, and

$$K_i = 1 + \left(\frac{k_f}{k_m} - 1\right)(1 - \phi)\alpha_i$$

$$(i = 1, 3) \tag{8}$$

$$G_i = 1 + \left(\frac{\mu_f}{\mu_m} - 1\right)(1 - \phi)\beta_i$$

α_i and β_i are functions of the aspect ratio ρ and Poisson's ratio ν_m of the matrix (see the Appendix).

A comparison of the longitudinal Young's modulus of a material filled with fibers aligned in the direction of the applied tensile load is given in Figure 2 for the Cox, Halpin-Tsai, and Chow equations. The materials are two filled and crystalline polymers that have a ratio E_f/E_m ranging from 21.2 (glass in epoxy resin), 100 (boron in epoxy resin), to 2400 (semicrystalline polyethylene) at constant $\phi = 0.40$. For lower values of E_f/E_m, there is little difference among the three relations. The discrepancy among them increases with increasing values of $E_f/E_m > 10^2$. Additional experimental verification of these equations is required for aspect ratios other than one and infinity over a wide range of E_f/E_m.

For spherical inclusions, the Kerner equation best predicts experimental data, as discussed by Kardos (Ref. 4) and Kenyon and Duffey (Ref. 30). If ρ is set equal to one, the results of Chow take

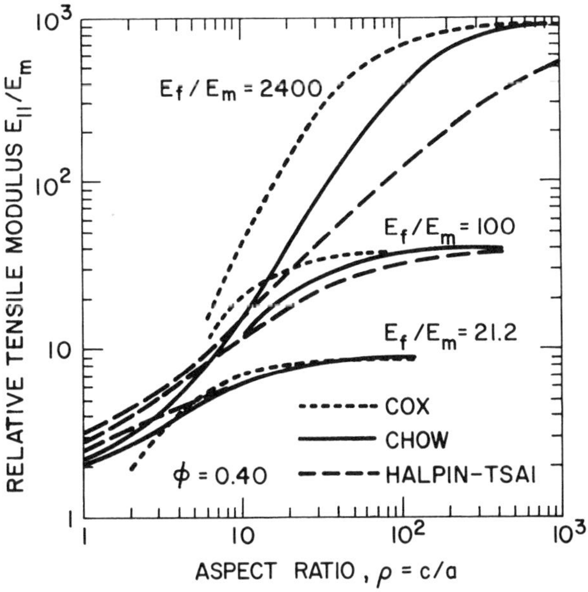

FIGURE 2 Comparison of the Cox, Chow, and Halpin-Tsai equations for three E_f/E_m ratios. (From Chow, Ref. 12.)

the same form as the Kerner equation. The longitudinal Young's modulus of a polymer filled with infinitely long fibers follows a mixing rule (Refs. 4, 16). When the aspect ratio approaches infinity, Eq. (7) becomes

$$\lim_{\rho \to \infty} \frac{E_{\parallel}}{E_m} = 1 + \phi \left(\frac{E_f}{E_m} - 1 \right) \tag{9}$$

which does have the form of a mixing rule. The summarizing point is that the Chow equations reduce to the Kerner equations for spherical fillers ($\rho = 1$) and the mixing rule for $\rho \to \infty$.

STRESS CONCENTRATION

An understanding of stress concentration in filled polymeric materials is critical to bounding their mechanical performance. Stress concentrations at fiber boundaries can cause premature failure if they are neglected. The stress concentration factor S is defined as the ratio of the maximum stress to the applied stress. When a uniform stress is applied parallel to the fiber direction, the maximum stress occurs at the ends of the fibers. The corresponding stress factor, developed in Reference 31, is:

$$S = \frac{(k_f/k_m)G_1 + 2(\mu_f/\mu_m)K_1}{2K_1G_3 + G_1K_3} \tag{10}$$

where the parameters are as defined in the preceding section. When $\phi \to 0$, and $\rho = 1$, Eq. (10) reduces to the classical Goodier equation.

For the case of rigid fibers at semidilute concentrations, Eqs. (9) and (10) become

$$S = [E_{\parallel}] = \frac{1}{1-\phi} \frac{\beta_1 + 2\alpha_1}{\beta_1\alpha_3 + 2\beta_3\alpha_1} \qquad (E_f/E_m \gg 1) \tag{11}$$

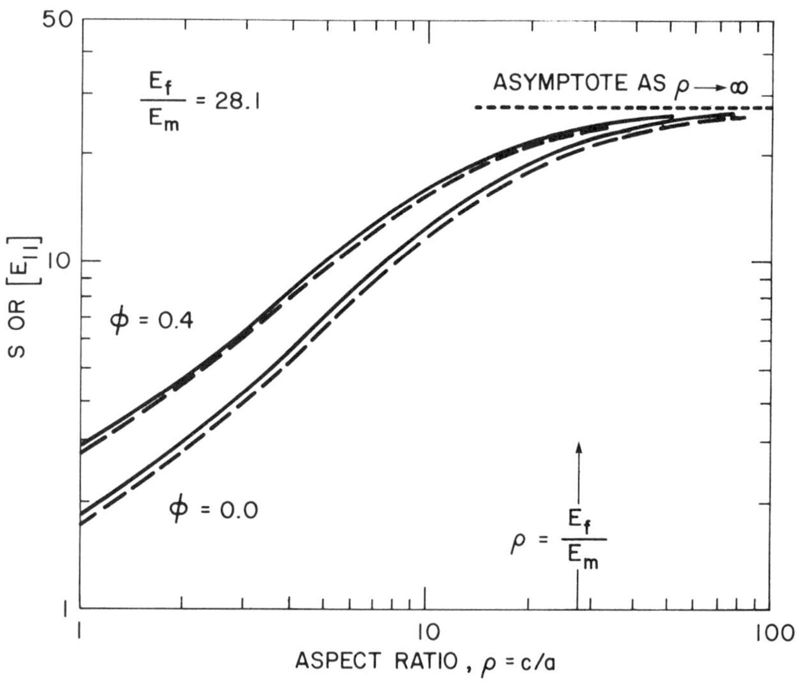

FIGURE 3 Comparison of the stress concentration factor and $(E_{\parallel} - E_m)/E_m\phi$ as a function of particle shape and filler content for glass in PPO: (—) stress concentration factor; (---) $(E_{\parallel} - E_m)/E_m\phi$. (From Chow, Ref. 36.)

where

$$[E_{\parallel}] = (E_f - E_m)/E_m\phi.$$

Equation (10) is plotted in Figure 3 for glass dispersed in polyphenylene oxide (PPO). The properties of the composite are $E_f = 73.1 \times 10^{10}$ dyn/cm^2, $\nu_f = 0.22$, $E_m = 2.6 \times 10^{10}$ dyn/cm^2, and $\nu_m = 0.35$. For $E_f/E_m \geqslant 28.1$ and using Eq. (11), $S \simeq [E_{\parallel}]$. This is usually true for most fiber-reinforced polymer composites. This indicates that Eq. (11) can provide a simple and reasonable approximation for high aspect ratios. When the aspect ratio approaches infinity, $E_{\parallel}$ reaches its asymptotic value $E_f/E_m - 1$. Figure 3 shows that the shape dependence of the stress concentration factor and the increase in tensile modulus starts to diminish for $\rho > E_f/E_m$. The diminishing effect of particle shape for $\rho > E_f/E_m$ is valid for values of E_f/E_m in the range from 10 to 10^3 (see Figure 2). This is an interesting observation to consider when processing similar composites. Advantage can be taken of this to optimize property-processing tradeoffs.

Stress concentrations cannot be used to determine the ultimate tensile strength of polymeric composites without consideration of the mode of fiber failure at the interface or in the matrix before fracture. It is unlikely that perfect bonding occurs or that flawless composites are made for evaluation. However, the stress concentration relations can be used to determine upper limits and perhaps to establish a predictive correlation with flawed composites.

ULTIMATE TENSILE STRENGTH

The ultimate tensile strength of a particulate-filled polymer is defined by Kelly (Ref. 16) as the maximum achievable stress or strain. This critical property has not received the same level of theoretical attention as the modulus of composites. This is primarily due to the complexity of fracture mechanisms. The prediction of the ultimate strength requires knowledge of the ultimate properties of the filler and matrix, the failure modes, the nature of the polymer-filler adhesion, perfection of alignment and spatial distribution of filler particles, stress concentration at the ends of the fibers, particle size and shape, and the brittle and ductile nature of the materials. These are discussed in detail by Kardos (Ref. 4), Nielsen (Ref. 5), Kelly (Ref. 16), and Chou and Kelly (Ref. 32). The basic concepts have been presented by Kelly (Ref. 16) and Nielsen (Ref. 5).

Kelly (Refs. 16, 32–34) applied the rule of mixtures and related the tensile strength in the fiber direction of composites containing discontinuous aligned fibers to the ultimate strength and volume fraction of fibers. These relationships do not include the role of filler end stress concentrations. If the adhesion between the rigid filler and matrix is good and if the fracture surface is not smooth but propagates from particle to particle, Nielsen (Ref. 5) predicted that the ultimate strain is a monotonic decreasing function of the volume fraction of filler.

Usually the primary failure mode of a filled polymer composite involves interfacial debonding, as discussed by Trachte and Di-Benedetto (Ref. 37). Based on Griffith's concept of fracture, discussed by Kelly (Ref. 16) and Davidge and Green (Ref. 35), a necessary, but not necessarily sufficient, condition for circumferential crack formation is that the strain energy stored in the two-phase medium must be equal to or greater than the energy required to create new surfaces, i.e.,

$$\Delta W \geqslant \gamma A \tag{12}$$

where $\gamma = \gamma_f + \gamma_m - \gamma_{fm}$ is the polymer-filler interfacial energy per unit area (work of adhesion) and A is the surface area produced. The strain energy associated with the formation of an equilibrium crack is due to the interaction between the elastic inhomogeneity and the uniformly applied stress σ_{ij}^A. From the generalized mean-field theory, Chow (Refs. 28, 29, 36) obtained the ultimate tensile strain

$$e_U \geqslant \left\{ \frac{12\gamma}{dE_m[E_\parallel](1 + [E_\parallel]\phi)} \right\}^{1/2} \tag{13}$$

The volume-to-surface ratio of the filler is characterized by

$$d = \frac{6V}{A} = \frac{2a}{f(\rho)} \tag{14}$$

where a is the radius of the minor axis of a spheroid and

$$f(\rho) = \frac{1}{\rho} \int_0^{\pi/2} \sin\theta[1 + (\rho^2 - 1)\sin^2\theta]^{1/2}\, d\theta \leqslant 1 \qquad (\rho \geqslant 1)$$

When $\phi \to 0$, it is expected that $e_U \to e_{U_m}$, the ultimate strain of the matrix, and

$$d \geqslant d_c \equiv \frac{12\gamma}{E_m e_{U_m}^2 [E_\parallel]_0} \tag{15}$$

where d_c is defined as the critical filler dimension and $[E_\parallel]_0 = \lim_{\phi \to 0}[E_\parallel]$ is the intrinsic tensile modulus. All materials contain inherent flaws, which act as stress concentrators, similar to fillers. Equation (15) indicates that the effect of $d > d_c$ is to introduce flaws larger than the inherent flaw size and so reduce the ultimate strain. It should also be noted that the critical filler size d_c is inversely proportional to the stress concentration factor, as shown by Eqs. (11) and (15).

The ultimate strain for glass in PPO is plotted in Figure 4. The ultimate strain decreases with an increase of ϕ or ρ and approaches its asymptote, which is a function of $[E_\parallel]$, for $\rho > E_f/E_m = 28.1$. There are very few data in the literature for short aligned

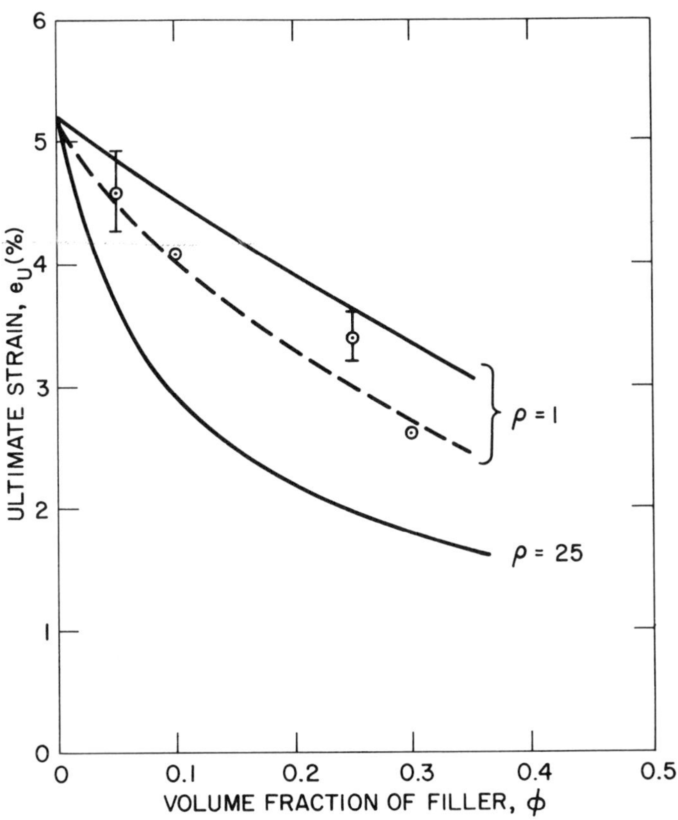

FIGURE 4 Effects of particle shape and filler content on the ultimate tensile strain of glass-filled PPO: (—) theory; (---) experiment. (From Trachte and DiBenedetto, Ref. 37.)

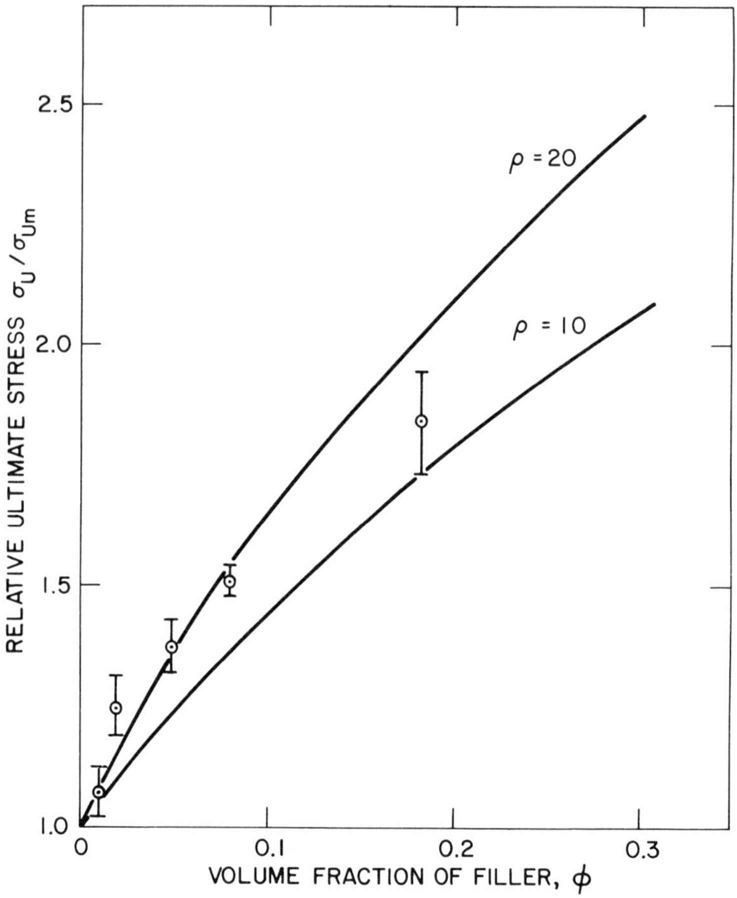

FIGURE 5 Effects of particle shape and filler content on the relative ultimate tensile stress of glass fiber-filled PMMA: (—) theory; (○) experimental data. (From Lee, Ref. 38.)

fibers, and Eq. (13) is, therefore, compared with the data of Trachte and DiBenedetto (Ref. 37) for $\rho = 1$. Equation (13) gives better agreement with the experimental data than Nielsen's Eq. (5), $e_U/e_{U_m} = 1 - \phi^{1/3}$. Assuming γ (glass bead/PPO) = 10^4 erg/cm^2, the critical diameter of the glass bead, from Eq. (15), is 9.2μ. The effect of particle size becomes an issue when it is greater than d_c.

The corresponding ultimate tensile stress is then

$$\sigma_U = \left[\left(\frac{12\gamma E_m}{d}\right)\left(\frac{1}{[E_{\parallel}]} + \phi\right)\right]^{1/2} \tag{16}$$

A comparison of Eq. (16) with Lee's experimental data (Ref. 38) for glass fibers in poly(methyl methacrylate) (PMMA) with E_m = 2.9×10^{10} dyn/cm^2 and $\nu_m = 0.35$ is made in Figure 5. The ultimate tensile stress increases with an increase of ϕ or ρ. This is

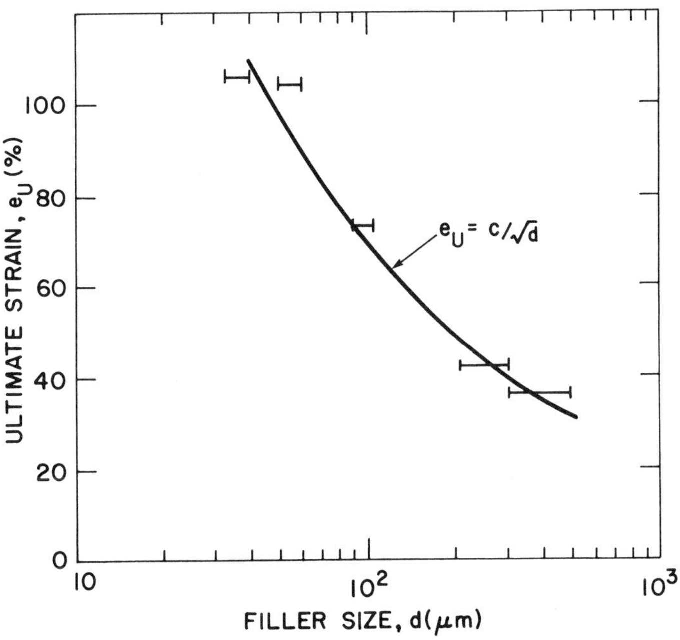

FIGURE 6 Dependence of ultimate tensile strain on particle size of sodium chloride in polyurethane rubbers: (—) experimental data. (From Schwarzl et al., Ref. 40.)

qualitatively in agreement with Kelly's equation (Ref. 16). Experimental data of Blumentritt et al. (Ref. 39) for aligned long fibers ($\rho > E_f/E_m$) are also consistent with the result that increased fiber concentration leads to increased tensile stress and to decreased ultimate strain. The theory discussed for short fiber-filled composites is based on interfacial debonding. Other failure mechanisms, including cohesive failure of the polymer matrix or fiber, can contribute to deviations between this approach and other data.

Both the ultimate tensile stress and the strain of particulate-filled polymers are inversely proportional to the square root of particle size—the ratio of volume to surfaces given by Eq. (14). Figure 6 validates this relationship for sodium chloride dispersed in polyurethane rubber, where $\rho = 1$ and $\phi \simeq 41\%$, data obtained from Schwarzl et al. (Ref. 40).

ORIENTATION EFFECTS

In general, an aligned composite will have nine independent elastic constants if it has three mutually perpendicular planes of symmetry. If the composite is isotropic in the plane normal to the longitudinal (fiber) axis, it has five independent elastic constants. This is known as a transverse isotropic composite.

Anisotropic Elastic Moduli

The elastic properties in the longitudinal (oriented) and transverse directions as a function of aspect ratio are illustrated in terms of intrinsic moduli. They are generally defined by

$$[\mu_{ij}]_0 = \lim_{\phi \to 0} \frac{\mu_{ij} - \mu_m}{\mu_m \phi} \tag{17}$$

Expressions for $[\mu_{12}]_0$, $[\mu_{13}]_0$, $[k]_0$, $[E_{\parallel}]_0$, and $[E_{\perp}]_0$ can be found in Reference 27. These relationships are the intrinsic shear moduli, the intrinsic bulk modulus, and the intrinsic Young's moduli. The intrinsic moduli are used to describe the effect of the shape of the filler particles and represent the change in moduli at dilute dispersions. The intrinsic shear moduli and longitudinal and transverse Young's moduli reflect the orientational effects of the fillers.

The five intrinsic elastic moduli are presented in Figures 7–9 as a function of ρ for $0.1 < \rho < 100$. The input parameters are given in Table 1 for glass/epoxy and boron/epoxy. The intrinsic bulk modulus is illustrated in Figure 7. The intrinsic longitudinal $[\mu_{13}]_0$ and transverse $[\mu_{12}]_0$ shear moduli are shown in Figure 8. Their asymtotic values were calculated from Eqs. (18) and (22) in the work

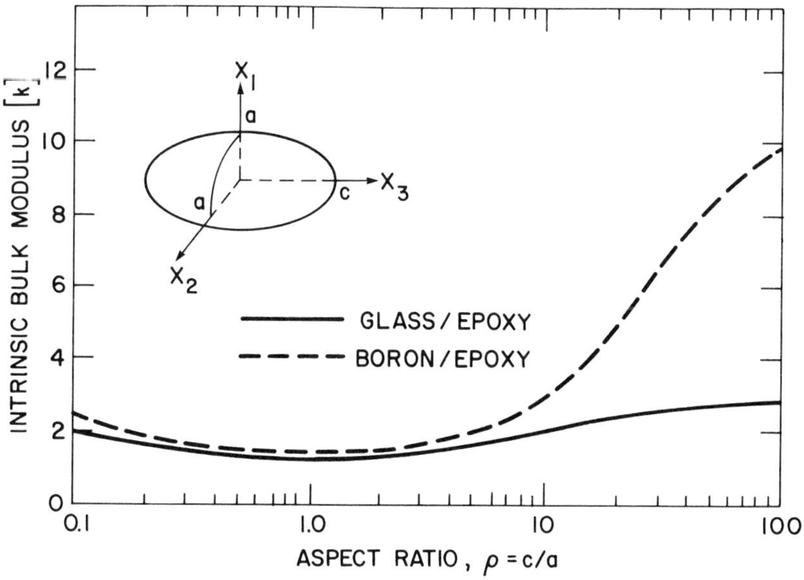

FIGURE 7 Intrinsic bulk moduli $[k]_0$. (From Chow, Ref. 27.)

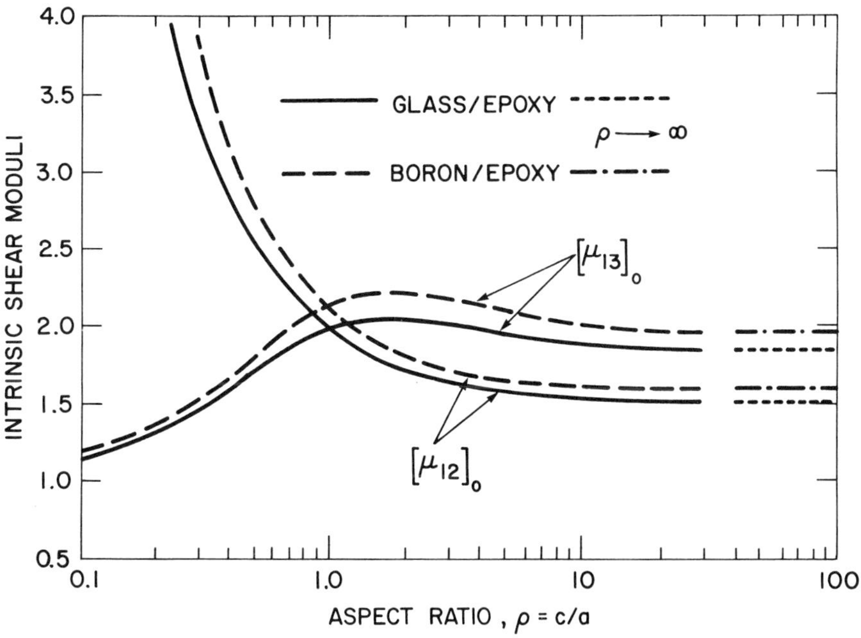

FIGURE 8 Intrinsic shear moduli $[\mu_{13}]_0$ and $[\mu_{12}]_0$. (From Chow, Ref. 27.)

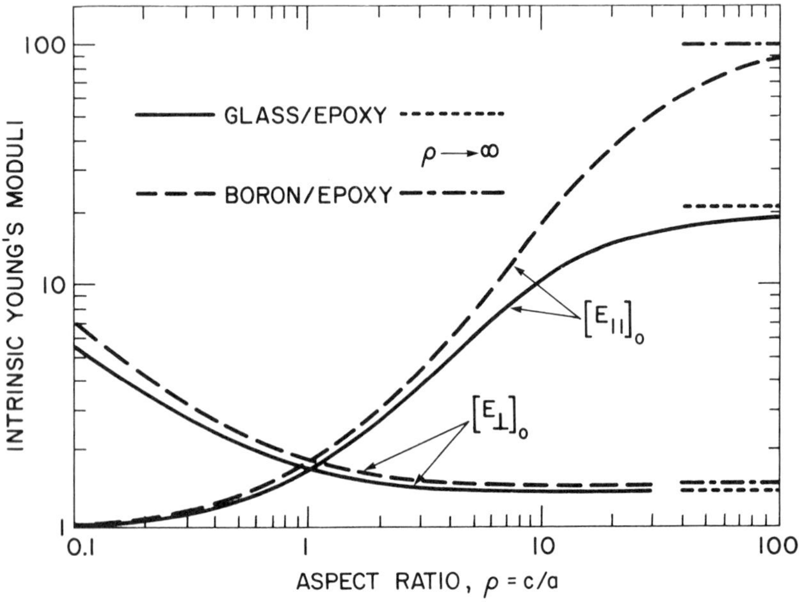

FIGURE 9 Intrinsic Young's moduli $[E_{\parallel}]_0$ and $[E_{\perp}]_0$. (From Chow, Ref. 27.)

TABLE 1 Moduli of the Filler and Matrix

Modulus	Glass in epoxy		Boron in epoxy	
	Filler (1)	Matrix (2)	Filler (1)	Matrix (2)
E ($\times 10^6$ psi)	10.6	0.50	60.0	0.60
ν	0.22	0.35	0.20	0.35
μ ($\times 10^6$ psi)	4.34	0.185	25.0	0.222
k ($\times 10^6$ psi)	6.31	0.555	33.0	0.666

of Chow and Hermans (Ref. 41). The intrinsic longitudinal $[E_{\parallel}]_0$ and transverse $[E_{\perp}]_0$ Young's moduli are shown in Figure 9. The values for $\rho \to \omega$ were calculated from relationships developed in Kelly's (Ref. 16) book. When $\rho = 1$, $\mu_{12} = \mu_{13}$, and $E_{\parallel} = E_{\perp}$, the shear and bulk moduli have the same value as predicted by the Kerner equation.

Tensile Moduli and Strength of Infinitely Long Fibers

The previous section dealt with orientation at 0 and 90°. McNally (Ref. 42) discusses results of mechanical measurements at other angles of orientation. In particular, the work of Dimmock and Abrahams (Ref. 43) for high-modulus carbon fibers in epoxy resin

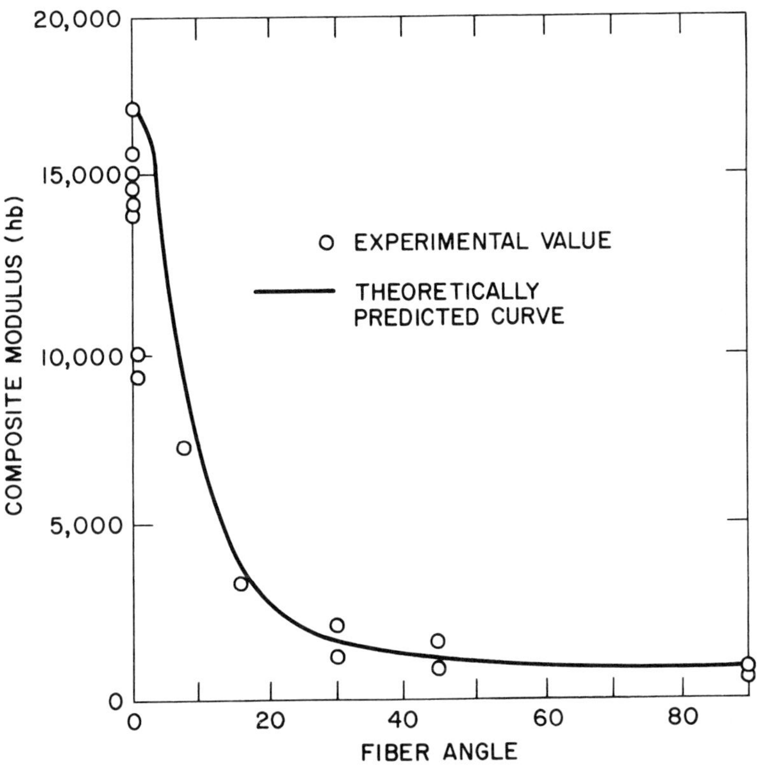

FIGURE 10 Angular dependence of modulus for a 43 vol % carbon fiber/epoxy composite (1 hb ≡ 1450 psi). (Adapted from McNally, Ref. 42.)

at 43 vol % is interesting. Although the computational details were not included, since it is for $\rho \to \infty$, their computations most likely follow Kelly's approach (Ref. 16) by including the angular dependence in the rule of mixtures. In Figure 10, the composite modulus is plotted as a function of orientation angle. At 0 and 90°, the values are equivalent to $E_{\parallel}$ and $E_{\perp}$ discussed earlier. The composite strength is given as a function of the angle between the fiber axis and the direction of applied stress in Figure 11. As the angle approaches 90°, the results are similar to Figure 10.

Orientation effects on thermal expansion will be discussed in a later section.

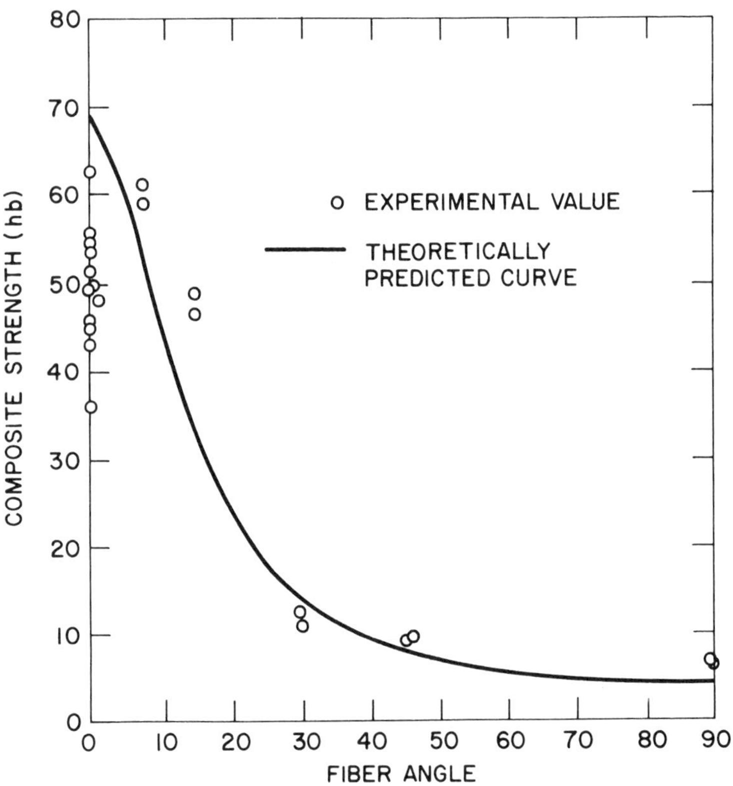

FIGURE 11 Angular dependence of composite strength for high-modulus carbon fibers in epoxy resin at 43% by volume (1 hb ≡ 1450 psi). (Adapted from McNally, Ref. 42.)

VISCOELASTIC PROPERTIES

Linear viscoelastic properties of composites can be obtained from extensions of the elastics models through straightforward application of the corrospondence principle. This has been done for many of the models discussed in previous sections. Application to the relaxation modulus and creep will now be discussed.

Relaxation Moduli

The correspondence principle has been applied to the Kerner, Takayanagi, and many other equations for spherical fillers and continuous fibers as discussed by Dickie (Ref. 44), Takayanagi et al. (Ref. 45), Schapery (Ref. 45), and Hashin (Ref. 47), respectively. In most cases, Poisson's ratio of the matrix is assumed to be real with $\nu_m'' = 0$ and $\nu^*_m = \nu'_m = \nu_m$. The effective complex shear modulus μ^* is described by the Kerner equation,

$$\frac{\mu^*}{\mu^*_m} = 1 + \frac{(\mu^*_f - \mu^*_m)\phi}{\mu^*_m + (\mu^*_f - \mu^*_m)(1 - \phi)\beta} \tag{18}$$

The elastic moduli are replaced by the corresponding shear moduli: μ^*, μ^*_m, and μ^*_f. Dickie (Ref. 44) has presented explicit expressions for the storage and loss moduli and for the loss tangent for Eq. (18). The filler is often assumed to be elastic for many applications, which simplifies Eq. (18) even further.

Viscoelastic properties typically vary by orders of magnitude in the region of the glass transition. The usual behavior of elastic fillers in a viscoelastic matrix is illustrated in Figures 12 and 13. Sodium chloride is dispersed in polyurethane rubber, as previously shown by Schwarzl et al. (Ref. 48). The elastic modulus of the sodium chloride is insensitive to temperature, but the shear modulus of the matrix drops significantly at its glass transition temperature T_g. That is,

$$\left(\frac{\mu_f}{\mu_m}\right)_{T<T_g} < \left(\frac{\mu_f}{\mu_m}\right)_{T>T_g} \tag{19}$$

Figure 12 shows that the fillers have a larger effect in raising the shear modulus above T_g than below it. This is consistent with Eq. (18). The effect of fillers on the dissipation of energy can be seen by observing the loss tangent as shown in Figure 13. Since the dissipation of elastic fillers is zero, which is particularly true for

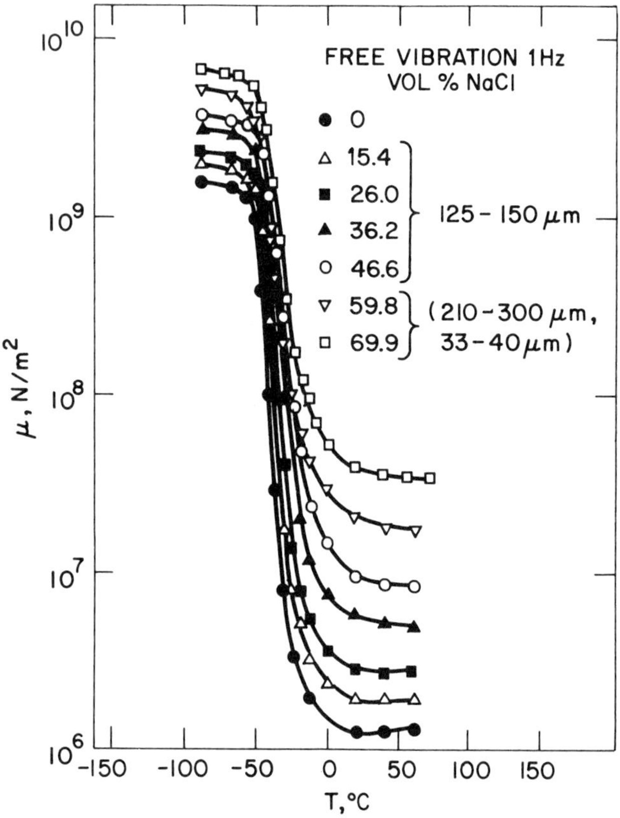

FIGURE 12 Shear modulus of polyurethane rubber filled with increasing amounts of sodium chloride. (From Schwarzl et al., Ref. 48.)

metal inclusions, the loss tangent of polymeric composites can be approximated, as discussed in Neilsen's book (Ref. 5), by

$$\tan \delta = \frac{\mu''}{\mu'} = (1 - \phi) \tan \delta_m \tag{20}$$

Equation (20) indicates why fillers often decrease the peaks of energy dissipation curves as illustrated in Figure 13. Fillers do not shift the glass transition temperature of the matrix polymer in two-phase mixtures. They do have the pronounced effect of broadening the transition region at higher filler concentration. This is due to the increase of the ratio μ''/μ' for $T > T_g$. For nonspherical particles, the basic approach to the problem is the same, as discussed in Reference 12.

Creep Elongation

Creep is deformation, usually under constant load, occurring over long periods of time. When a constant tensile stress is applied to a composite specimen under isothermal conditions, the relative creep of filled [e(t)] and unfilled [$e_m(t)$] polymers can be determined, as discussed by Nielsen (Ref. 5), by

$$\frac{e(t)}{e_m(t)} = \frac{E_m(t)}{E_{\parallel}(t)} \tag{21}$$

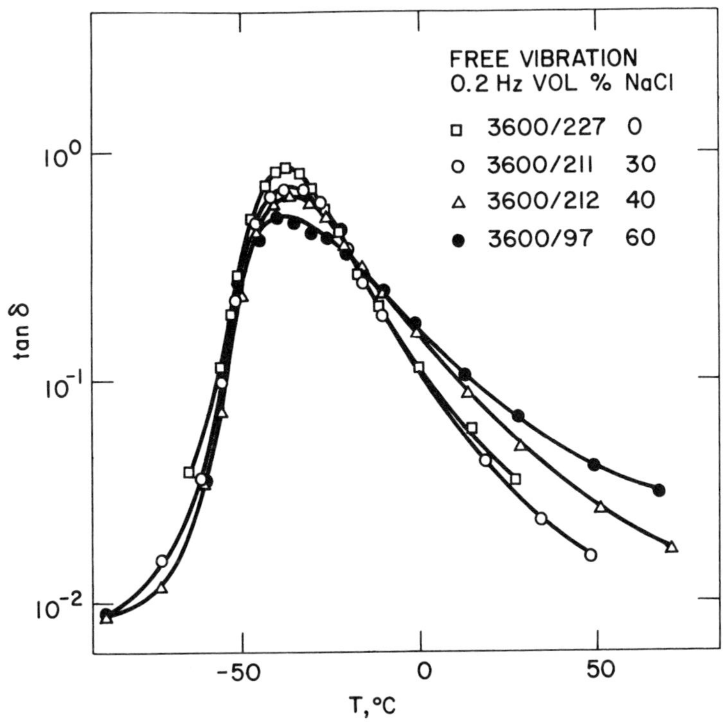

FIGURE 13 Loss tangent of polyurethane rubber filled with increasing amounts of sodium chloride. (From Schwarzl et al., Ref. 48.)

The effective relaxation modulus, $E_{\parallel}(t)$, can be evaluated by the quasi-elastic method discussed by Schapery (Ref. 49). A visco-elastic solution is approximated by an elastic solution with all the elastic constants replaced by time-dependent relaxation moduli. Applying the approximate quasi-elastic method to Eq. (7) for $E_{\parallel}$, the creep of filled polymers can be calculated from the creep of the unfilled polymer $[1/E_m(t)]$. Figure 14 indicates that the application of Eq. (21) works well for polyethylene filled with kaolin, as shown by Nielsen (Ref. 50). The elastic filler actually decreases the creep if there is no appreciable dewetting of the particles.

In nonisothermal situations, creep is affected by temperature changes mainly through thermal expansion (γ_f, γ_m) and stress relaxation. Usually, both undergo sudden changes in the glass transition region. This effect can be considered by looking at an extension of Eq. (21), discussed in Reference 12.

$$e(t) = \frac{\sigma_0}{E_{\parallel}(t)} + \int_0^t \theta_{33}(t - s)\mathring{T}(s)\, ds \qquad (22)$$

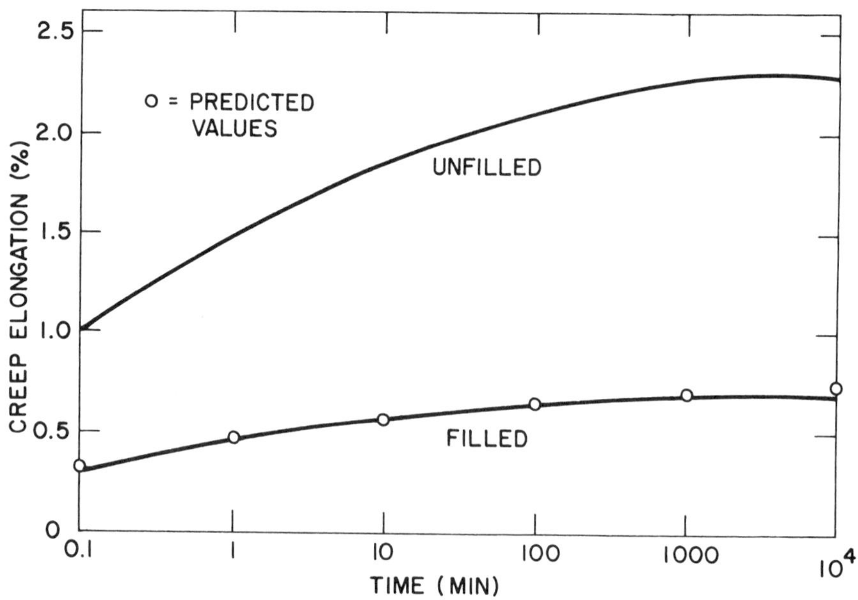

FIGURE 14 Creep of polyethylene unfilled and filled with kaolin (ϕ = 0.20, T = 60°C). The calculations were made using Eq. (21). (From Nielsen, Ref. 5.)

where σ_0 is the constant stress applied. The longitudinal thermal function $\theta_{33}(t)$ can be evaluated in the same way as $E_{\parallel}(t)$. Based on the quasi-elastic method

$$\theta_{33} = \theta_m + \frac{k_f}{k_m(t)} \frac{(\gamma_f - \gamma_m)\tilde{G}_1(t)\phi}{2\tilde{K}_1(t)\tilde{G}_3(t) + \tilde{G}_1(t)\tilde{K}_3(t)} \tag{23}$$

where

$$\tilde{K}_i(t) = 1 + \left[\frac{k_f}{k_m(t)} - 1\right][(1-\phi)\alpha_i + \phi] \qquad (i = 1,3) \tag{24}$$

$$\tilde{G}_i(t) = 1 + \left[\frac{\mu_f}{\mu_m(t)} - 1\right][(1-\phi)\beta_i + \phi]$$

A typical elastic filler and viscoelastic matrix with a real ν_m and constant γ_f and γ_m are assumed in the above relations. The value of γ_m depends on whether the temperature is above or below the glass transition of the matrix material. There are very few experimental data available for comparison with these results.

THERMAL EXPANSION

The thermal expansion of polymeric materials can be dramatically affected by fillers. Recent work on the particle shape effect on the thermal expansion of polymeric composites filled with aligned ellipsoidal particles at finite concentration will be discussed. Included will be an illustration of the extension of the "mean-field theory," discussed in the previous sections by application of the generalized approach of Eshelby, for elastic moduli and tensile strength to the thermal expansion of two-phase materials.

Predicting the effective thermal expansion of a filled polymer in terms of phase geometry, orientation, and constituent material properties has been discussed a great deal in the past 30 years. The discussions range from empirical to theoretical analyses. Composites with dispersed spheres, discussed by Schapery (Ref. 49), Kerner (Ref. 51), Turner (Ref. 52), and Wang and Kwei (Ref. 53), represent most of work. The topic has also been reviewed by Sendeckyi (Ref. 3) and Holliday and Robinson (Ref. 19).

The work discussed here will focus on composites filled with ellipsoidal particles. Considering a uniaxially oriented structure as discussed previously, the effective volumetric thermal expansion coefficient γ is related to the effective longitudinal (θ_{33}) and transverse ($\theta_{11} = \theta_{22}$) linear thermal expansion coefficients by

$$\gamma = 2\theta_{11} + \theta_{33} \tag{25}$$

The effective longitudinal, linear thermal expansion coefficient, derived by Chow (Ref. 29), is

$$\theta_{33} = \theta_m + \frac{k_f}{k_m} \frac{(\gamma_f - \gamma_m)\tilde{G}_1 \phi}{2\tilde{K}_1\tilde{G}_3 + \tilde{G}_1\tilde{K}_3} \tag{26}$$

and the effective transverse, linear thermal expansion coefficient is

$$\theta_{11} = \theta_m + \frac{k_f}{k_m} \frac{(\gamma_f - \gamma_m)\tilde{G}_3 \phi}{2\tilde{K}_1\tilde{G}_3 + \tilde{G}_1\tilde{K}_3} \tag{27}$$

The bulk modulus k and $\tilde{G}_i$ and $\tilde{K}_i$ are as given by Eq. (24) in the time-independent forms.

The volumetric thermal expansion coefficient is given by

$$\gamma = \gamma_m + \frac{k_f}{k_m} \frac{(\gamma_f - \gamma_m)(\tilde{G}_1 + 2\tilde{G}_3)\phi}{2\tilde{K}_1\tilde{G}_3 + \tilde{G}_1\tilde{K}_3} \tag{28}$$

When $\rho = 1$, $\beta_1 = \beta_3$, $\alpha_1 = \alpha_3 = \alpha$ (see Appendix), and $\theta_{11} = \theta_{33} = \theta$, Eqs. (26)–(28) reduce to

$$\begin{aligned} \gamma &= 3\theta \\ &= \gamma_m + \frac{k_f}{k_m} \frac{(\gamma_f - \gamma_m)\phi}{1 + (k_f/k_m - 1)[(1 - \phi)\alpha + \phi]} \end{aligned} \tag{29}$$

which is the well-known Kerner equation.

The derived relations for the linear and volumetric thermal expansion coefficients of a filled polymer are illustrated in Figures 15 and 16 for a glass-filled epoxy resin. The longitudinal and transverse linear thermal expansion coefficients, shown in Figure 15, indicate a strong anisotropic dependence on particle shape. Figure 16 shows that the volumetric thermal expansion coefficient is not very sensitive to the particle shape. In both figures, it is apparent that the effect of particle shape diminishes with increasing filler concentration. Again, the Kerner equation, shown by Nielsen (Ref. 9) to be consistent with experimental data, is a limiting case of the results in this section.

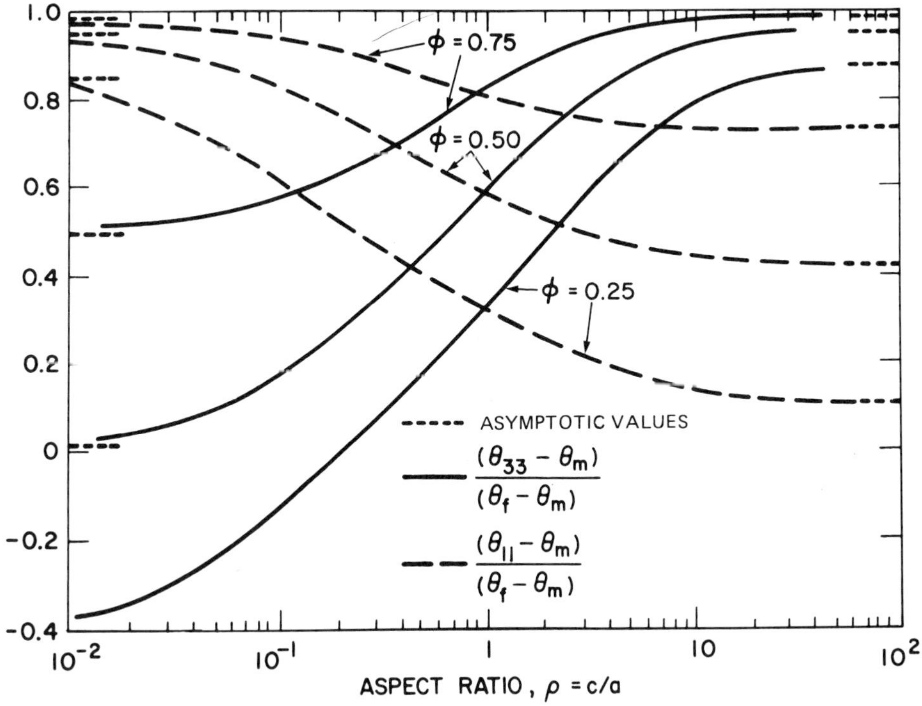

FIGURE 15 Coefficients of linear thermal expansion. (From Chow, Ref. 28.)

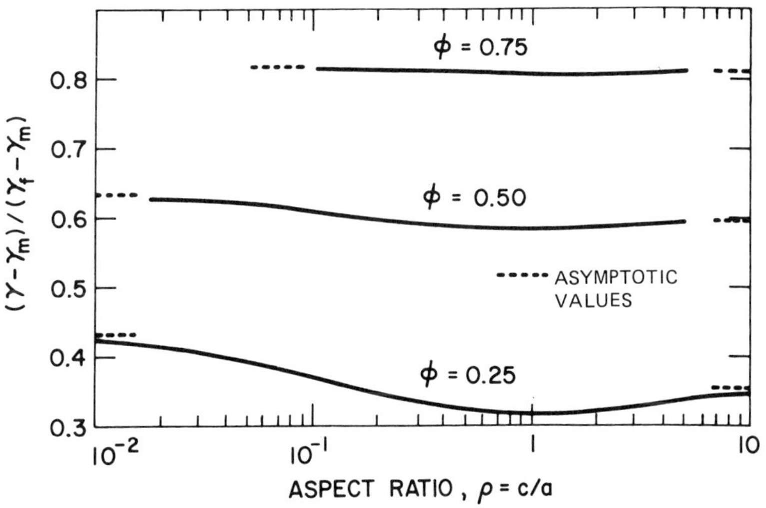

FIGURE 16 Coefficient of volumetric thermal expansion. (From Chow, Ref. 28.)

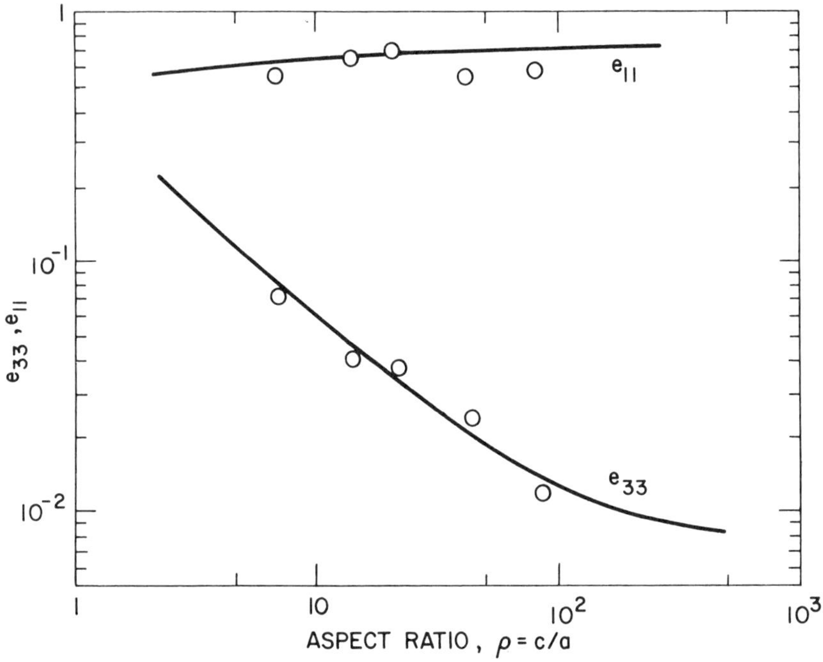

FIGURE 17 Aspect ratio dependence of expansion strains in parallel and perpendicular orientations for a nylon fiber/rubber composite. (Adapted from McNally, Ref. 42.)

An alternative approach using Schapery's (Ref. 49) and Halpin-Tsai's (Ref. 17) equations is duscussed by McNally (Ref. 42) and is illustrated in Figure 17. Figure 17 also gives data for longitudinal and transverse expansion strains for an oriented nylon fiber-elastomer system as a function of aspect ratio and for $\phi = 0.35$. The expansion strains are given by $e_{ii} = \theta_{ii}\Delta T$. Noting that $\theta_f < \theta_m$ in Figure 15, the results in Figure 17 are consistent with those given in Figure 15. Similar to that in Figure 16, the volume expansion is not sensitive to the aspect ratio. Papanicolaou et al. (Ref. 55) review theories existing at that time with their work on metal-epoxy composites.

THERMAL CONDUCTIVITY

In addition to desirable mechanical properties obtained by adding fillers to polymeric matrices, improved thermal characteristics are of interest. During the past several decades, there has been a great deal of activity in deriving methods for predicting the thermal conductivity of polymeric composites. Progelhof et al. (Ref. 6) and Bhattacharya and Chaklader (Ref. 15) have written recent reviews on the topic. A myriad of empirical, semiempirical, and theoretical approaches occur in the literature. One thing hampering their continuing evaluation is the lack of good experimental data encompassing sample preparation, range, and measurement technique. In the literature to date, the approach of Lewis and Nielsen appears to be as adequate as any and is, therefore, chosen for discussion.

Lewis and Nielsen modified the Halpin-Tsai equation [see Eq. (6)] for the elastic modulus and applied it to thermal conductivity by including the effect of the maximum packing fraction v_f of the filler. The modified equations, given by Nielsen (Ref. 56), are

$$\frac{\kappa}{\kappa_m} = \frac{1 + AB\phi}{1 - BC\phi} \tag{30}$$

and

$$A = k_E - 1 \tag{31}$$

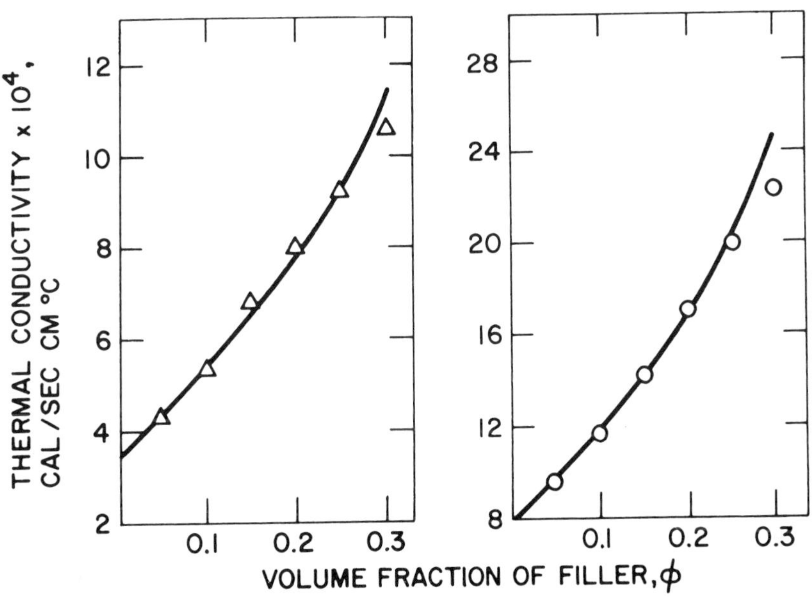

FIGURE 18 Thermal conductivity versus volume fraction of filler. (—) theoretical; (△) MgO in polystyrene; (○) MgO in polyethylene. (Adapted from Nielsen, Ref. 56.)

$$B = \frac{\kappa_f/\kappa_m - 1}{\kappa_f/\kappa_m + A} \tag{32}$$

$$C = 1 + \frac{1 - v_f}{v_f^2} \phi \tag{33}$$

where κ is the composite thermal conductivity, A is related to the generalized Einstein coefficient k_E, and B is related to the conductivity of the components. Values of A and v_f are given for many geometric shapes and orientations in the work of Progelhof et al. (Ref. 6).

The Lewis and Nielsen model is compared with experimental data of Sundstrom and Lee (Ref. 57) in Figure 18. The data represent MgO in polystyrene and polyethylene. The calculations are based on $v_f = 0.64$ and $A = 3.0$. Other work reported by Bhattacharya and Chaklader (Ref. 15) supports the general observations that thermal conductivity increases with increasing filler content, with increasing temperature, and with particle size.

CONCLUSIONS AND SUMMARY

Theoretical work concerning mechanical and thermal properties of filled polymeric composites is of continuing interest and leads the availability of good experimental data. Good data, including particle distribution and orientation, adhesion between constituents, and thermal effects, are very difficult to obtain.

Stress-strain behavior and tensile modulus are more easily measured than most mechanical properties. Still, there is not a great deal of data available. Generally, as the filler loading increases, the composite modulus increases, the ultimate strain decreases, and the ultimate strength increases. Theoretical models abound, but none has been proved to hold over wide ranges of aspect ratio and filler-to-matrix modulus ratios.

The stress concentration model of Chow can be calculated from experimental data. When $\phi \to 0$ and $\rho = 1$, it reduces to the classical Goodier equation. A plot of the equation for glass in PPO shows that $S \simeq [E_\parallel]$ for $E_f/E_m > 28.1$. Furthermore, the shape dependence of the stress concentration factor and tensile modulus becomes small when $\rho > E_f/E_m$. This is true for values of E_f/E_m from 10 to 1000. This information is worthwhile when considering processing trade-offs. Because of the difficulty in establishing the fiber failure mode at the interface or within the matrix, the ultimate tensile strength of polymeric composites cannot be directly determined from stress concentration factors.

If interfacial debonding between the filler and matrix is assumed to be the primary failure mode, critical filler dimensions can be determined. This reveals that for $d \geqslant d_c$, flaws are introduced that reduce the ultimate strain. Both the ultimate stress and strain of a particulate-filled polymer are inversely proportional to the square root of particle size. Because of the number of uncontrollable parameters, good experimental data are difficult to obtain.

Composite viscoelastic properties are obtained by the application of the correspondence principle to extend elastic models. It is predicted and observed that fillers have a large effect in raising the shear modulus above the glass transition region than below it. Fillers often decrease energy dissipation peaks, do not shift the glass transition temperature, but do broaden the region at high filler concentrations. Elastic fillers reduce creep if no great amount of dewetting of particles occurs.

Recent progress in modeling the thermal expansion of polymeric composites substantiates experimental observation. While the longitudinal and transverse linear thermal expansion coefficients have a strong anisotropic dependence on particle shape, the volumetric thermal expansion coefficient does not. The effect of particle shape diminishes with increasing filler concentration.

Finally, the extension of elastic models to predict thermal conductivity, as well as several other approaches, leads to reasonable agreement with the very limited experimental data. The most obvious trend observed is the increase in thermal conductivity with increasing volume fraction of filler.

Efforts to replace metals with polymeric composites will continue for weight reduction, cost reduction, and improved function. Ongoing theoretical and experimental efforts in predicting the mechanical and thermal properties of these composites are essential.

APPENDIX

The parameters for Eqs. (8) and (24) are as follows:

$$\alpha_1 = \frac{4\pi Q}{3} - 2(2\pi - I)R$$

$$\alpha_3 = \frac{4\pi Q}{3} - 4(I - \pi)R$$

$$\beta_1 = \left(\frac{4\pi}{3} - \frac{4\pi - 3I}{1 - \rho^2}\right) - 4(I - 2\pi)R$$

$$\beta_3 = \left[\frac{4\pi}{3} - \frac{(4\pi - 3I)\rho^2}{1 - \rho^2}\right] Q + (4\pi - I)R$$

where

$$Q = \frac{3}{8\pi}\frac{1}{1 - \nu_m}, \qquad R = \frac{1}{8\pi}\frac{1 - 2\nu_m}{1 - \nu_m}$$

and

$$I = \begin{cases} [2\pi\rho/(1-\rho^2)^{3/2}][\cos^{-1}\rho - \rho(1-\rho^2)^{1/2}] & (\rho < 1) \\ [2\pi\rho/(\rho^2-1)^{3/2}][\rho(\rho^2-1)^{1/2} - \cosh^{-1}\rho] & (\rho > 1) \end{cases}$$

When $\rho \to 1$, we have

$$\alpha_1 = \alpha_3 = \alpha = \frac{1}{3}\left(\frac{1 + \nu_m}{1 - \nu_m}\right)$$

$$\beta_1 = \beta_3 = \beta = \frac{2}{15}\left(\frac{4 - 5\nu_m}{1 - \nu_m}\right)$$

REFERENCES

1. R. M. Christensen, *Mechanics of Composite Materials*, Wiley, New York (1979).
2. J. A. Manson and L. H. Sperling, *Polymer Blends and Composites*, Academic, New York (1976), chap. 12.
3. G. P. Sendeckyi, *Composite Materials*, Academic, New York (1974), vol. 2, chaps. 3 and 4.
4. J. L. Kardos, Crit. Rev. Solid State Sci., *3*:419 (1973).
5. L. E. Nielsen, *Mechanical Properties of Polymers and Composites*, Dekker, New York (1974), vol. 2.
6. R. C. Progelhof, J. L. Throne, and R. R. Ruetsch, Polym. Eng. Sci., *16*:615 (1976).
7. J. L. Kardos, W. L. McDonnel, and J. Raisoni, J. Macromol. Sci. Phys., *B6*:397 (1972).
8. K. Sato, Prog. Organic Coatings, *4*:271 (1976).
9. L. E. Nielsen, J. Compos. Mater., *1*:100 (1967).
10. Y. S. Lipatov, Adv. Polym. Sci., *22*:1 (1977).
11. R. A. Dickie, in *Polymer Blends* (D. R. Paul and S. Newman, eds.), Academic, New York (1978), vol. 1, chap. 8.

12. T. S. Chow, J. Mater. Sci., *15*:1873 (1980).

13. T. S. Chow, C. A. Liu, and R. C. Penwell, J. Polym. Sci., Polym. Phys. Ed., *14*:1311 (1976).

14. T. S. Chow, R. C. Penwell, and C. A. Liu, Polym. Eng. Sci., *17*:436 (1977).

15. S. K. Bhattacharya and A. C. D. Chaklader, Polym. Plast. Technol. Eng., *20*:35 (1983).

16. A. Kelly, *Strong Solids*, 2d ed., Clarendon, Oxford (1973), chap. 5.

17. J. E. Ashton, J. C. Halpin, and P. H. Petit, *Primer on Composite Materials: Analysis*, Technomic, Stamford, Conn., (1969), chap. 5.

18. L. J. Broutman and R. H. Krock, *Modern Composite Materials*, Addison-Wesley, Reading, Mass. (1967).

19. L. Holliday and J. Robinson, J. Mater. Sci., *8*:301 (1973).

20. A. R. Von Hippel, *Molecular Science and Engineering*, MIT Press, Cambridge (1959).

21. M. Takayanagi, H. Harima, and Y. Iwata, Rep. Prog. Polym. Phys., (Jpn.), *6*:121 (1963).

22. H. L. Cox, Br. J. Appl. Phys., *3*:72 (1952).

23. P. J. Barham and R. G. C. Arridge, J. Polym. Sci., Polym. Phys. Ed., *15*:1177 (1977).

24. R. Hill, J. Mech. Phys. Solids, *13*:213 (1965).

25. J. J. Hermans, Proc. R. Acad. (Amsterdam), *B70*:1 (1967).

26. N. Law and R. McLaughlin, J. Mech. Phys. Solids, *27*:1 (1979).

27. T. S. Chow, J. Appl. Phys., *48*:4072 (1977).

28. T. S. Chow, J. Polym. Sci., Polym. Phys. Ed., *16*:959 (1978).

29. T. S. Chow, J. Polym. Sci., Polym. Phys. Ed., *16*:967 (1978).

30. A. S. Kenyon and H. J. Duffey, Polym. Eng. Sci., *7*:1 (1970).

31. T. S. Chow, Polymer, *20*:1576 (1979).

32. T. W. Chou and A. Kelly, Annu. Rev. Mater. Sci., *10*:229 (1980).

33. W. H. Bowyer and M. G. Bader, J. Mater. Sci., *7*:1315 (1972).

34. P. T. Curtis, M. G. Bader, and J. E. Bailey, J. Mater. Sci., *13*:377 (1978).

35. R. W. Davidge and T. J. Green, J. Mater. Sci., *3*:629 (1968).

36. T. S. Chow, J. Polym. Sci., Polym. Phys. Ed., *20*:2103 (1982).

37. K. L. Trachte and A. D. DiBenedetto, Int. J. Polym. Mater., *1*:75 (1971).

38. J. K. Lee, Polym. Eng. Sci., *8*:195 (1968).

39. B. F. Blumentritt, B. T. Vu, and S. L. Cooper, Polym. Eng. Sci., *14*:633 (1974).

40. F. R. Schwarzl, H. W. Bree, and C. J. Nederveen, in *Proc. Fourth Int. Congr. Rheology* (1965), part 15, p. 241.

41. T. S. Chow and J. J. Hermans, J. Compos. Mater., *3*:382 (1969).

42. D. McNally, Polym. Plast. Technol. Eng., *8(2)*:101 (1977).

43. J. Dimmock and M. Abrahams, Composites, *1*:87 (1969).

44. R. A. Dickie, J. Appl. Polym. Sci., *17*:45 (1973).

45. M. Takayanagi, S. Nemura, and S. Minami, J. Polym. Sci., *C5*: 113 (1964).

46. R. A. Schapery, J. Compos. Mater., *1*:228 (1967).

47. Z. Hashin, Int. J. Solid Struct., *6*:539 (1970).

48. F. R. Schwarzl, H. W. Bree, C. J. Nederveen, G. A. Schwippert, L. C. E. Struik, and C. W. Van der Wal, Rheol. Acta, 5: 270 (1966).

49. R. A. Schapery, J. Compos. Mater., *2*:380 (1968).

50. L. E. Nielsen, Trans. Soc. Rheol., *13*:141 (1969).

51. E. H. Kerner, Proc. Phys. Soc. (London), *B69*:808 (1956).

52. P. S. Turner, J. Res. Natl. Bur. Stand., *37*:239 (1946).

53. T. T. Wang and T. K. Kwei, J. Polym. Sci., *A2, 7*:889 (1969).

54. J. C. Halpin, J. Compos. Mater., *3*:732 (1969).

55. G. C. Papanicolaou, S. A. Paipetis, and P. S. Theocaris, J. Appl. Polym. Sci., *21*:689 (1977).

56. L. E. Nielsen, J. Appl. Polym. Sci., *17*:3819 (1973).

57. D. W. Sundstrom and Y. D. Lee, J. Appl. Polym. Sci., *16*: 3159 (1972).

5

The Concept of Mesophase
Mechanisms of Crack Propagation

P. S. THEOCARIS

Department of Engineering Science
Athens National Technical University
Athens, Greece

While theories up to now have considered both fiber composites and particulates as two-phase materials constructed by embedding a system of parallel fibers or uniformly arranged spherical particles inside a matrix, with a perfect bonding between phases, experience has shown that it is imperative to introduce a third hybrid phase, developed in between the main phase, which plays a significant role in the mechanical behavior of the entity.

In this study further evidence will be added on the significance of this phase in the composite by analyzing the system in its microstructure. That is, very thin layers of the representative volume element (RVE) of a composite are assumed to be cut out from the bulk, incorporating either a radial thin layer of a fiber and its adjacent matrix or a central thin layer of the spherical particle and its surrounding material. A crack arbitrarily incident to the interface of the system was considered propagating in the unit and an experimental study of its behavior as the crack approaches this boundary gave positive proof of the existence and importance of this *mesophase*.

This study is accompanied by a whole-field study of models for both types of materials to bring out the overall influence of the mesophase on the mechanical and other physical properties of the system.

The mechanism of fracture of a specific fiber or particulate can be accurately simulated by studying the dynamic behavior of a crack approaching a bimaterial interface, where one phase stands for the matrix and the other for the fiber or particulate material.

By varying the mechanical properties of the two phases and the relative inclination of the interface to the initial crack direction, one can find possible angles where a crack splits an inclusion, including the case where the crack propagates momentarily along the interface.

Interesting conclusions concerning the real stress state and crack velocity at the mesophase were derived, including rapid velocity changes and a diffusion of the stress singularity in a thin layer surrounding the interface, thus giving evidence for the existence of the mesophase and defining qualitatively the role of this pseudophase in strengthening the composite.

INTRODUCTION

The progressive failure of a composite is in general very complex and it is not possible to describe its successive steps analytically. If composites are to be effective in high-performance applications, it is imperative to have accurate methods for describing their failure behavior. Because of the complicated internal structure of composites, analysis of the mechanical behavior of such materials is much more difficult than that of homogeneous substances.

It is clear that the stress field in composite systems is three-dimensional. For an approximate solution extensive use has been made of the variational principles of the potential and complementary energies to develop the governing differential equations. Special consideration must be given to *macroscopic discontinuities* such as holes, inclusions, and especially cracks embedded in the composite, since these additional perturbations of the stress field aggravate the overall state of stress. Then the progressive failure process in a composite is extremely complex and each successive step cannot be described in detail.

Besides the macroscopic discontinuities in a composite, its own structure, which consists of at least two phases, introduces insurmountable problems for the study of these materials on the microscale; it is not possible to make an exact study of the state of stresses and strains in the alternating phases, compatible with the principles of continuity and equilibrium.

In this chapter models will be used which simulate the behavior of actual two-phase systems in a meridional section of a composite (particularly a fiber composite). The behavior of two successive phases of the composite, when a crack in one phase approaches the interface between the two phases, passes it, and continues to propagate in the second phase, will be studied. Such models cannot give a quantitative prediction of the mechanical behavior of real composites; however, they reveal the mechanisms hindering the abrupt and catastrophic propagation of a crack initiated somewhere in a homogeneous material.

Characterization of a composite system is a difficult task because of the great number of parameters affecting its mechanical behavior. Such parameters are the shape and size of the reinforcing inclusions, the physicochemical properties of its constituents, the volume fraction of filler, the bonding between the phases, the type of dispersion, and the amount of particle agglomeration.

Among these factors, the volume of filler fraction is of cardinal importance. This factor is characterized by the so-called *optimum concentration*, which is regarded as the *limit of saturation* of the adsorption centers on the surface of the filler by the macromolecules. For filler concentrations greater than this optimum value, the continuity of the network structure is upset, and this leads to a deterioration rather than to an improvement in the mechanical properties with filling. This reversal of the reinforcement is explained by the difference in the coefficients of thermal expansion of the filler and the polymeric matrix and by the occurrence around the filler particles of a stress concentration zone, which greatly weakens the adhesion bonding of polymer and filler.

Moreover, the filler per se does not exhibit any resistance to crack growth, and it reduces the content of the continuous phase, which is responsible for the strength of the composite.

However, the adhesion bonding between polymer and filler is an important factor, greatly affecting the strength of the composite. By *adhesion* we mean the energy of interaction of phases at the interface. For instance, proper surface treatment of the filler particles can lead to a better adhesion between polymer and filler and, consequently, to a change in the overall mechanical behavior of the composite system. Thus, the condition of the mesophase layer is a dominant factor regulating the overall strength of the composite, since microdefects at the mesophase lead to localized stress concentrations, which may be greater than the average stress in the bulk of the material. If the localized stresses are sufficiently high, they may lead to defect growth and to failure of the material. Thus, the rate of crack growth depends strongly on the degree of inhomogeneity of the material, not only macroscopically, but also microscopically, because the presence of filler influences the behavior of the composite.

With stressing of a cracked composite, there is always an additional stretching in the region surrounding the crack tip and a reorientation of the polymer macromolecules, which leads to strengthening. At the same time, the polymer material close to the fillers is characterized by strains, which are greater than the average elongation of the whole specimen. The value of this greater elongation is proportional to the volume fraction of filler. Moreover, the degree of additional orientation of the macromolecules at the place of rupture depends greatly on the volume fraction of filler. On the other hand, the formation of a third phase, at the interface of matrix and eventually the filler, reduces the capacity of deformation of the bound macromolecules.

From the observations above it is clear that the volume fraction of filler plays an important role in the mechanism of crack propagation in composites.

Extensive experimental evidence indicates that various fracture mechanisms dominate in composites, depending on the amount of inclusions incorporated in the substance. Thus, for low filler concentrations, cracks appear in the matrix material first and the fillers tend to constrain the growth of the cracks, thus increasing the toughness of the composite. For high filler concentrations, cracks are developed and propagate in high densities of fillers following interfaces.

The study of crack propagation mechanisms gives insight into the influence of structure and interaction of phases in a composite system. This is because the structure of a composite material is characterized by inhomogeneities, such as voids and microcracks, which may be studied qualitatively by electron and optical microscopy, while their effect on the strength of the composite system may be studied both qualitatively and quantitatively by the well-known method of caustics along with high-speed photography. Moreover, from the theoretical point of view, one can use models describing the structure and the mechanical behavior of composite materials on both the macroscopic and microscopic scales.

Theoretical confrontation of this formidable problem is impossible. Therefore, we seek experimental means of studying the mode of propagation of a crack in the microstructure of a composite. In order to study the behavior of composites from a fracture-mechanics perspective, we must isolate from the whole composite system an elementary area consisting of the two component materials (matrix and filler) connected along an interface.

Then consider a thin specimen corresponding to a meridional section of a unit consisting of part of the matrix and its adjacent inclusion. Since we are interested in a qualitative study of the mechanism of propagation of cracks in composites, the model is satisfactory. Furthermore, the meridional section of the cylindrical inclusion corresponds to the most difficult case, and therefore we know that we are studying an extremum in the mode of propagation of the crack. Another extremum is derived from the study of an equatorial section with a circular inclusion in the matrix. Thus, in spite of the scaling problem, this model simulates adequately extreme situations in the composite material in many respects.

Moreover, for the study of crack propagation in these biphase specimens it is assumed that, outside the interface layer, the two materials are homogeneous and isotropic. It is known that the stress singularities change as the crack approaches the interface, due to the constraints imposed by one phase on the other. This change of the order of singularity also affects the behavior of the crack as it crosses the mesophase.

The fracture behavior of such biphase specimens was studied by the method of dynamic caustics. The caustic created around the tip of a stationary or running crack carries all the information necessary to evaluate the instantaneous crack velocity, the mode of deformation, and the intensity of the stress field at the vicinity of the crack tip. Since the caustics may be recorded with a high-speed camera, regardless of the opaqueness or transparency of the material, the method can be applied directly to the materials being studied.

From this experimental study of the mode of crack propagation along meridional and equatorial sections of fiber and particulate composites by the method of caustics, we give a detailed assessment of the mode of fracture and discuss its influence in relation to improving the properties of an initial isotropic elastic or viscoelastic material (the matrix) by reinforcement with another material with appropriate properties.

MODELS FOR COMPOSITES

Literature Review

The procedure for defining a geometric model representing the macroscopic mechanical behavior of a composite involves the analysis of a representative volume element, or volume cell, of the material. The models introduced by Hashin (Ref. 1) for particulates and by Hashin and Rosen (Ref. 2) for fiber composites are generally accepted. They assume in both cases representative volume elements (RVEs) of the materials consisting of a gradation of sizes of cells, corresponding to the volume filling configuration, existing in the particular material studied. Moreover, a fixed ratio of radii between the spheres or cylinders of the inclusions (r_f) and the respective quantities for the matrix (r_m) is defined, in order to analyze a single RVE representing the behavior of the entire composite.

Other types of models, besides these simple geometric ones include the *self-consistent model*, according to which the average values for stresses and strains in either phase are determined by solving two separate problems, whose superposition yields the final configuration of the model. The solution of the two individual problems allows evaluation of the average properties of the composite by knowing the respective properties of either phase. The self-consistent model was applied to composites by Budiansky (Ref. 3) and by Hill (Ref. 4).

An important variation of the self-consistent model is the *three-phase model* introduced by Kerner (Ref. 5). In this model the inclusion is enveloped by a matrix layer, which, in turn, is embedded in an infinite medium with the unknown macroscopic properties of the composite.

From these typical models and many others in the literature which are not mentioned here, those introduced by Kerner (Ref. 5) and Van der Poel (Ref. 6) give satisfactory results and can be extended to multiphase models. The difference between these two models lies in their boundary conditions. The Kerner model considers the problem as an *average elastic problem* and finds the conditions between moduli and volume contents between phases in a generic spherical grain of the body without explicitly giving the states of stress and strain in the composite, which is an extremely difficult task. Van der Poel considers the same problem, but applies the solution of Godier (Ref. 7) and Fröhlich and Sack (Ref. 8) for a spherical cavity subjected to a hydrostatic loading or a pure shear loading to the case of the RVE of a typical particulate.

The initial Van der Poel model, which is valid for hard fillers and incompressible matrices, has been extended by Schwarzl and Eikhoff (Ref. 9) to incorporate a description of particulates in a wide temperature range. Further use of the Van der Poel model and the Schwarzl and Eikhoff ideas was made by Maurer in his interesting dissertation (Ref. 10).

Similar attempts to study three-phase particulate composites were made by Sagalaev and Simonov-Emiljanov (Ref. 11) with the Kerner-Kerner model, where the classical Kerner model was used twice to cover successively by pairs the phases of a three-phase composite (Ref. 12), by Kudykina and Pervak (Ref. 13), by Takano and Sakanishi (Ref. 14), and finally by Spathis et al. (Ref. 15). All these models yield the moduli of the particulate composites in terms of the moduli and respective volume fractions of three constituent phases of the composite, without considering, except for the last model, that the middle phase is derived from the interaction of the two main phases and depends on their degree of adhesion. All these models were compared previously in Reference 16.

Finally, the model of Takayanagi et al. (Ref. 17) should be mentioned, in which the influence of the filler modulus, E_f, is taken into account in evaluating the respective modulus of the composite, E_c. In this model the filler volume fraction, $\underline{v}_f = \lambda\varphi$, was combined with a part $(1 - \varphi)\lambda\underline{v}_m$ of the matrix volume fraction, $\underline{v}_m$, which had a different influence on E_c than the rest of the matrix volume fraction $(1 - \lambda)\underline{v}_m$. The factors λ and φ of the model can be defined in different empirical ways.

While the model of Takayanagi et al. was designed mainly for two-phase materials, it could be extended to incorporate the influence of the mesophase (Ref. 18). A further extension of the concept of using mixed-mode connections, that is, in series and in parallel, for the elements representing the mechanical properties of the particulate seems very promising, since it makes the model highly flexible, so that it can be adapted to the real behavior of the substance.

Concept of Mesophase, Role of Adhesion

The novel element in the Hashin and Hashin-Rosen models (Refs. 1, 2) is the introduction of a third phase which lies between the two main phases (inclusions and matrix) and contributes to the progressive unfolding of the properties of the inclusions to those of the matrix without discontinuities. These models incorporate all transition properties of a thin boundary layer of the matrix near the inclusions. Thus, this pseudophase characterizes the effectiveness of the bonding between phases and defines an *adhesion factor* of the composite.

Around an inclusion embedded in a polymeric matrix a rather complex situation develops, with areas of imperfect bonding, permanent stresses due to shrinkage, high stress gradients, or even stress singularities, due to the geometry of the inclusions, voids, microcracks, etc. Moreover, the interaction of the surface of the filler with the matrix is usually much more complicated than a simple mechanical effect. The presence of a filler actually restricts the segmental and molecular mobility of the polymeric matrix, as adsorption-interaction in polymer surface layers into filler inclusions occurs. Then the polymeric matrix, cast around the inclusions, creates phenomena of physical and chemical adsorption.

Physisorbed (physically adsorbed) layers of the matrix contribute, in general, to weak mesophases. However, the physical interpenetration of the boundary layer of the matrix in cavities and other rough regions of the surfaces of solid inclusions, together with the restrained development of the molecular structure of the polymeric chains there and any other structural variations of the adjacent layers, creates an *intermixing* and *interpenetrating phenomenon*, which considerably influences the molecular structure of the *mesophase*, resulting in variations in its mechanical strength. Thus, the mechanical properties of the matrix films and layers close to the interfaces, are strongly depending on the physical situation of this boundary layer.

On the other hand, *chemisorbed* (chemically adsorbed) molecules on the interfaces create structural variations by developing *beaded structures* of *caged molecules* or *ladderlike molecules*. All these types of chemisorbed elements near inclusions lead to rapid

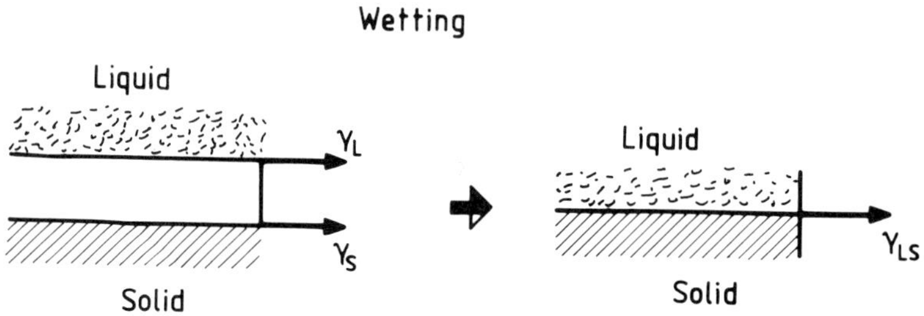

FIGURE 1 Wetting of the liquid phase when cast on the solid one. $\Delta F_W = \gamma_{SL} - (\gamma_S + \gamma_L) < 0$.

variations of the properties and mechanical strength of the interface layer close to the surfaces of inclusions.

This physical and chemical bonding between the main phases of the composite depends on a series of factors related to the individual phases and their affinity. The surface characteristics of the phases, including their chemical properties, topology, and homogeneity, as well as their coupling possibilities, are the main factors influencing the extent of the mesophase.

Equally important is the degree of *wetting* of the liquid phase when cast on the solid one. This depends primarily on the chemical characteristics of the phases and the degree of affinity between them (Figure 1). It is characterized by a *coefficient of wetting* that depends on the surface free energies of liquid, γ_L, and solid, γ_S, phases. The wetting coefficient, $\Delta F_W = \gamma_{SL} - (\gamma_S + \gamma_L)$, regulates the degree of spreading of the liquid phase and its penetrating properties on the solid surface by lowering the surface tension.

The degree of wetting and spreading of the liquid phase influences the adhesion characteristics between the phases and the amount of their diffusion (Figure 2a and b). However, the speed of spreading of the liquid phase also depends on the state of the solid surface and especially on its roughness.

On the other hand, the roughness of the solid surface may later contribute beneficially to the extent of the physisorbed layer, thus increasing the adhesion by physical means (Fig. 2c). In this way, *mechanical grafting* is developed, which adds to the amount of adhesion, but it is also a source of stress concentrations and singularities.

Finally, phenomena of migration of the curing agent of the liquid polymeric phase have been detected at the vicinity of the interfaces, which lead to a weak boundary layer (Fig. 2d); this layer, being less brittle than the surrounding matrix material, creates a bumper between phases, thus altering the fracture mode of the composite.

Although chemical bonding successfully explains a great many of the phenomena developed at the mesophase, it seems not to be sufficient to give a complete explanation for all the phenomena appearing at the boundary layers between phases.

Contamination of the surface by entrapped air bubbles and other impurities in the liquid, not yet set, polymeric matrix, which gather around the inclusions, and incomplete coverage of the interface are also responsible for the extent of mesophase and the overall strength of the composite. Moreover, restrained polymerization of the boundary layer, due to the presence of the inert inclusions, also contributes significantly to the extent of mesophase.

Interpenetration phenomena between phases, due to diffusion and mechanical grafting, are important factors in reinforcement mechanisms (Ref. 19). There are two kinds of intermixing procedures: penetration of the matrix polymer into the chemisorbed layers of the mesophase, and migration of the physisorbed molecules of the mesophase into the matrix phase. Both types of structures contribute to reinforcement of the composite.

The structural situation at the mesophase between an inert fiber and a polymeric matrix may be represented as shown in Figure 3. The chemisorbed part of the mesophase is indicated by open squares. Between conglomerations of such material there are ladderlike penetrations of physisorbed material, represented by open circles. This

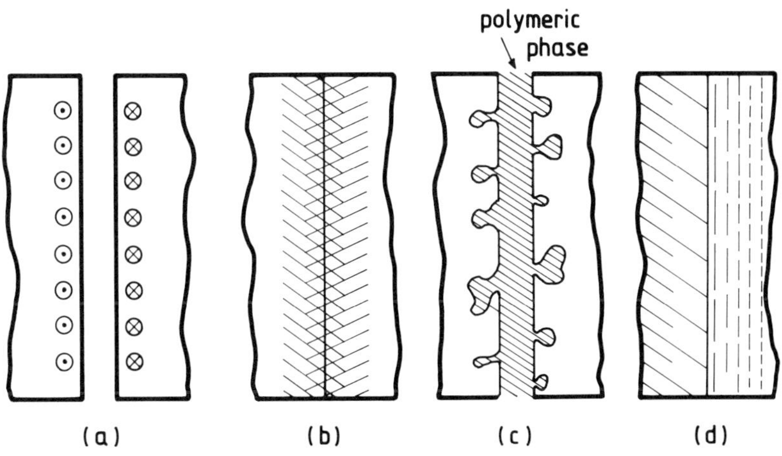

FIGURE 2 (a) Wetting, (b) spreading, (c) mechanical grafting, and (d) weak boundary layer developed when the liquid phase is cast on the solid one.

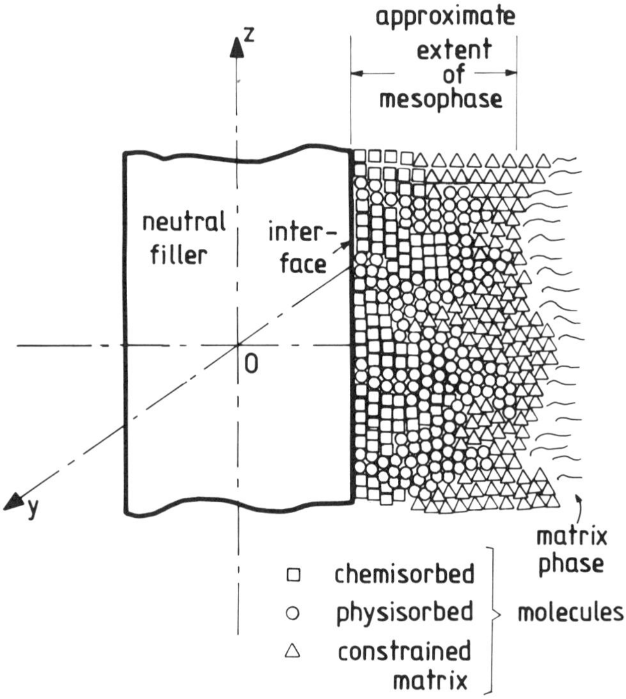

FIGURE 3 Structural situation at the mesophase between an inert fiber and a polymeric matrix indicating the chemisorbed, physisorbed, and constrained matrix layers lying between the fiber and the matrix.

material also extends outward, penetrating into the pure matrix material, and it is represented by open triangles. Therefore, in addition to the chemical adsorption phenomena which constitute the backbone of adhesion between the main phases, there are further secondary bonds which play an important role in the overall strength of the composite.

The phenomena happening discussed above, occurring around an inclusion, create an *intermediate boundary layer* of variable thickness. This zone extends beyond the thin layer including the phenomena of *physisorption* and *chemisorption*, and it incorporates the zones of imperfect bonding and shrinkage stresses, high stress gradients, or even stress singularities due to geometric discontinuities of the surfaces of inclusions, to the concentration of voids, and to the impurities, microcracks, and other anomalies.

The majority of reported values for the thickness of the matrix mesophase fall in the range 10 to 10,000 Å. However, it should be understood that this layer represents only the part which includes all physical and chemical alterations of the matrix material. Beyond this inner layer there exists another more extended layer which is influenced by the existence of mesophase, but its properties can be assumed to be close to the mechanical properties of the bulk matrix.

Although the density of mesophase and its mechanical properties can be assumed constant, layers of the pseudophase close to the interface may have dramatic property variations in order to accommodate the differences between the properties of the inclusions and the matrix. Structural gradients within the mesophase layer have been measured by ultrasonic techniques (Ref. 20). Although the ultrasonic measurements were concerned with the influence of the weak boundary layer close to the interface, where in some cases a reduction of the shear modulus of the matrix was measured, in general, a progressive variation of the mechanical properties of the material of the mesophase through its thickness is expected, to accommodate different states of mechanical and physical behavior between the extreme values of these quantities in the matrix and the inclusions.

In this chapter the existence of the boundary layer constituting the mesophase, and developed between the two main phases of a two-phase composite, is taken into account in developing a model describing the mode of propagation of cracks in unidirectional fiber composites or particulate composites. The influence of this layer as well as of the inclusions on the retardation and hindrance of propagation of cracks in composites is qualitatively studied by an appropriate model, and important insight is gained in the overall behavior of the composites and improvement of their mechanical and physical properties relative to those of their constituent materials.

Unfolding Models Introducing Mesophase

Figure 4 presents principal sections of the models of Hashin (Ref. 1) and Hashin and Rosen (Ref. 2) and the respective improvements in our models for particulate composites and fiber-reinforced composites. In the Hashin-Rosen models the matrix phases (m) are concentric spheres or cylinders and the respective mesophases (i) disappear.

These models for particulate- and fiber-reinforced composites were introduced previously and are described in References 21–25. However, a brief description of both models is given here. Figure 5 represents the RVE for a particulate (a) and a fiber composite (b) surrounded by an elementary unit cube of the composite material.

For the particulate composites the component phases are interconnected through consecutive spherical interfaces of the filler, mesophase, and matrix. The dominant transverse interconnection through

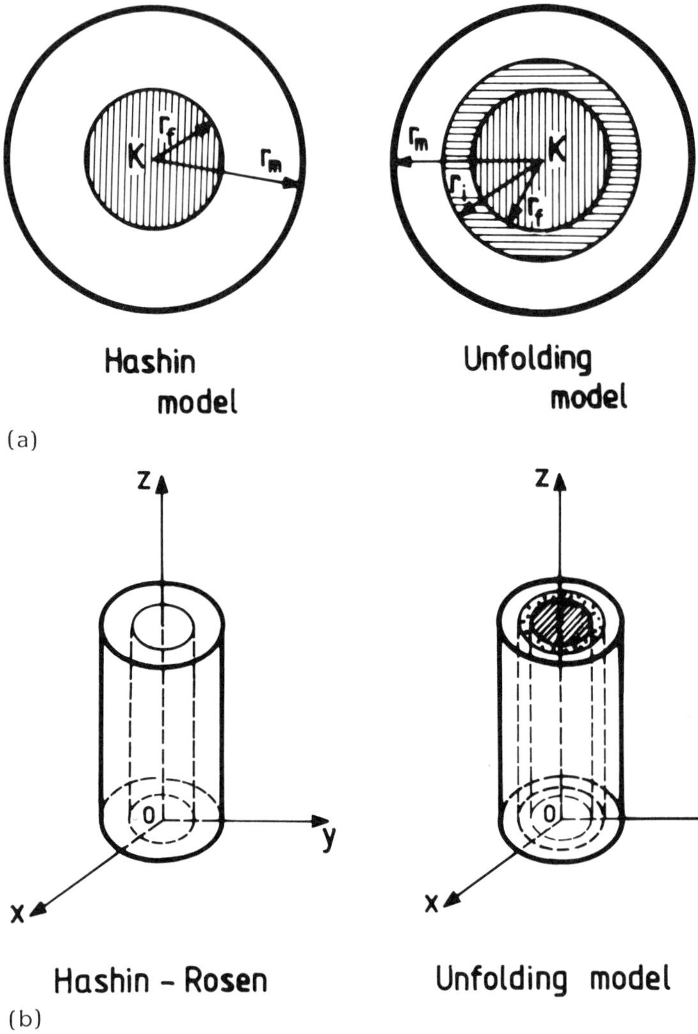

FIGURE 4 (a) Principal sections (particulates) and (b) axonometric views (fiber-reinforced composites) of the Hashin-Rosen two-phase models and their respective three-layer unfolding models (m, i, and f denote the matrix, mesophase, and filler, respectively).

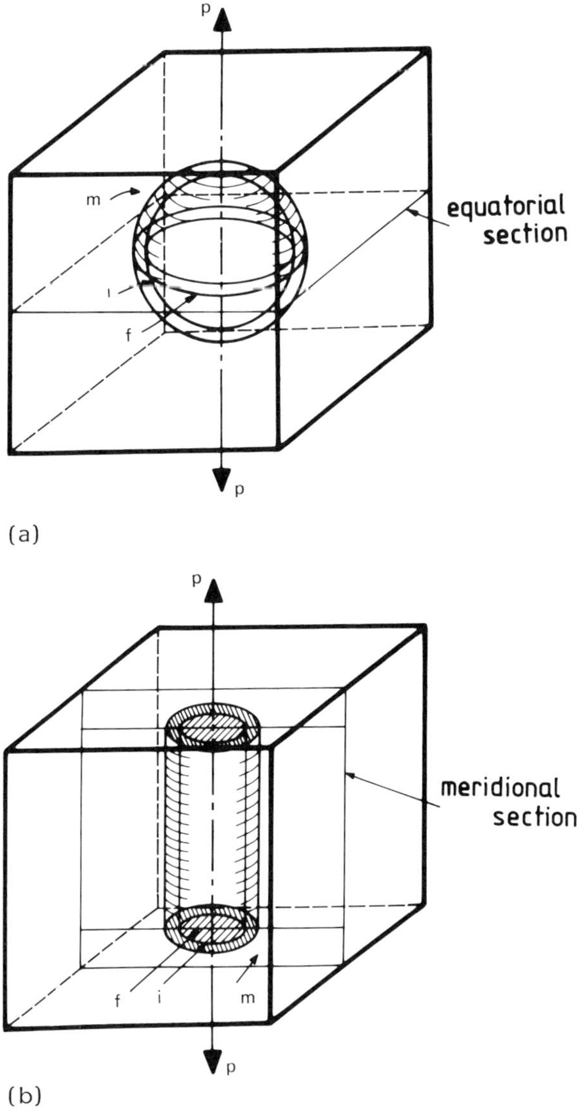

FIGURE 5 Configurations of the representative volume elements for (a) particulates and (b) fiber-reinforced composites (m, i, and f denote the matrix, mesophase, and filler, respectively).

shear, which holds for the fiber composites, is insignificant in this case, adhesion now being achieved by a continuously varying combination of shear and normal forces at the interfaces. Thus, variable components of the hydrostatic and shearing loading were developed between these surfaces. Then, it is closer to the real force distribution situation to assume that the contributions of phases are superposed in serial layers of the phases, placing one on top of the other in the direction of application of the external load.

In this case, the compliances, instead of the moduli, should be added to derive the compliance of the composite, since now the capacitances of the electrical analogy, corresponding to the moduli of the elements, are connected in series. This argument explains the necessity of adding the compliances of phases, instead of moduli, in this case.

However, since the approximation of superposed flat layers is again far from reality, because of the average sphericity of the layers, the expressions for the bulk compliance of the composite should be given by complicated expressions for the constituent compliances.

In the case of a unidirectional fiber composite, a simple improved law of mixtures (Figure 5b), where the contribution of the mesophase is taken into account, may be considered satisfatory. This is expressed by

$$E_c = E_f \underline{v}_f + E_i \underline{v}_i + E_m \underline{v}_m \tag{1}$$

where the subscripts c, f, i, and m denote composite, filler, mesophase, and matrix, respectively, and the $\underline{v}$'s are their volume fractions. In Eq. (1) the mesophase modulus, E_i (r), according to the three-term unfolding model (Refs. 21–25), is expressed by

$$E_i\,(r) = E_m + E_f\left(\frac{r_f}{r}\right)^{\eta_1} - E_m\left(\frac{r_f}{r}\right)^{\eta_2} \tag{2}$$

where the r's are the polar distances from the axis of the filler. The exponents η_1 and η_2 define the steepness of each term and may be calculated by measuring the thickness of the mesophase, r_i, and the modulus of the composite, E_c.

It can be readily shown from the boundary conditions between mesophase and matrix (for $r = r_i$) that the following relation holds between η_1 and η_2:

$$\frac{E_f}{E_m} = \left(\frac{r_i}{r_f}\right)^{(\eta_1 - \eta_2)} \tag{3}$$

This relation, combined with Eq. (1), where Eq. (2) for the mesophase modulus $E_i(r)$ is inserted, defines the values for η_1 and η_2 if E_c and r_i are measured.

The two term model is expressed by

$$E_i(r) = E_f\left(\frac{r_f}{r}\right)^{2\eta} + \left[E_m - E_f\left(\frac{r_f}{r_i}\right)^{2\eta}\right]\left(\frac{r - r_f}{r_i - r_f}\right) \qquad (4)$$

In this model the only unknown exponent, 2η, is defined by introducing Eq. (4) into Eq. (1) and inserting the values of E_c and r_i, respectively.

To now define the thickness of the mesophase, Δr_i, Lipatov's law was considered, which relates this thickness to the jump in heat capacity values at the glass transition temperature of a polymer. According to this law, the difference in heat capacity jump between the pure polymer of the matrix and the composite at their respective glass transition zones defines the extent of mesophase. Measurements with suitable samples of the matrix polymer and the composite in a differential scanning calorimetry apparatus then define accurately the values of Δr_i.

Lipatov (Ref. 26) showed that a relation holds between a *weight constant*, λ, defining the mesophase volume fraction v_i, and the jumps of the heat capacity ΔC_p^f of the filled composite and ΔC_p^0 of the unfilled polymer:

$$\lambda = 1 - \frac{\Delta C_p^f}{\Delta C_p^0} \qquad (5)$$

where λ multiplies the filler volume fraction $\underline{v}_f$.

In order to define the volume fraction, $\underline{v}_i$, of the mesophase for the particular composite studied, a series of dilatometric measurements must be made in a differential scanning calorimeter over a range of temperatures including the glass transition temperatures of the matrix and the composite, containing different amounts of fillers. All $C_p = f(T)$ curves plotted for pure polymers and composites presented typical jumps, ΔC_p, in heat capacity in the glass transition zone. For particulate composites Lipatov's formula is expressed by

$$\frac{(r_f + \Delta r)^3}{r_f^3} - 1 = \frac{\lambda v_f}{1 - v_f} \qquad (6)$$

For unidirectional fiber-reinforced composites this relation was modified (Ref. 23) and is expressed by

$$\frac{(r_f + \Delta r)^2}{r_f^2} - 1 = \frac{\lambda \underline{v}_f}{1 - \underline{v}_f} = \frac{\underline{v}_f + \underline{v}_i}{\underline{v}_f} - 1 \tag{7}$$

Equations (6) and (7) yield the radius $r_i = (r_f + \Delta r)$ of the mesophase from the experimental values of ΔC_p^0 and ΔC_p^f, respectively.

On the other hand, accurate experimental evaluation of the modulus of the composite, E_c, yields another quantity necessary for the solution of the two versions of the improved law of mixtures considering the influence of mesophase. The solution of these equations

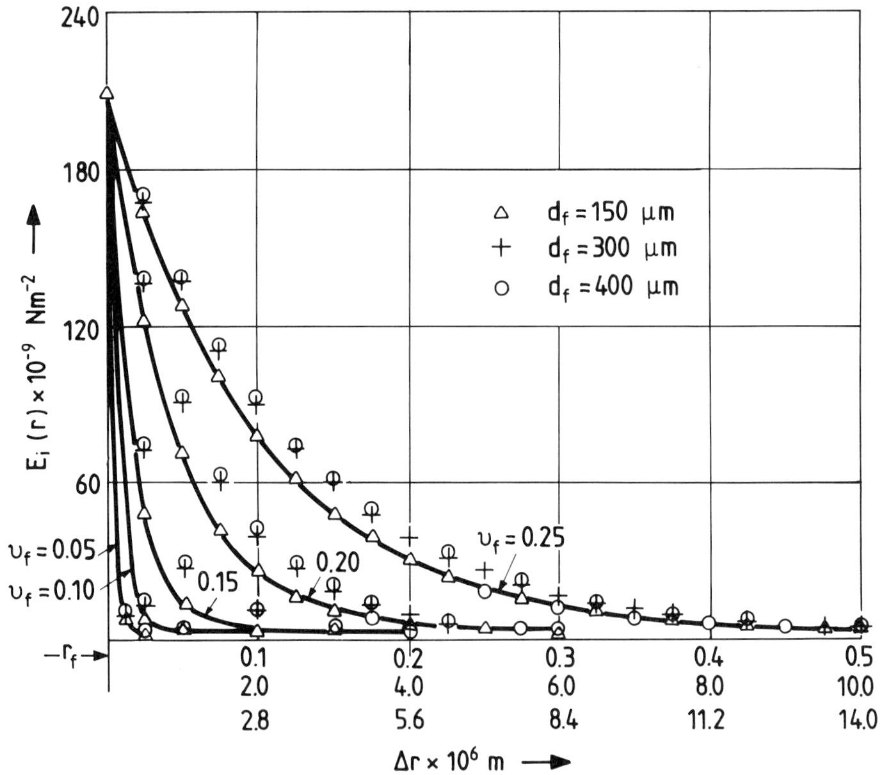

FIGURE 6 Variation of the mesophase moduli with polar distance r for different filler volume contents $\underline{v}_f$ for the three-term unfolding model.

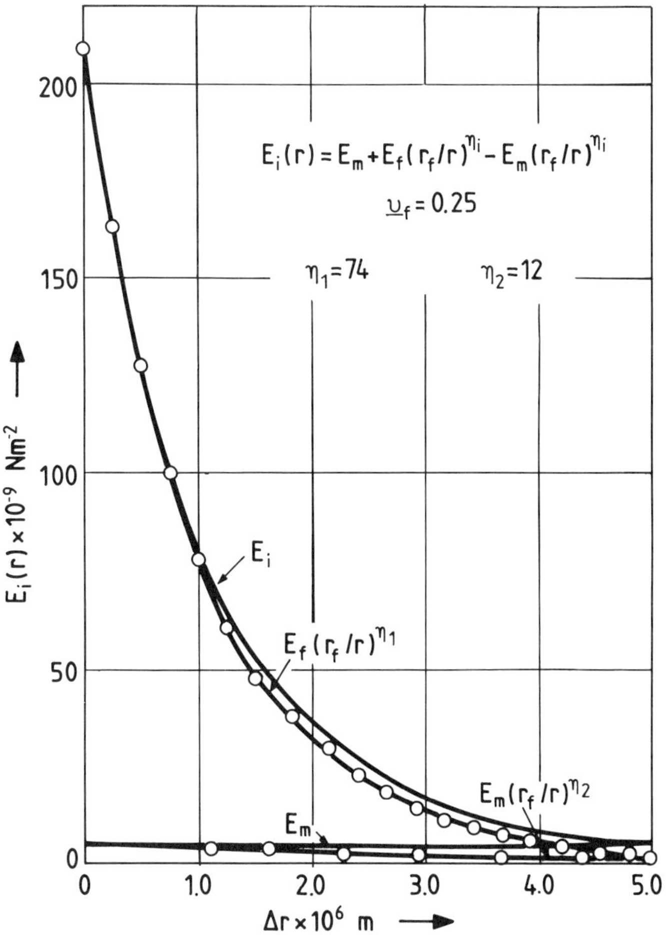

FIGURE 7 Mode of evolution of the variable terms $E_f(r_f/r)^{\eta_1}$ and $E_m(r_f/r)^{\eta_2}$, contributing to the definition of the mesophase modulus, versus polar distance r from the filler boundary for a 25% iron-epoxy particulate composite.

yields the exponents of negative-power expressions for the variation of the mesophase modulus, $E_i(r)$, and constitute an effective means for defining the adhesion quality of the bonding between phases in the composite.

The models have been applied to iron-particulate composites with filler volume contents up to 25% with satisfactory results. Figure 6 presents the variation of the mesophase moduli, $E_i(r)$, with the polar distance, r, of the mesophase for different filler volume fractions, $\underline{v}_f$, corresponding to both unfolding moduli. Normalization of these moduli to the modulus variation for the higher filler volume content indicates the coincidence of all these $E_i(r)$ versus r curves, proving the uniformity of the results for the same kind and quality of preparation of these iron-epoxy particulate composites.

Figure 7 presents the variation of the η_1, η_2, and 2η exponents for the different $\underline{v}_f$'s. Various values of the difference $A = (\eta_1 - \eta_2)$ of the exponents or of the 2η exponent indicate a different quality of adhesion.

Both versions of the unfolding model gave reasonable thicknesses for the mesophase layer in its proper sense. Moreover, the mode of variation of the mesophase modulus in the layer indicated the role of transition played by this layer in adapting and smoothing out the large differences in mechanical properties between the phases.

The values of the thicknesses Δr_i of the mesophase defined by the models may be considered reasonable, extending up to a few micrometers for the highest filler volume contents and the worst adhesion quality, as is the case with the iron particulates of Figures 6 and 7. For satisfactory adhesions their range is up to 2000 to 5000 Å for high filler volume fractions (see examples with glass fiber-epoxy composites in References 22 to 24, where for $\underline{v}_f = 70\%$ $\Delta r_i = 2000$ Å). It is worth mentioning that this layer was calculated at the critical zone of the glass transition of the composite, where a resonance of disparities between phases should occur. Furthermore, since this layer embraces all the factors contributing to its existence, and not just the physisorption and chemisorption phenomena of macromolecules in this zone, as in the evaluation of such layers in other cases, a thicker mesophase layer should be expected than that evaluated from single chemisorption phenomena. The thickness, Δr_i, of the mesophase layer is then a quantity depending directly on the quality of adhesion and yields a simple and sure means for estimating this parameter.

PROPAGATION OF CRACKS IN FIBER COMPOSITES UNDER STATIC LOADING

For the dynamic study of mesophase phenomena we used a series of composite epoxy resin plane specimens with a length of 0.30 m, a width of 0.10 m, and a thickness of 0.002 m. The specimens consisted of two full-length phases of equal width bonded together along their common interface (Ref. 27).

The mechanical properties of the two phases of the composite were varied by adding different amounts of plasticizer (Thiokol LP3) to the pure epoxy polymer (Epikote 828, Shell Co.). The two phases were bonded together without any other adhesive. This was achieved by casting the material of phase II along the longitudinal boundary of phase I, which was already prepared and appropriately placed in the mold. After consolidation of phase II, the composite specimen was thermally treated for 1 week with small temperature gradients to ensure total polymerization of the plates and stress-free boundaries at the interfaces.

Phase I always contained an initial transverse slit of constant length a_0 = 0.020 m, with a lip opening less than 0.0003 m. This slit was always used to initiate the crack at the same position in phase I.

A number of different specimens were prepared by using five different types of compositions in all possible combinations as phase I or II. The different compositions contained amounts of plasticizer increasing by 10% increments for each composition from 0 up to 40%. Thus, five groups of composite specimens were studied, each of which had phase I (slotted) with a constant percentage C of plasticizer, and phase II (plain strip) with varying amounts of plasticizer. Furthermore, another series of tests was performed with both phases of the same composition (0 and 10% plasticizer) in order to reveal the influence of the mesophase in identical duplex composites.

Addition of plasticizer in the epoxy prepolymer reduced the modulus of elasticity of the substance and made it more rubbery and therefore more ductile. Interesting results were obtained by arranging the tests in two groups, i.e., those in which the crack propagates from a hard phase I to a soft phase II, and those in which the crack moves from a soft phase I to a hard phase II. Table 1 gives the static mechanical and optical properties of the various plasticized polymers used in the tests.

The applied dynamic load was of the falling-weight type with a constant stress rate equal to $\dot{\sigma} = 10 \times 10^3$ N/cm^2-sec. For the study of the crack propagation process a Cranz-Schardin high-speed camera was used, disposing 24 sparks with a maximum frequency of 10^6 frames per second. With the help of an external generator the sparks can be set with any desired frequency up to the limit.

The optical part of the experimental arrangement was very simple. the light beam from each spark reflected on a spherical mirror of great reflecitivity and, after passing through the specimen, was focused on the lens of the camera (see Figures 8 and 9).

The first phenomenon studied was crack arrest at the mesophase. The crack propagated in phase I with a constant maximum velocity,

TABLE 1 Mechanical and Optical Constants of Pure and Plasticized Epoxy Polymers

Plasticizer (C%)	Elastic modulus, E (N/cm^2)	Poisson's ratio (ν)	Stress optical coefficient, C_t ($m^2 N^{-1} \times 10^{-2}$)	Coefficient of optical anisotropy, ξ
0	324,000	0.338	17.03	0.1401
10	316,000	0.340	20.03	0.1161
20	292,000	0.344	20.79	0.1080
30	245,000	0.358	21.66	0.0983
40	179,000	0.430	30.78	0.0859

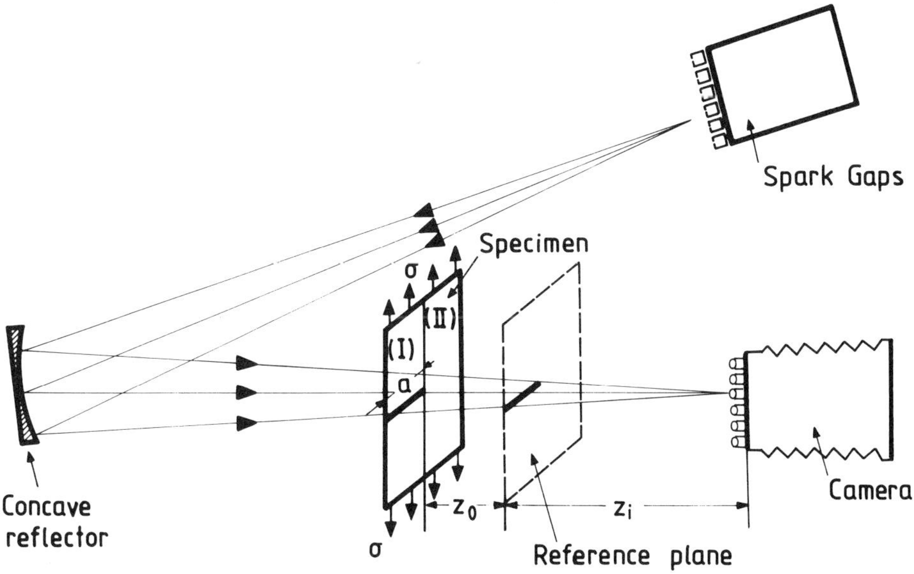

FIGURE 8 Optical part of the experimental arrangement.

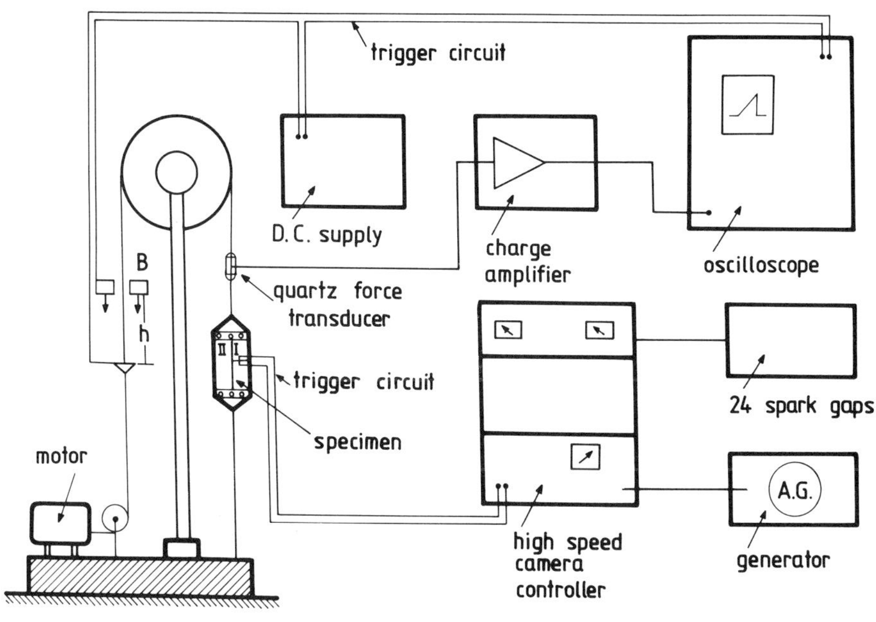

FIGURE 9 Mechanical and electrical part of the experimental arrangement.

which was maintained up to a very short distance from the interface. The crack starts decelerating within the mesophase zone and finally stops completely at the interface.

The process of crack deceleration, crack arrest at the interface, and crack acceleration in phase II is characterized by the gradual disappearance of the caustic in phase I and its subsequent appearance in phase II close to the interface.

As the crack tip and therefore the caustic are entering the mesophase, the forefront of the caustic is continuously cut off, while the rest of the caustic retains its initial shape of a nondistorted generalized epicycloid. This phenomenon continues after the crack arrest, that is, after the crack tip has reached the interface, until the whole area covered by the caustic disappears, submerged at the interface.

Furthermore, during this progressive disappearance of the caustic in phase I, there are no signs that a new caustic is in the process of development in phase II.

Subsequently, and while the crack remains in the arrested condition, a new caustic is gradually appearing in phase II. The new caustic initially has the form of an arc of a circle, until it emerges in a few microseconds as a full circle.

In Figure 10 a series of photographs is presented which shows the crack propagation in a composite with a soft phase I (20% plasticizer) and hard phase II (10% plasticizer). The mode of crack propagation from a hard phase I (unplasticized) to a soft phase II (20% plasticizer) is presented in Figure 11.

It is worth noting that all caustics approaching the interface are progressively truncated without any shape distortion, whereas no caustics appear in phase II during the whole period of crack deceleration and arrest, although the crack tip practically touches the interface.

The final semicircular form of the caustic in the crack arrest period coincides with a typical form of a static caustic, and it can directly yield the value of the respective stress intensity factor. We assume that the partial-arc forms correspond to semicircular caustics with a transverse diameter $D_t = 2l/\pi$, where l is the apparent arc length.

The continual disappearance of the caustic in phase I near the interface and its subsequent emergence in phase II show that the caustic and therefore the stress-strain field at the crack tip propagate through the interface in a wavelike manner. It can be derived that the stress field at the crack tip is directly correlated with the stress waves which emanate from the tip of the propagating crack and which are observed by high-speed photography in all experiments performed with transparent materials, such as Plexiglas and polycarbonate.

Moreover, it has been frequently observed that stress waves emanating from the crack tip are reflected at the opposite free

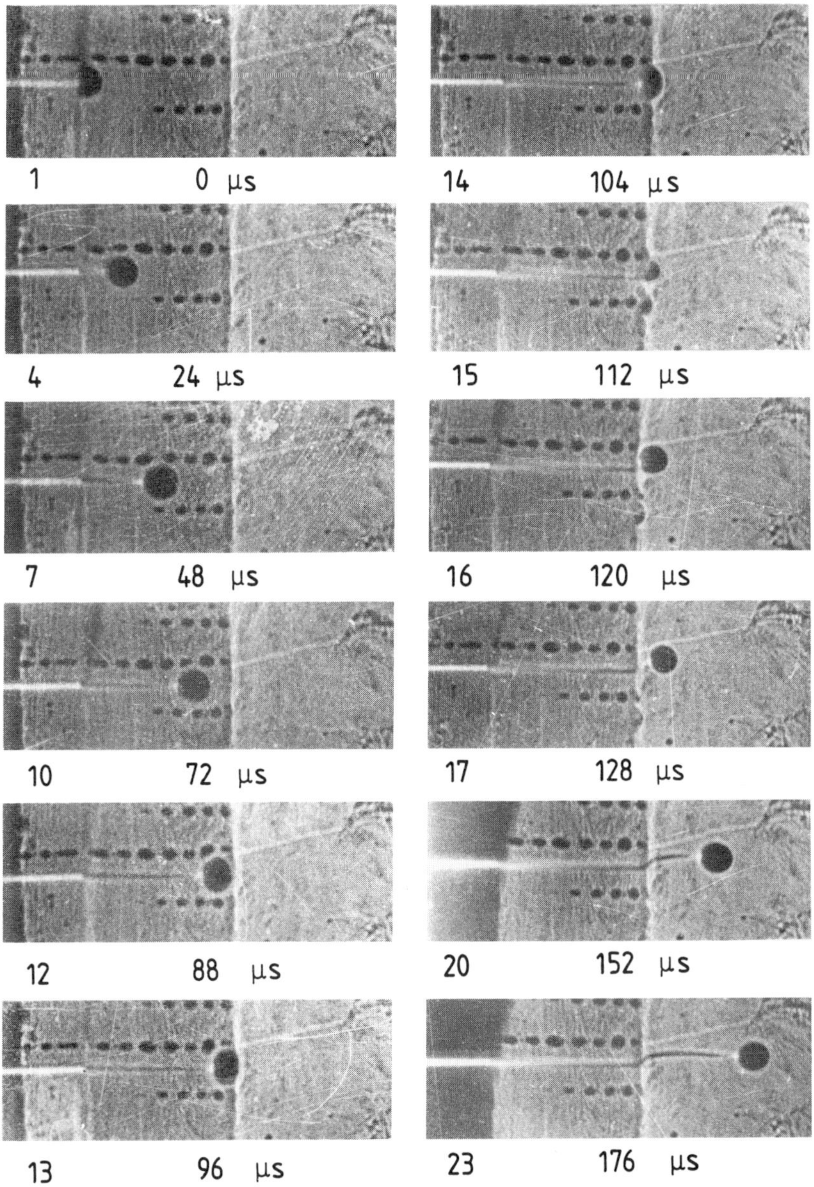

FIGURE 10 Series of photographs showing crack propagation in a composite (phase I, 20% plasticizer; phase II, 10% plasticizer).

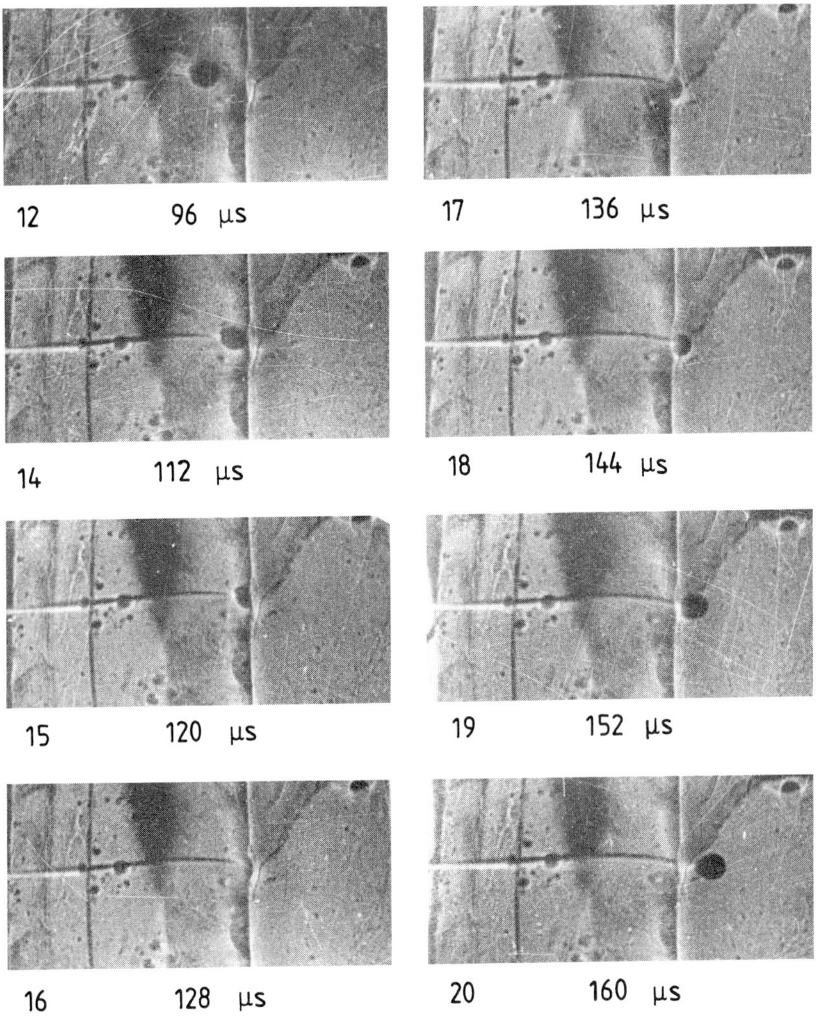

FIGURE 11 Series of photographs showing crack propagation through the interface of a composite (phase I, 0% plasticizer; phase II, 20% plasticizer).

boundary of the specimen, after complete rupture of the specimen. There they create a new caustic, which travels backward along the lips of the existing crack, which separated the specimen in two pieces, together with the reflected wave and with a velocity equal to the shear wave velocity of the substance of the specimen.

From all these phenomena we assume that the disappearance and reappearance of the caustic at the interface are due to the distortion of the stress field close to the crack tip, because of the reflection and refraction of the stress waves accompanying the propagating crack tip at the not yet disrupted interface. This means that the stress-strain field is not transmitted uniformly from one phase to the other.

When the crack is initiated in phase II, it accelerates and propagates in the composite in the same typical process of crack acceleration, up to a new maximum constant propagation velocity. The constant velocity of the propagating crack in phase II is higher than the constant crack propagation velocity in phase I. This phenomenon has been observed in any combination of materials of the phases, and it also appears in duplex specimens, where both phases are made of the same material.

A plausible explanation may be that the new fracture begins after a crack arrest period and under almost the same load conditions that prevailed during the first initiation of crack propagation in phase I. However, this new fracture process takes place with a much longer initial crack than the initial slit in phase I.

The crack propagation velocities in the phases also depend on the material properties of each phase. The effect of these material properties is superimposed on the effect due to crack arrest at the interface.

The effect of the material characteristics of the phases may be studied systematically by taking into account the phase classification of the specimens. For the identical phase II classification, that is, for the group of specimens in which phase II is kept the same, the stress intensity factor in this phase increases continuously with increasing ductility of phase I. The same is true of the crack propagation velocity. As the ductility of phase I increases, the corresponding crack propagation velocity in phase II increases.

In contrast, in the tests with identical phase I we could not determine any definite change of the crack velocity or the stress intensity factor in phase I with changing ductility of phase II. The crack propagation velocity of each material in phase I converges to a mean value characteristic for each material.

Another important phenomenon of fracture in composite plates is that for hard phase I the crack passes through the mesophase without any change in direction. Figure 11 shows the path of the crack through the composite specimen (hard phase I, unplasticized; soft phase II, 20% plasticized) for a typical case. For the inverse arrangement, when the crack starts from a soft phase I and enters a hard phase II, the crack deviates from its original direction as it enters phase II of the composite, forming an angle of about 60–70° with the interface. After entering the second phase and at a relatively short distance from the interface, it again takes its K_1-dominated path. Figure 10 (frames 20–23) exemplifies this case.

These phenomena may be explained by assuming that when the crack starts from the soft phase and enters the hard phase, it creates shear stresses at the interface, which reaches a high stress concentration. These increased shear stresses in some cases caused a disruption of the interface at the region near the area of crack propagation, which resulted in a bifurcation of the crack into two branches. Figure 12 shows the fracture process for a composite specimen with soft phase I (30% plasticizer) and hard phase II (unplasticized). The crack propagated in phase I with a constant velocity and was arrested at the interface. The increased shear stress at the interface and, of course, some eventual additional imperfections in this region (for instance, weak bonding) made the crack propagate along the interface before entering phase II.

The appearance of two caustics in phase II surrounding the tips of the propagating crack along the interface is illustrated in Figure 12, which clearly shows the splitting of the single crack in phase I into two branches, starting at the interface. The two branches propagate along the interface, and at some distance from the point of initial bifurcation they have lost some of their strain energy, since they move along the interface in a complex K_I and K_{II} mode in which, eventually, they meet some points of a stronger interface bonding, which create insuperable obstacles to the crack propagation. At these points a new distribution of stresses develops in the composite; phase I assumes less and less of the load as the crack propagates along the interface, which causes an arrest of both branches. At these points, both branches enter phase II and progressively are oriented in a direction normal to the direction of the applied load.

It is worth noting that the angle subtended by the crack axes and the longitudinal interface when the two crack branches enter phase II is originally approximately 70° and then progressively increases to 90°. However, the phenomenon of crack bifurcation and branching at the interface is complicated and warrants an exhaustive quantitative study.

Figure 13 presents a series of photographs from a group of tests with duplex epoxy resin plates, both with 20% plasticizer. Immediate branching of the crack arriving at the interface is apparent as soon as the crack enters phase II, which is identical to phase I. However, again the angle of inclination of the branched cracks is of the order of 60° with the interface.

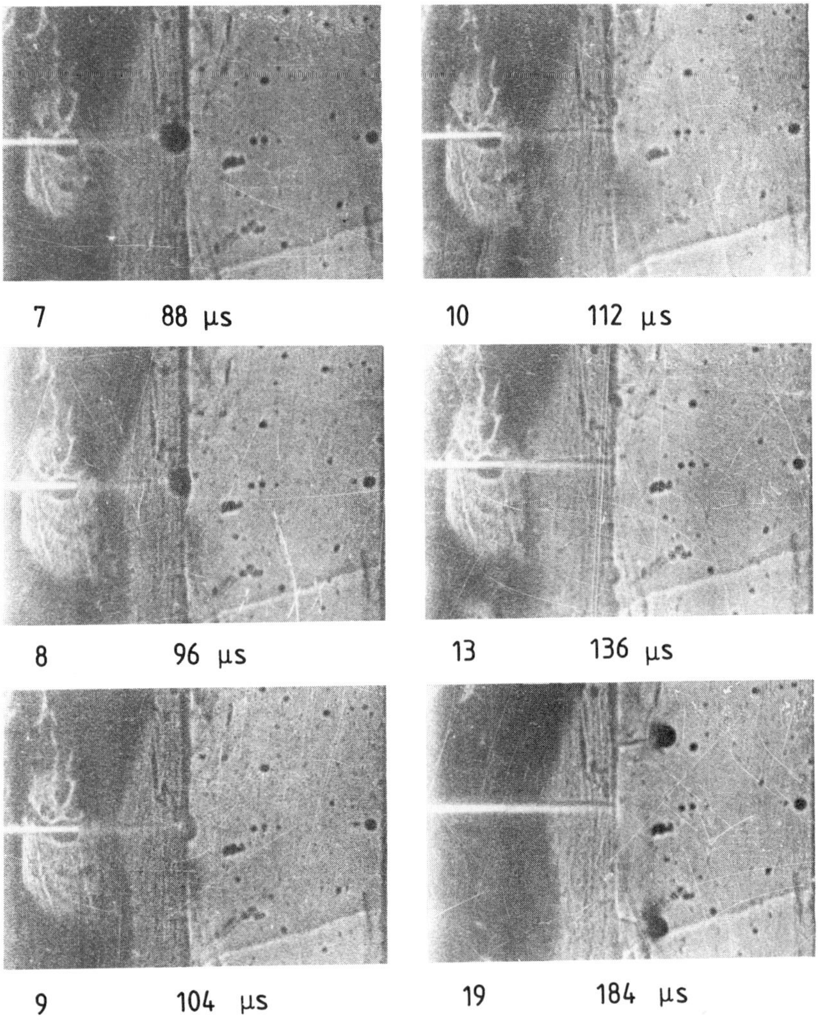

FIGURE 12 Series of photographs showing crack bifurcation in a specimen with soft phase I (30% plasticizer) and hard phase II (0% plasticizer).

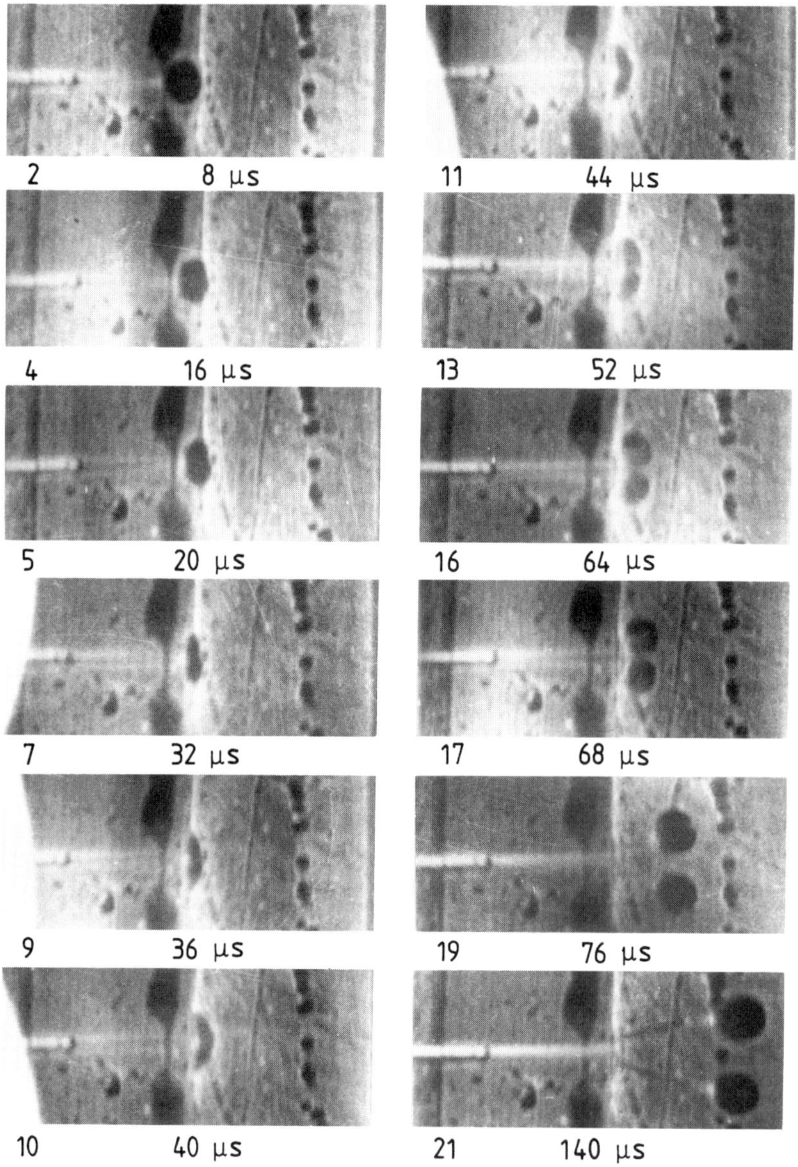

FIGURE 13 Series of photographs showing crack propagation in a duplex plate in which both phases were prepared from a cold-setting epoxy resin plasticized by adding 20% Thiokol LP3 plasticizer.

When the crack splits into two branches, the energy release per crack surface is also split into two parts, and therefore the crack velocities of the branches are lower than that expected for single-crack propagation. This phenomenon is related to the stress distribution around the two crack tips.

It is now worth studying the shear stresses that develop along the interface of the composite specimen and tend to tear it apart, that is, in a direction different from the initial direction of crack propagation. The reason for the development of these shear stresses along the interface is the unequal distribution of tensile stresses in the two phases of the composite, which have different mechanical properties.

Assuming a composite specimen with perfect bonding of phases, it is readily concluded that the relation of the stresses in the two phases of the composite is directly proportional to the relation of their elastic moduli. More concretely, for $\varepsilon_I = \varepsilon_{II} = \text{const}$, it is valid that $\sigma_I/E_I = \sigma_{II}/E_{II} = \text{const}$ and therefore $E_I/E_{II} = \sigma_I/\sigma_{II}$. The difference $\Delta\sigma$ between σ_I and σ_{II} results in a shear stress along the interface ($\tau = \Delta\sigma/4$). These shear stresses superimpose on the singular stresses around the crack tip when the crack reaches the interface. At this instant there is a danger that the interface will be disrupted. Therefore, the phenomena described above can be explained by considering the stress distributions in the two phases as functions of the elastic characteristics of the different materials.

Since all composite specimens were subjected to uniform elongation, it was noted in the case of soft phase I that the elongation at fracture of the composite is higher than in the case of hard phase I. This is because the elastic modulus of the soft phase is lower than the modulus of the hard phase, whereas stable interface bonding requires equal elongation and strain on both phases. Therefore, a hard phase II is forced to sustain a higher elongation and strain than in the case where it is used as phase I. This higher strain corresponds to a larger value of fracture stress. The increased stresses in the brittle material of phase II explain the tendency for a further increase of the crack propagation velocity.

The stress intensification of a hard phase II also means a higher absolute value of the difference between the phase stresses at the interface, that is, a higher shear stress at the interface. The relation of the phase stresses for two given materials (excluding the effects due to the crack tip singularity) is always the same, independent of which material is used as phase I. This means that the ratio σ_h/σ_s of the stresses when the hard material is used as phase I is equal to the ratio σ'_h/σ'_s when the soft material is used as phase I, where σ_h, σ_s denote the fracture stresses for hard phase I and σ'_h, σ'_s denote the fracture stresses for soft phase I.

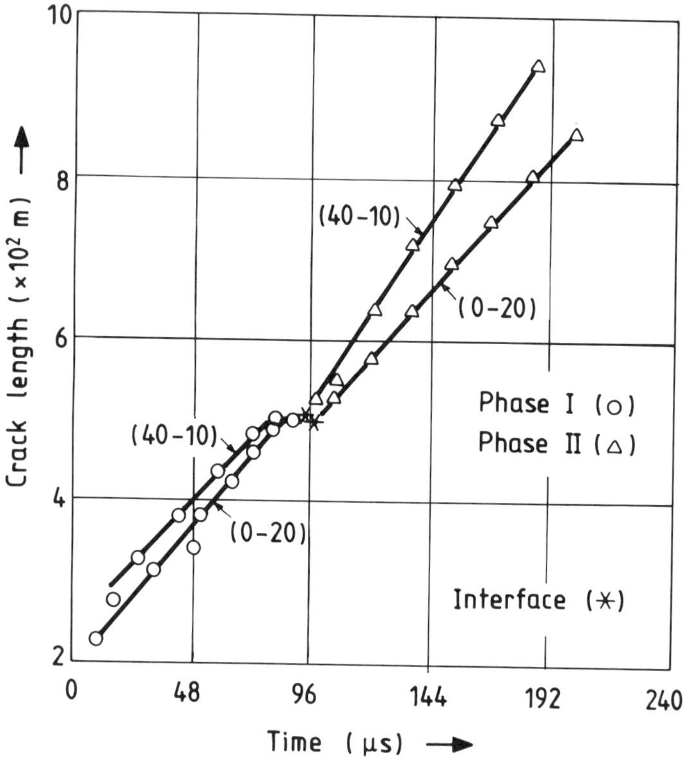

FIGURE 14 Crack length versus crack propagation time for a composite with phase I 40 or 0% plasticizer and phase II 10 or 20% plasticizer.

From the agruments developed previously, it can be shown that $\sigma'_h > \sigma_h$ and $\sigma'_s > \sigma_s$ is always valid. Based on these two inequalities, as well as on the equality of the ratios $\sigma_h/\sigma_s = \sigma'_h/\sigma'_s$, it is readily found that $(\sigma'_h - \sigma'_s) > (\sigma_h - \sigma_s)$. Thus, in the case of soft phase I, the increased shear stresses at the interface cause a local deviation of the direction of the crack axis, or a bifurcation of the propagating crack. In the case of a hard phase I, the effect of the shear stresses at the interface is much less.

The shortening of the crack arrest time in the case of a hard phase II can be explained in terms of the increased elastic strain energy stored in this phase of the specimen. This effect was also determined by Ishikawa et al. (Ref. 28) for crack arrest at a small hole lying ahead of the propagating crack.

Figure 14 presents the variation of the crack length with time for two characteristic compound specimens. The open circles correspond to soft (40% plasticized) and hard (unplasticized) phases I, while open triangles to hard (10% plasticized) and soft (20% plasticized) phases II. The slopes of these curves clearly indicate constant average velocities for the cracks in either phase. For the first case the crack velocities are $\underline{v}_I$ = 314 m/sec and $\underline{v}_{II}$ = 461 m/sec, while for the second they are $\underline{v}_I$ = 331 m/sec and $\underline{v}_{II}$ = 350 m/sec. In between the two constant velocities in the respective phases, there is a small time interval for the crack arrest period at the mesophase indicated with an asterisk. This time interval was estimated to be of the order of 8 to 10 μsec.

The marked difference between the crack velocities in phase II in the two examples is mainly attributed to the influence of the mechanical properties of phases I on the crack propagation in phases II. For a detailed quantitative discussion of this influence see Reference 27.

Figure 15 presents the variation of the instantaneous crack length with time for the case of fracture of the group of specimens indicated

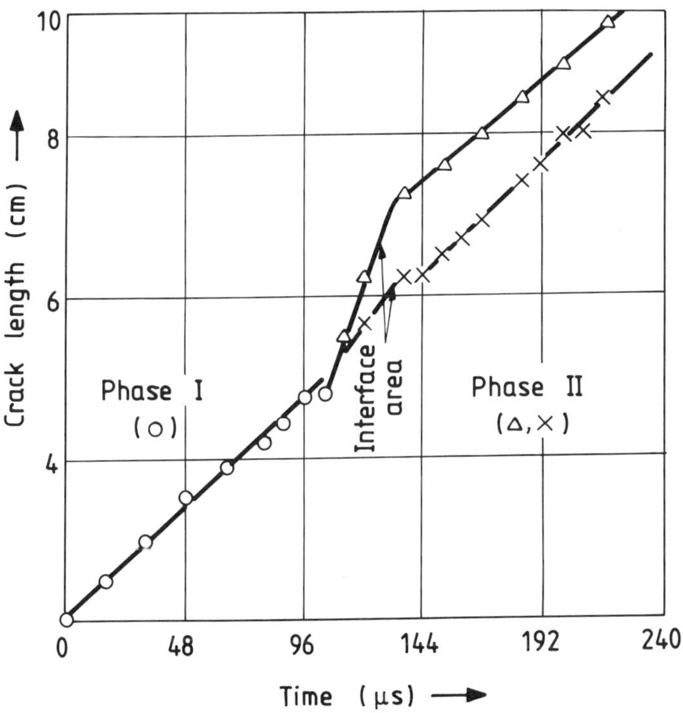

FIGURE 15 Crack length versus crack propagation time for a specimen with phase I (30% plasticizer) and phase II (0% plasticizer). Notice the two crack arrest processes on reaching the interface (before bifurcation) and before entering phase II (after bifurcation).

in Figure 12 with a soft phase I (30% plasticized) and a hard phase II (unplasticized). In this case the velocities in phase II were $\underline{v}^u_{II}$ = 298 m/sec and $\underline{v}^d_{II}$ = 257 m/sec, whereas the crack velocity in phase I was $\underline{v}_I$ = 262 m/sec. The almost equal values of the velocities in both phases are explained by the branching of the crack and splitting of the strain energy into two parts for either crack propagating in phase II. If branching did not take place the crack velocity in phase II should be much higher, as it is with the first case of Figure 14.

We now study the variation of the stress intensity factor at the tip of the propagating crack in both phases. The stress intensity factor remains constant after the crack initiation in phase I and within a certain distance from the interface. As the crack tip reaches the zone influenced by the mesophase the stress intensity factor starts to increase, and at the interface it reaches a value

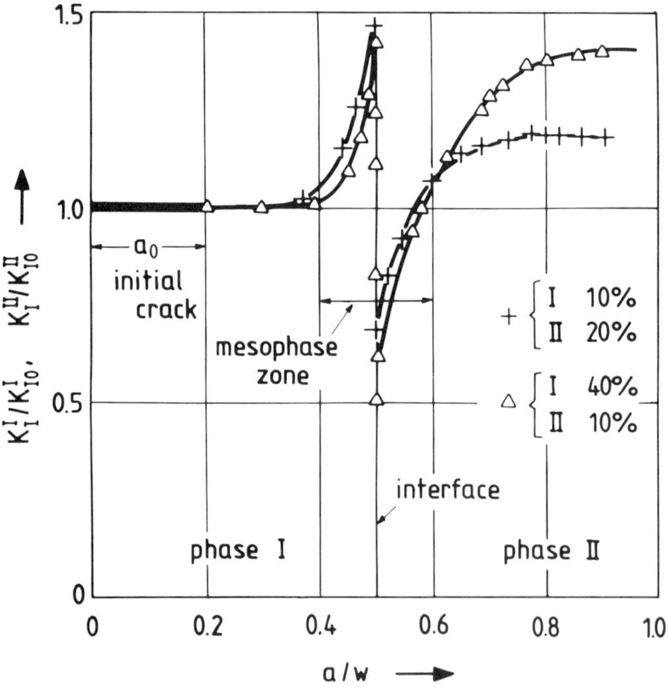

FIGURE 16 Variation of the K_I stress intensity factor at either phase of the composite for (a) a hard-soft sequence and (b) a soft-hard sequence of phases.

30–50% higher than its average constant value. This increase is due to the existence of both the mesophase, which plays the role of a transverse boundary, and phase II, with its different material characteristics. Although the contribution of phase II to the increase of the value of K_I is considerable, much greater than that of the mesophase, its mechanical properties seem not to influence the variation of K_I and therefore phase II may be considered as a reinforcing element of the interface. This leads to the conclusion that the increased values of the stress intensity factor depend mainly on the stable interface.

The singular stress field at the tip of the crack arrested at the interface also influences the phase II stress distribution, so that, shortly after the initiation of the crack in phase II, the stress intensity factor takes values considerably larger than the values of K_I in phase II during crack initiation; this increase is of the order of 20 to 50% of the initial value of K_I in phase II. This phenomenon of the progressive increase of K_I as the crack propagates in phase II is apparent from the successive size of caustics. These caustics reach a final stable size at the end of the zone influence by the mesophase. Figure 16 shows the variation of the stress intensity factor at each phase of the composite, normalized to its initial value at the initiation of the crack at either phase, versus the crack length.

PROPAGATION OF CRACKS IN FIBER COMPOSITES UNDER DYNAMIC LOADING

The previous paragraph was limited to cases where the external dynamic loading was kept constant and the investigation was concerned only with the influence of the two main factors, that is, the mechanical properties of either phase of the plate and the order of succession of the phases during the initiation and propagation of cracks. We now study a third important factor, that is, the influence of the rate of application of the external loading, in combination with the other two factors (Refs. 29–31). Again, interesting conclusions may be derived, yielding an insight into the mechanism of fracture in multiphase materials.

Thus, we have investigated and compared the crack propagation velocities in bimaterial plates under both static loading conditions and dynamic loading conditions at different strain rates (Ref. 29).

Similar investigations of velocity under both static and dynamic load conditions were performed by, among others, Katsamanis et al. (Ref. 32), using cracked plates made of polymeric substances. These authors proved that the constant crack propagation velocity depends on the critical stress in fracture for both static and

TABLE 2 Mean Crack Propagation Velocities in Phase I of the Biphase Specimens

	v_I mean (m/sec)		
	$\dot{\varepsilon}_1 = 0.5\ (\text{sec})^{-1}$	$\dot{\varepsilon}_2 = 0.2\ (\text{sec})^{-1}$	$\dot{\varepsilon}_3 = 0.6 \times 10^{-4}\ (\text{sec})^{-1}$
v_I^{0}	391	363	307
v_I^{10}	368	351	284
v_I^{20}	352	337	258
v_I^{30}	335	304	243
v_I^{40}	318	283	235

dynamic loading. They also showed that for the same fracture stress the constant crack propagation velocity reaches higher values when fracture takes place under dynamic loading with higher strain rates. On the other hand, the constant crack propagation velocity tends to increase more abruptly with an increase of fracture stress for the case of static loading, but never reaches the fracture velocity values of the corresponding dynamic case. The validity of these results is extended here for the case of biphase specimens.

For this purpose, a series of tests similar to those described above were executed, the only difference being that the externally applied loads were changing from static to dynamic. Three loading velocities were applied, creating different strain rates, $\dot{\varepsilon}$, for each test. These strain rates are given in Table 2.

As a first result from these experiments it may be stated that all general qualitative characteistics of the crack propagation process in bimaterial plates were proved valid also for dynamic loadings. Tests showed primarily that the crack propagation velocity in phase I, which depends only on the material characteristics of phase I and the loading conditions of the biphase plate, increases with increasing strain rate.

Table 2 presents the mean crack propagation velocities in the five different materials used as phase I for the three different loading and deformation rates.

The velocities in phase II of the bimaterial plates are highly dependent on the material characteristics of phase I. Furthermore, they all reach higher values than the corresponding phase I velocities, due to the crack arrest effect.

Specimens of the same composition fractured under two different dynamic loadings showed a uniform crack velocity distribution. The phase II crack propagation velocity of the specimen fractured under a higher strain rate exceeded that of the same specimen fractured under a lower dynamic strain rate by approximately the same fraction as the corresponding phase I velocities. This means that the effect of crack arrest influences the crack propagation process in a similar way for both cases of dynamic loading.

A significant discrepancy in the above model occurs, however, in the case of static loading. The crack propagation velocity starts with a relatively low value in phase I of all bimaterial plates, but after the crack arrest period reaches a value of the same order as the fracture velocity in phase II under the higher strain rate dynamic loading. This velocity is significantly higher than the corresponding fracture velocity for the same phase II material under the lower strain rate dynamic load. Therefore, the crack arrest effect results in a much higher velocity increase in phase II, for the case of static loading. This result could not be predicted from previous experimental investigations.

The phenomenon of high velocity increases developed in phase II due to crack arrest becomes clearer when a soft material occupies phase I. This arrangement of phases imposes such velocity increases, after the period of crack arrest, that crack bifurcation may occur at the interface area.

The results described above are summarized in Figure 17, which shows the dependence of the crack propagation velocity in phase II of biphase plates (with identical material for phase II, corresponding to a 20% plasticized epoxy polymer) on the material characteristics of phase I for the three loading velocities. It is worth noting again the increased values of the v_{II} crack propagation velocity for the specimens fractured under static load.

From this series of experiments we conclude that the *crack arrest effect* plays a more significant role in the fracture process of biphase plates subjected to static load then in the case of dynamic fractures. This can be explained only if we take into consideration the different stress distributions during the fracture process. The crack arrest separates the fracture process of the specimen into two different parts. Either of these parts acquires the characteristics of a distinct fracture process. When there is a type of fracture of the specimen corresponding to static load, the applied load remains constant for the whole fracture period; that is, the fracture process in phase II takes place under the influence of the maximum (constant) external load.

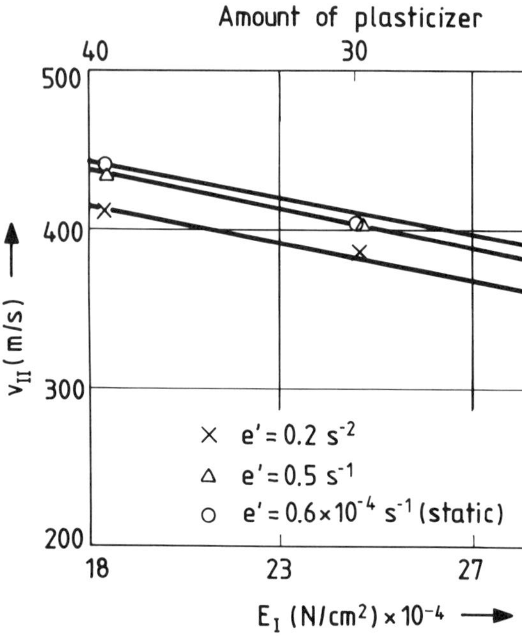

FIGURE 17 Dependence of phase II crack propagation velocity on the modulus of elasticity of phase I for both static and dynamic load conditions (phase II, 20% plasticized).

Notice that the reinitiation of the crack at the interface and the fracture of phase II of the specimen take place under a lower external load, since the initial crack is now as long as the whole width of the strip of phase I. This is indeed the case when fracture takes place under the influence of a dynamic load.

To complete this study it is necessary to show the dependence of the crack propagation mode through the interface on the state of stress dominating at the mesophase.

In order to eliminate the influence of the mechanical properties of the constituent materials of the phases, we performed a series of tests with identical materials in both phases. In this way all other influences were minimized and the effect of the mesophase was dominant.

This investigation showed that the process of crack arrest at the interface of duplex specimens depends primarily on the mesophase and secondarily on the abrupt change in mechanical properties of the materials of the phases. Therefore, the dynamic problem of a crack reaching the mesophase of a biphase plate cannot be appropriately approached by using only the results of the static analysis.

It is well known (Ref. 33) that fracture of an elastic medium is accompanied by emission of stress waves from the tip of the propagating crack. This emission is associated with a reduction of the stresses at the crack tip from a stress level equal to the tensile strength of the material down to zero (Ref. 34).

Experimental approaches to the problem have shown (Refs. 35, 36) that, at both edges of a growing crack, all three types of stress waves are excited (longitudinal, transverse, Rayleigh) and that the characteristics of the Rayleigh stress waves are related to the stored elastic energy at the crack tip before fracture (Refs. 37, 38).

Thus, it becomes clear that the relation between fracture stress waves and crack propagation can, in principle, be approached by means of average effects, that is, by investigating the overall characteristics of the stress wave field emanating from the crack tip and the influence it can have on the fracture process itself.

It has already been mentioned that, for duplex epoxy specimens subjected to dynamic tensile loads (Refs. 27, 29), continuous stress wave propagation ahead of the crack tip is inhibited when the crack approaches the mesophase, which results in a disturbance or even in destruction of the singular stress field at the crack tip and forces the crack to stop. The external load, when it is still acting on the specimen during the crack arrest time, is designed to create a new singular stress field in order to cause crack reinitiation.

For this series of tests with duplex specimens the types of specimens and the experimental technique were the same as those used previously with plates of two different phases (Ref. 29). Furthermore, the same phenomena appeared during fracture of the duplex specimens. Indeed, the cracks, after propagating with a constant maximum velocity in phase I of the duplex specimen, decelerated abruptly within the mesophase and finally completely stopped when they reached the interface.

When the cracks approached the mesophase a progressive disappearance of the caustic in phase I took place, whereas the remaining parts of the caustics were left unaffected, preserving their initial shapes and dimensions of a generalized epicycloid. Furthermore, during this progressive disappearance of the caustics in phase I, there were no signs that new caustics were in the process of development in phase II.

Subsequently, after the arrest periods at the interfaces, new caustics emerged in phase II in exactly the same way as with the biphase plates.

The experiments in duplex specimens showed that all the qualitative characteristics of the crack arrest process described above remained unchanged for all possible material compositions of the bimaterial specimens. These results imply that the crack arrest process is, to a great extent, independent of the mechanical properties of the phases and is mainly related to the existence of the mesophase.

Figure 13 shows the crack propagation in a 20–20% epoxy polymer duplex specimen. Frames 5 to 11 show the period of crack arrest at the interface, since more than half of the caustic has been submerged at the interface. Frames 4 to 8 show the submergence of the caustic at the interface. Notice the significant deformation of the interface as it absorbs the kinetic energy of the crack and the strain energy stored at the crack tip.

Frames 9 to 11 illustrate the second period of the crack arrest interval. The specimen was fractured at a relatively high strain rate, which resulted in considerable deformation of the interface (frames 4, 5, and 8), and a crack bifurcation, after crack arrest, is developing as indicated in Reference 29. In frames 12–21 the development of bifurcation is apparent.

Figure 18 shows the fracture process in a duplex Plexiglas specimen. The crack arrest at the interface and the subsequent submergence-emergence of the caustic, the deformation of the interface (frames 3, 4, and 6), and the emission of fracture stress waves immediately after crack reinitiation (frame 8) are apparent. Notice also the much larger (than in the case of epoxy resin specimens) deformation of the interface, to the right as the crack arrives at it (frame 6) and to the left as the crack leaves it (frame 7); this deformation is also apparent when a crack arrives at the free end of a Plexiglas specimen.

From the tests with duplex epoxy or Plexiglas plates it is evident that the influence of the mechanical properties of the phase is simply superimposed on the phenomena due to the intrinsic properties of the mesophase isolated in the fracture process of duplex specimens. Thus, a reinforcing phenomenon appears to be due to the mesophase, which either increases the crack propagation velocity (soft phase I) or counteracts it (hard phase I).

All the experimental evidence described above shows that the main factor which determines the dynamic crack propagation process through a bimaterial plate is not the abrupt change in mechanical properties of the material from phase I to phase II, but the existence of the mesophase.

Therefore, in a fiber-reinforced composite, the mesophase constitutes a definite material discontinuity. It acts first as a *buffer* and then as a *crack arrester*. Although the mechanism of crack arrest depends on the sequence of the phases (soft-hard or hard-soft), the macroscopic mechanism of crack propagation delay is in principle the same. This mechanism may be broadly described by stating that the kinetic and strain energies deposited at the crack

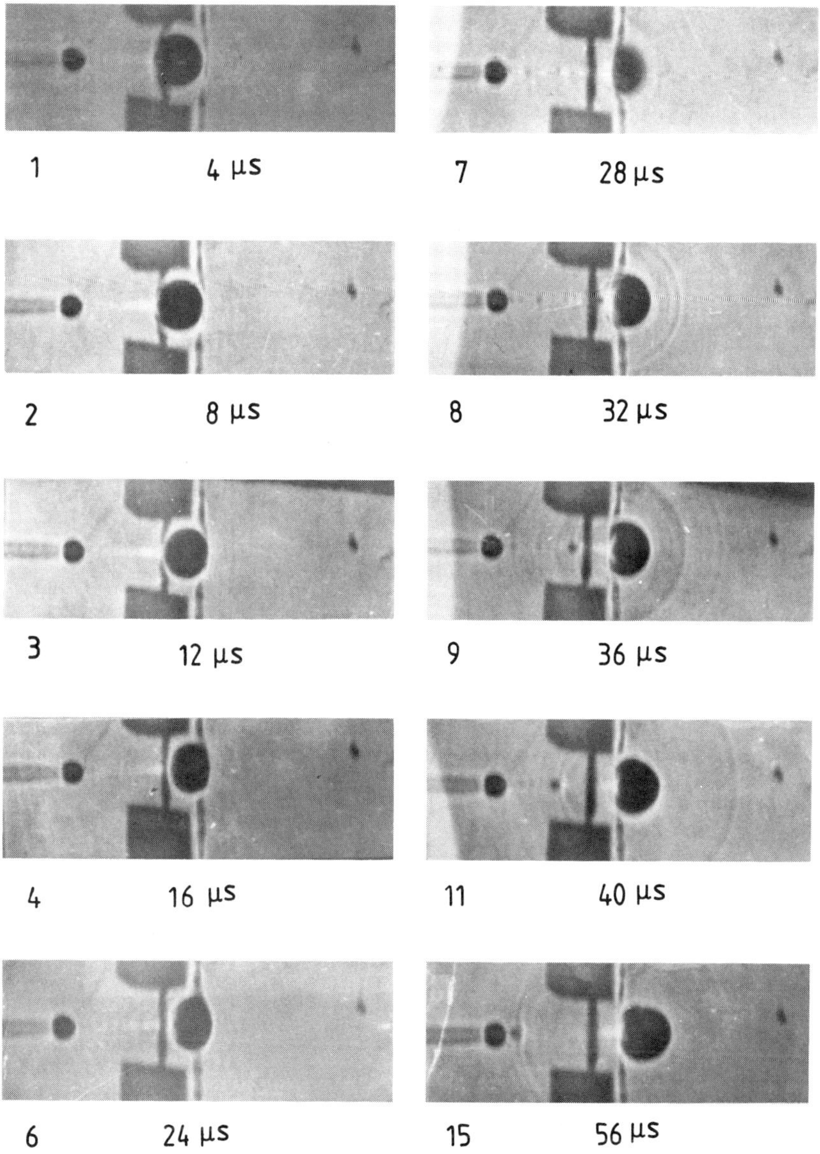

FIGURE 18 Series of photographs showing crack propagation through the interface of a Plexiglas duplex specimen.

tip are mainly converted into energies of deformation of the boundaries of the discontinuity. Then a delay phenomenon appears in which, although the external load is still acting on the specimen, it does not create any singular stress concentration at the tip of the crack. However, in a few microseconds, during the second period of crack arrest, a new singular stress concentration is created, this time beyond the barrier of the mesophase, which causes crack reinitiation in phase II. The process of crack reinitiation is governed by the material properties of phase II, the crack length, the external load, etc.

Besides its main role as a buffer of the propagating crack and a barrier to its path, the mesophase also acts as a reflector of energy. We already stated that part of the strain energy at the propagating crack tip is reflected backward when the crack is arrested at the interface. A small caustic is traveling in the opposite direction with the shear wave velocity of the substance of the phase due to the reflected shear wave at the intersection of the crack axis and the interface. The caustic, although small, is perceptible and corresponds to some amount of energy traveling backward, thus relieving the material by a small amount.

The third important role of mesophase is to act as a *splitter* and *deviator* of the crack path and therefore of the energy accompanying it.

Indeed, when the crack is traveling from a brittle to a ductile phase and when its velocity and energy are high enough, deviation of the path of the crack is very common. The crack is forced either to follow the direction of the interface, thus deviating from the minimum energy path of the K_I mode, or to enter directly into the soft phase but after a bifurcation or multifurcation. The new oblique paths are dominated by both K_I and K_{II} stress intensity factors and therefore consume much more energy than the single K_I path.

Moreover, splitting of the crack into a number of cracks results in a division of the reserve of the energy into as many parts as the number of new cracks. The effect of the mesophase acting as a splitter and deviator is a very rapid and abrupt reduction of the level of energy stored in the propagating crack, which may cause an immediate arrest of all secondary cracks and prevent complete failure of the structure.

THE DYNAMIC STRESS INTENSITY FACTOR IN FRACTURING COMPOSITES

The role of the mesophase in dynamic crack propagation will be better illustrated by studying the details of the behavior of the stress intensity factor during the fracture process.

The instantaneous values of the K_I and K_{II} components of the stress intensity factor can be readily found from the respective diameters of the caustics and the orientation of their axes of symmetry with respect to the crack axis (Ref. 39). Figure 16 presents the variation of the K_I stress intensity factor in either phase of the composite for (a) a hard-soft sequence and (b) a soft-hard succession of phases. In the hard-soft case we used as phase I a 10% plasticized epoxy resin, while phase II was a 20% plasticized polymer. In the second case the phases were 40% and 10% plasticized epoxies, respectively.

For all bimaterial specimens, the stress intensity factor remained almost constant after crack initiation in phase I and within a certain distance from the interface. As the crack reached the zone of influence of the mesophase, which extends on both sides beyond its width, the stress intensity factor started to increase, and at the mesophase it reached a value of 30 to 50% higher than its constant value far away from it.

Similarly, in all bimaterial plates tested, after the crack arrest period, the crack started propagating in phase II with a relatively low transient value of the stress intensity factor, later reaching a considerably larger stable value. The same phenomena were also apparent during the fracture of one-material duplex specimens.

This increase of the K_I value is a result of the interaction of the moving crack with the mesophase. Although the method of caustics, like all other experimental methods for determining the stress intensity factor, was developed in relation to the near-tip solution of the singular stress field for a crack in an infinite medium, it has been shown (Ref. 39) that it can also be used in cases of cracks approaching boundaries and other discontinuities better and closer to these discontinuities, since they are engendered by an infinitesimal circle around the crack tip. However, it seems that as the crack approaches the interface, the relation between K_I and the rate of crack length increase, $\dot{a}$, cannot be considered linear, and therefore, some error may be introduced in evaluating K.

It has been already mentioned that when the crack approaches the mesophase in phase I, and the crack arrest period starts, where the caustic, following the running crack, submerges in the interface, its arresting mode depends only on the presence of the mesophase; phase II does not significantly participate. We may therefore assume that during this period the singularity at the crack tip retains its previous value of $p = 0.5$. The caustic formed during the crack arrest period consists of an arc of the previous entire caustic, which is now being progressively submerged in the interface. Then it is reasonable to assume that these caustic arcs correspond to entire caustics with a transverse diameter of $D_t = 2l/\pi$, where l is the length of the arc of the respective entire caustic, and we have therefore calculated the stress intensity factor accordingly. In the period of reemergence of the crack in phase II the new caustic resembles the typical caustics for static fields.

During this period a new order of singularity, different from p = 0.5, is developed. This singularity is noncomplex only in the case where the crack meets the interface at right angles (Ref. 40) and, for the cases of biphase plates studied here, deviates only insignificantly from the value of p = 0.5 (Ref. 41). Extreme values for the singularities were calculated for specimens with a hard phase I and a soft phase II (40% plasticized). For this case it was equal to p = 0.56. For the inverse case, where the soft phase I consisted of a 40% plasticized epoxy and the hard phase II was unplasticized, this order was evaluated to be p = 0.475. Again, for this second period of crack arrest, the stress intensity factor for the final caustic form was calculated according to Reference 41, since this form corresponded with great accuracy to the respective static caustic at the tip of a crack meeting a bimaterial interface at a right angle.

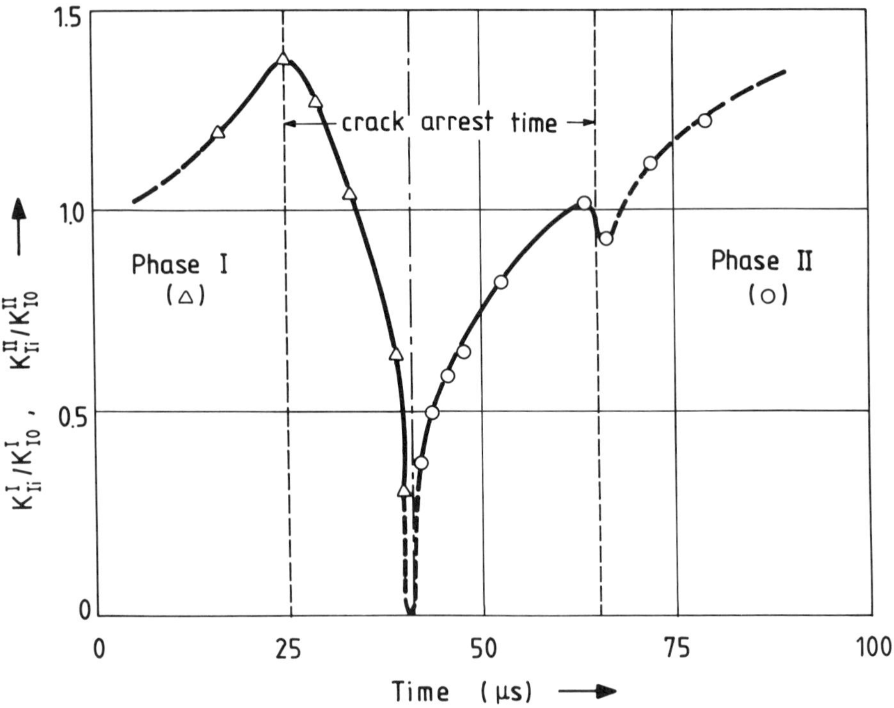

FIGURE 19 Variation of the stress intensity factor with time during crack arrest, normalized to its initial value at the initiation of the crack at each phase, for a (0–40) bimaterial specimen (phase I, 0% plasticizer; phase II, 40% plasticizer).

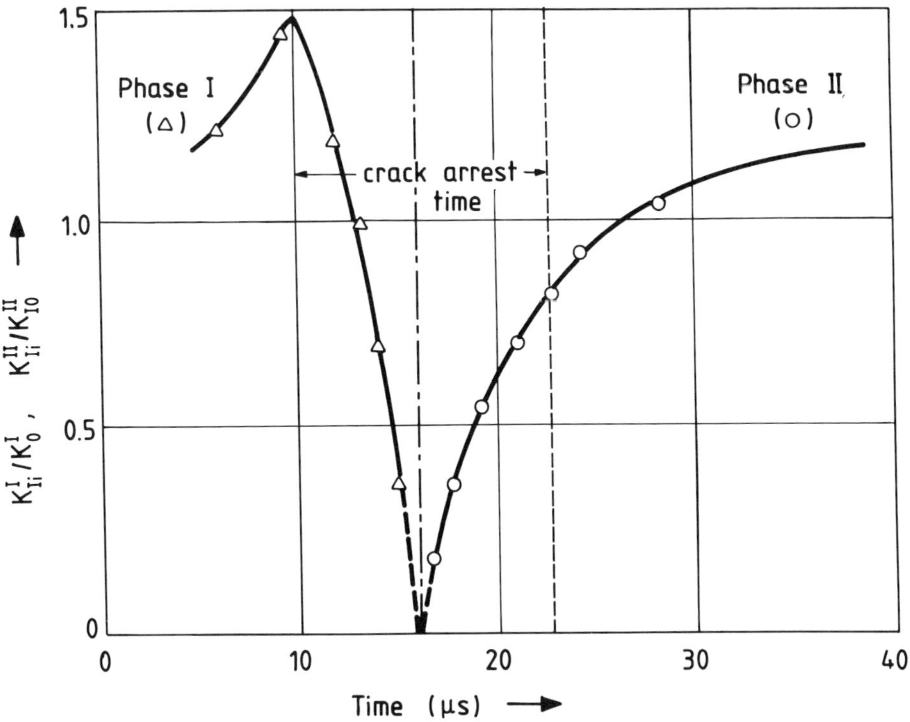

FIGURE 20 Variation of the stress intensity factor with time during crack arrest, normalized to its initial value at the initiation of the crack at each phase, for a duplex Plexiglas specimen.

At the same time we assumed that the preliminary arc forms of the emerging caustic corresponded to complete caustics with a transverse diameter of $D_t = 2l/\pi$, where l is the length of the arc of the respective part of the caustic.

This rule for the calculation of K_I, from preliminary caustic forms, is in accordance with the experimental evidence. The stress intensity factor can be considered, to a good approximation, as a function of the half-perimeter s of the caustic ($2s \simeq \pi D_t$). At the same time the boundary conditions for the initial and final instants of crack arrest [$K_I = f(D_t)$] are satisfied (Ref. 41). Finally, after the crack reinitiation in phase II we assumed that the typical order of singularity p = 0.50 was again attained.

Figures 19 and 20 show the variation of the stress intensity factor with time, normalized to its initial value at the initiation of the crack at each phase, for a 0–40% biphase specimen (p = 0.56) and a Plexiglas duplex specimen (p = 0.50).

During the first period of crack arrest the caustic, appearing only in phase I, submerges in the interface, until it completely disappears, which means that the stress intensity factor takes continuously lower values up to zero. Subsequently, the caustic emerges in phase II of the specimen, continuously taking larger values, up to the critical value of crack reinitiation.

However, the different order of singularity from the typical $p = 0.5$ influences the variation of the K_I value only insignificantly in the second period of crack arrest in phase II.

A CRITERION FOR THE DISRUPTION AND BIFURCATION OF PROPAGATING CRACKS

A quantitative criterion will be established which is based on the T criterion (Ref. 52) and relates the mechanical properties of the interface with the mechanical properties of the subsequent phase of penetration of the crack for the critical condition of crack propagation either along the interface or along the second phase (Ref. 42).

It is well known that the process of crack propagation in composites is often accompanied by a simultaneous process of debonding of the structural parts of the composite. A similar problem was first approached by Cook and Gordon (Ref. 43), who studied the possibility of interface disruption ahead of an elliptical hole in a flat plate. Subsequent investigations dealt with the problem of delamination in layered composites (Ref. 44), as well as the problem of fiber debonding and pull-out in fiber-reinforced specimens (Refs. 45–48).

The problem of interface disruption of biphase specimens seems to be considerably different from the already investigated problems of fiber debonding, or of delamination of a layered composite, since in these cases the crack continues to propagate normal to the fibers and the externally applied load, while debonding appears only as a secondary phenomenon, accompanying the main crack propagation process. This major difference between fiber-debonding and pull-out or delamination of a composite specimen, and interface disruption of a biphase specimen, makes the new investigation necessary.

It is, for example, obvious that, while in the case of a biphase specimen all fracture energy is dissipated for the interface disruption, in the case of a fiber-reinforced specimen the main amount of fracture energy is dissipated in crack propagation of the through-the-thickness crack, normal to the fibers, and only a minor fraction of this energy is associated with fiber debonding, fiber pull-out, and postdebonding frictional work. Thus, the critical strength of a biphase specimen interface is expected to be much larger than the critical strength of the fiber-matrix interface of a fiber composite.

The present study can be used to approach the problem of crack propagation parallel to the fibers, which has not been previously investigated. As stated in the literature (Ref. 49), fracture of a notched fiber-reinforced composite, loaded parallel to unidirectional fibers and normal to the crack, can sometimes take place parallel to the fibers by not only delaminating the fiber-matrix interface but also splitting the matrix itself. The mechanical properties of the composite as a whole ahead of the crack tip should be considered, whereas in the normal direction from the crack tip, the mechanical properties of the matrix should be taken into account.

The fact that the stress concentration at the crack tip continues to be of mode I during the whole period of crack propagation in phase I, crack coalescence, and crack arrest at the interface allows the conclusion that the crack propagation normal to the arrested mode I crack, that is, along the interface depends mainly on the difficulty in overcoming the obstacle of the mesophase and on the weakness of the interface region due to local debonding, impurities, inhomogeneities, microcracks, etc.

To strengthen this evidence, crack propagation experiments were undertaken in duplex PMMA specimens with weak interfaces. These duplex specimens were composed of two strips of identical dimensions ($0.30 \times 0.05 \times 0.004$ m^3), bonded together along their common longitudinal boundaries by means of a Plexiglas powder-chloroform adhesive suspension. Their phase I was initially slotted and the fracture process took place under the same load conditions as in the case of the biphase specimens. The experiments showed that, again, the coalescence of the moving transverse crack tip with the longitudinal interface results, in most cases, in crack propagation along the longitudinal interface, but now the new crack propagation process is initiated before the crack tip reaches the interface and is arrested there. The propagating transverse crack initiates another crack propagation process along the interface, as it approaches the interface, thus creating a longitudinal crack ahead of its path of propagation.

Figure 21 shows this process of early-interface disruption, occurring as the transverse crack approaches the interface of a duplex PMMA specimen, due to weak interface bonding. In frame 6 of Figure 21, where the moving crack is still in phase I, the early disruption of the interface is initiated, resulting in the formation of a new small crack lying ahead of the propagating transverse crack and oriented normal to its path of propagation. At both tips of this new longitudinal crack, two small caustics are formed, which indicate that singular stress concentrations develop in these areas. The longitudinal crack propagates in both directions along the interface, as its tips absorb more and more energy due to their coalescence with the moving transverse crack. Thus, in frame 6, the ratio of

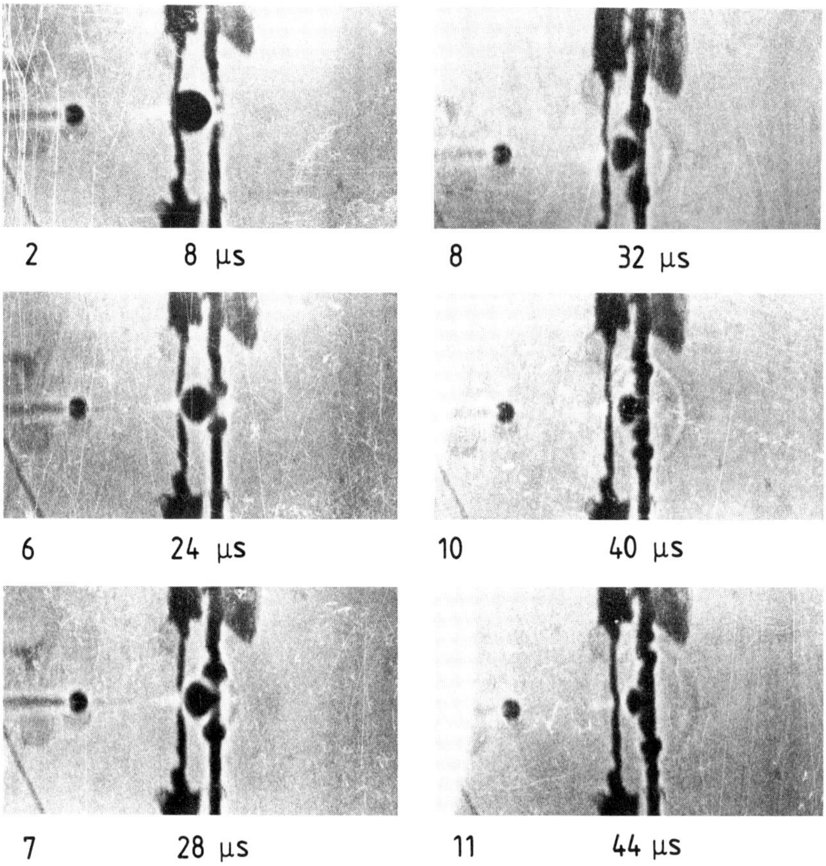

FIGURE 21 Process of an early interface disruption in a duplex PMMA specimen with an interface parallel to the loading axis of the plate.

the stress intensity factor at the tip of the main transverse crack, K_I, to the stress intensity factors at the upper or lower tips of the longitudinal crack, K_u^*, K_ℓ^*, is approximately $K_I/K_{u,\ell}^* = 4.35$, whereas this ratio in frame 8 is reduced to approximately $K_I/K_{u,\ell}^* = 2.05$.

We now investigate the critical conditions under which the interface disruption takes place, which are identical to the critical conditions for a stable interface, that is, the critical strength of the interface bonding for crack propagation through the interface without any disruption phenomena.

Similar problems were investigated theoretically for a transverse crack terminating normally at a bimaterial interface (Ref. 50) or crossing an interface (Ref. 51). These investigations are of little interest for our case, however, not only because they deal with the static problem, and therefore cannot approach the dynamic phenomena of crack arrest, but mainly because they concentrate on the theoretical evaluation of the stress intensity factor at the crack tip, which in our case is determined experimentally by the method of caustics.

On the other hand, the previous investigation and solution by Cook and Gordon (Ref. 43) refers only to a stress distribution ahead of an elliptical hole (that is, on the extension of the major axis of the hole, $\theta = 0°$) with a given relation between the principal axes of the ellipse (for example, $a/b = 100$) and a finite tip radius ($\rho = b^2/a$). It seems, therefore, inadequate to handle the problem of an infinite crack terminating at right angles to the interface of two dissimilar media, which corresponds to this investigation.

We consider a transverse crack, loaded under mode I conditions, which has been already arrested at the interface, while the singular stress field has been already rebuilt around the crack tip, and we investigate the conditions under which it will not enter phase I of the plate, but will initiate a new crack along the interface. For this purpose we shall consider the T criterion (Ref. 52), which enables us to correlate directly the mechanical properties of the materials used with the critical situation under investigation.

The T criterion is formulated as follows:

1. The crack propagates in the direction of maximum *dilatational energy density*, T_V.
2. The crack initiates when the maximum value of T_V exceeds a critical value $T_{V,0}$, which is considered as a material constant.
3. The angular distribution of T_V around the crack tip is evaluated along the initial or any subsequent elastic-plastic boundary as defined by the Mises yield condition (Ref. 53), where the distortional component of the elastic energy density, T_D, is constant.

This criterion implies that (Ref. 52):

$$T_V(r,\theta_0) \geqq T_{V,0} = \text{const.}$$

$$\left.\frac{\partial T_V(r,\theta)}{\partial \theta}\right|_{\theta=\theta_0} = 0, \qquad \left.\frac{\partial^2 T_V(r,\theta)}{\partial \theta^2}\right|_{\theta=0} < 0 \tag{8}$$

$$T_D(r,\theta) = T_{D,0} = \text{const.}$$

For plane-stress problems, the total energy density T and its dilatational and distortional components are given by Ref. 53:

$$T = \frac{1}{2E}[\sigma_x^2 + \sigma_y^2 - 2\nu\sigma_x\sigma_y + 2(1+\nu)\tau_{xy}^2] \tag{9}$$

$$T_V = \frac{1-2\nu}{6E}(\sigma_x + \sigma_y)^2 \tag{10}$$

$$T_D = \frac{1}{6G}(\sigma_x^2 + \sigma_y^2 - \sigma_x\sigma_y + 3\tau_{xy}^2) \tag{11}$$

We assume that the longitudinal interface sustains the singular stresses in the direction $\theta = 90°$ to the crack axis. Moreover, we assume that both phases of the specimen are brittle, with different degrees of brittleness, and we can therefore assume that the initial Mises yield locus is small enough to be inscribed inside a circle of radius r_0 around the crack tip equal to the radius of the respective *initial curve* of the caustic (Ref. 54).

Since we have a mode I deformation when the crack tip is in phase I, the T_V density has a maximum in the direction $\theta = 0°$, that is, $T_{v,max} = T_{v,\theta=0°}$. During the second period of the crack arrest at the interface the stress concentration of the crack tip increases constantly until $T_{v,\theta=0°}$ exceeds the value of $T_{v,II}$, which corresponds to the material resistance of phase II, lying ahead of the crack tip; that is,

$$T_{v,\theta=0°} \geqq T_{v,II} \tag{12}$$

However, since the crack tip lies at the interface and therefore the stress intensity factor is influenced by the neighborhood of phase II in the biphase plate, the region of the crack tip constitutes in general a region of reduced strength. It is, therefore, possible that T_v along the interface ($T_{v,\theta=90°}$) exceeds the value $T_{v,i}$, corresponding to the interface strength, before Eq. (12) is satisfied.

The critical condition for the disruption of the interface may be therefore expressed by

$$\left.\begin{aligned} T_{v,\theta=0°} &\leqq T_{v,II} \\ T_{v,\theta=90°} &\geqq T_{v,i} \end{aligned}\right\} \quad \frac{T_{v,\theta=0°}}{T_{v,\theta=90°}} \leqq \frac{T_{v,II}}{T_{v,i}} \tag{13}$$

For the case of a crack terminating at the interface of the two phases, the stress in the close vicinity of the crack tip and at a distance $r = r_0$ are given by (Ref. 50):

$$\sigma_x = \frac{K_I}{(2\pi r_0)^p} \frac{\cos\theta}{2(1 - \sin\theta/2 \sin 3\theta/2)}$$

$$\sigma_y = \frac{K_I}{(2\pi r_0)^p} \frac{\cos\theta}{2(1 + \sin\theta/2 \sin 3\theta/2)} \tag{14}$$

$$\tau_{xy} = \frac{K_I}{(2\pi r_0)^p}$$

where the exponent p ($0 < p < 1$) is the order of singularity at the tip of the crack, which normally intersects the interface.

From Eqs. (10) and (14) and for the two directions of interest, that is, $\theta = 0°$ and $\theta = 90°$, we obtain:

$$T_{v,\theta=0°} = 4 \frac{K_I^2}{(2\pi r_0)^{2p}} \frac{1 - 2\nu_{II}}{6E_{II}} \tag{15}$$

$$T_{v,\theta=90°} = 2 \frac{K_I^2}{(2\pi r_0)^{2p}} \frac{1 - 2\nu_i}{6E_i} \tag{16}$$

where the indices II and i stand for phase II and interface i, respectively.

On the other hand, the values of the quantities $T_{v,II}$ and $T_{v,i}$ can be expressed in terms of the critical values of the stress intensity factors K_{Ic} for crack initiation under mode I conditions ($\theta = 0°$) in a material having the mechanical properties of phase II or of the interface, respectively. Thus, we have

$$T_{v,II} = \frac{1 - 2\nu_{II}}{6E_{II}} \quad 4\,\frac{(K_{Ic}^{II})^2}{2\pi r_{0,II}} \tag{17}$$

$$T_{v,I} = \frac{1 - 2\nu_{i}}{6E_{i}} \quad 4\,\frac{(K_{Ic}^{I})^2}{2\pi r_{0,i}} \tag{18}$$

The critical condition (13) for interface disruption now becomes

$$2\,\frac{(1 - 2\nu_{II})E_i}{(1 - 2\nu_i)E_{II}} \leqq \frac{(1 - 2\nu_{II})E_i(K_{Ic}^{II})^2 r_{0,i}}{(1 - 2\nu_i)E_{II}(K_{Ic}^{i})^2 r_{0,II}}$$

which yields

$$2\,\frac{r_{0,II}}{r_{o,i}} \leqq \left(\frac{K_{Ic}^{II}}{K_{Ic}^{i}}\right)$$

If we substitute in these relations the expressions for the radii of the two initial curves, $r_{0,II}$ and $r_{0,i}$, given by (Ref. 55):

$$r_{0,II} = \left[\frac{(K_{I,c}^{II})^2}{12\pi G_{II} T_{D,0}^{II}}\right] f_{II}(\theta) \qquad \text{and} \qquad r_{0,i} = \left[\frac{(K_{I,c}^{i})^2}{12\pi G_i T_{D,0}^{i}}\right] f_i(\theta)$$

and assume further that the correction functions $f_{II,i}(\theta)$ are approximately equally to unity for $\theta = 0°$, we derive

$$\frac{T_{D,0}^{i}}{T_{D,0}^{II}} \leqq 0.5\,\frac{G_{II}}{G_i}$$

or

$$\frac{2T^{i}_{D,0}G_{i}}{T^{II}_{D,0}G_{II}} \leqq 1 \tag{19}$$

This is a necessary condition for the disruption of the interface and therefore the branching of the crack approaching it.

But from Eq. (11) and $\theta = 0°$ we have that

$$T_{D,II}G_{II} = \frac{\sigma^2_{0,II}}{6} \qquad \text{and} \qquad T_{D,i}G_{i} = \frac{\sigma^2_{0,i}}{6}$$

Then Eq. (19) may be written

$$\sigma_{0,i} \leqq 0.707\,\sigma_{0,II} \tag{20}$$

Equation (20) indicates that when the yield or fracture stress for the mesophase is smaller than or equal to the yield or fracture stress of phase II divided by the square root of 2, there is a possibility of disruption of the interface.

Equation (20) is also supported by the series of experimental results, since the tendency for interface disruption appears only in the cases of a hard phase II and a soft phase I, where the $\sigma_{0,II}$ stress for the brittle phase II is always higher than the respective value of $\sigma_{0,I}$ for phase I. Then, if the yield or fracture strength of the mesophase is rather weak, approaching that of the matrix, Eq. (20) may be satisfied and branching of the crack along the interface may occur.

However, for the inverse case of a hard phase I and a soft phase II, Eq. (20) can be satisfied for very poor adhesion, since in this case the value of $\sigma_{0,II}$ is very low.

For practical application of the new criterion, the material properties G and $T_{D,0}$ for both the mesophase region and phase II must be known. First we evaluate the E_i modulus of the mesophase by using its average value from the three-term or two-term unfolding models by knowing the moduli of the phases and the extent of mesophase (Ref. 22). The extent of mesophase may be found by using Lipatov's theory (Ref. 26) and differential scanning calorimetry measurements of the matrix material and the composite.

To evaluate the $T_{D,0}$ component of energy at the mesophase, we may assume that the initial curve of the caustic for parallel-light incidence is approximately equal to the initial Mises yield locus. Moreover, the Poisson ratio of mesophase mat be taken equal to the Poisson ratio of the matrix, since the mesophase is derived from the matrix material. Finally, for perfect wetting, the ratio of the elastic moduli E_i/E_{II} may be taken equal to the ratio of the respective tensile fracture stresses. That is,

$$\frac{E_i}{E_{II}} = \frac{\sigma_{0,i}}{\sigma_{0,II}}$$

Then Eq. (20) becomes:

$$E_i^a \leqq 0.707\ E_{II}$$

For a soft-hard mesophase and assuming an approximate value for $E_i^a = (2E_I + E_{II})/3$, we can readily deduce that for

$$E_m \leqq 0.25\ E_f \tag{21}$$

a disruption of the interface should happen.

All these assumptions are justified by extensive experimental evidence concerning the behavior of various types of epoxy composite materials. For this purpose a number of biphase epoxy specimens were examined. Some of these specimens were constructed out of a hard phase II and a soft phase I, whereas the rest were constructed with a hard phase I and soft phase II. Phase I was initially slotted and the specimens were fractured under the influence of dynamic loading. In all these tests only two disruption phenomena occurred, which were due to an accumulation of imperfections in the interface region. These results can be predicted by the criterion developed above. For the more critical combination of materials, that is, for the specimens having as phase I a 40% plasticized epoxy and an unplasticized phase II, we have from Table 1 that

$$\frac{E_I}{E_{II}} = \frac{182}{330} = 0.551$$

Then the E_i^a modulus may be approximated by the center of gravity of a parabola, whose apex has an ordinate E_I and passes through E_{II}. This good approximation yields

$$E_i^a = \frac{2E_I + E_{II}}{3} = 2.27 \times 10^9 \text{ N/m}^2$$

Introducing these values for E_{II} and E_i^a in Eq. (21), we obtain

$$2.27 \times 10^9 \text{ N/m}^2 \leqq 2.289 \times 10^9 \text{ N/m}^2$$

The approximate equality of the two sides of this inequality explains the uncertainty of disruptions in the tests.

INFLUENCE OF THE ANGLE OF INCLINATION OF THE INTERFACE

Up to now we have extensively examined biphase specimens in which a propagating crack approaches the mesophase zone and touches the interface, in the case where this interface is oriented parallel to the externally applied load and normal to the propagating crack under conditions of mode I deformation. However, since the filaments in the composite are in general arranged in different orientations and the externally applied loads are more complicated than simple tension, it is not seldom that cracks in the composites meet interfaces at an oblique angle. We shall therefore now study the influence of the slant of the interface (Ref. 56). The angles of slant of interfaces studied were $\theta = 30$, 45, and 60° to the transverse direction of the specimens. All other characteristics of the biphase specimens were similar to those in the previous tests with interfaces parallel to the loading axis.

The series of tests involving bimaterial plates with slanted interfaces indicated a clear similarity to the phenomena appearing in cracks propagating in bimaterial plates with longitudinal interfaces. However, important differences in the attitude of the propagating crack and the behavior of the materials of the phases were also detected. Figure 22 presents a series of photographs indicating the propagation mode of an initially transverse crack in a biphase plate with a soft phase I (30% plasticizer) and a hard phase II (10% plasticizer to the epoxy). The interface of the two phases subtended an angle $\theta = 60°$ with the transverse axis of the plate. The plate was submitted to a dynamic loading creating an average strain rate $\dot{\varepsilon} = 4 \text{ sec}^{-1}$.

A first and important observation is that the crack propagates in both phases under approximate mode I conditions. This remark is based on the fact that the caustic senses the existence of mixed-mode

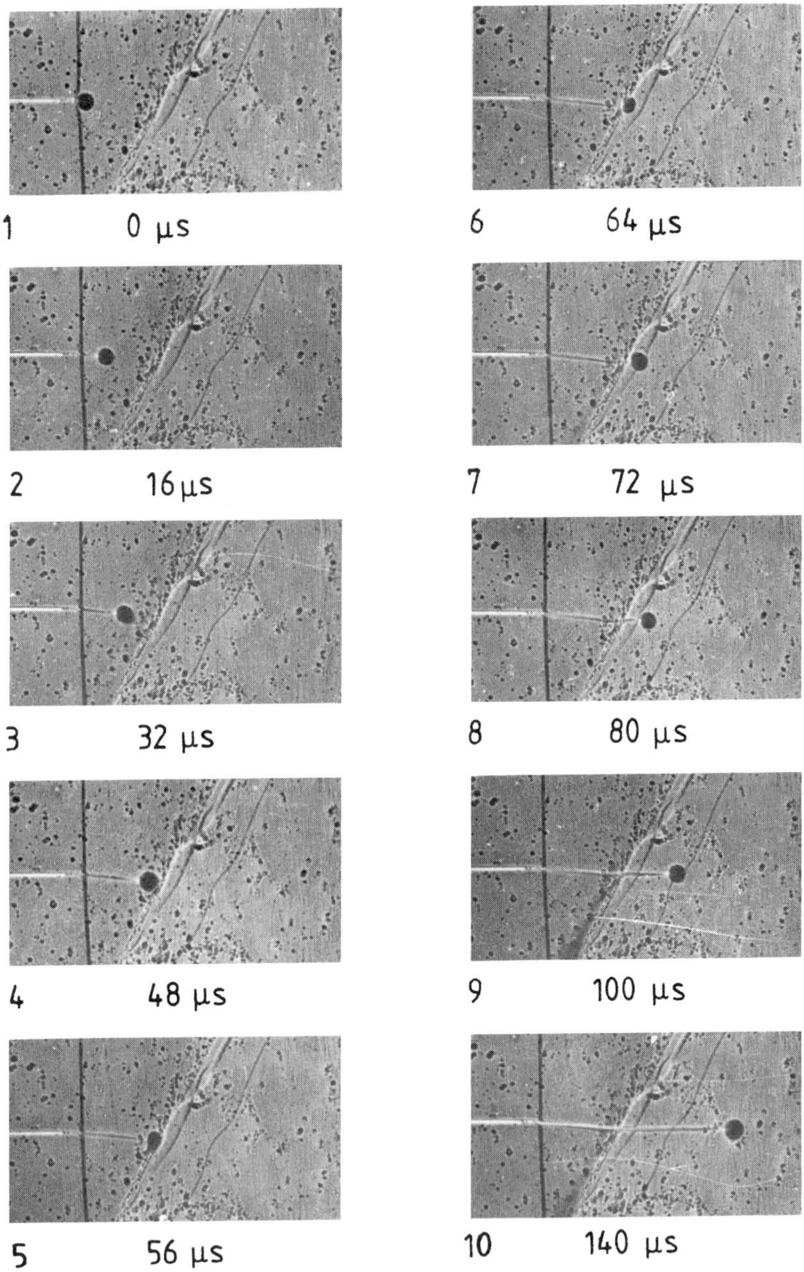

FIGURE 22 Series of photographs showing crack propagation in a bimaterial specimen (phase I, 30%; phase II, 10% plasticized) with an oblique interface.

deformation by its rotation about the crack tip, and such a phenomenon also did not appear in this series of tests.

For the case of slant interfaces fractured dynamically the following remarks may be made:

1. Crack propagation in phase I takes place with a variable velocity, which increases continuously from its initial value up to a maximum value. The maximum value occurs at some critical distance between the crack tip and the axis of the interface.
2. After passing this maximum, the propagating crack decelerates continuously to a minimum value of velocity, attained when the crack tip reaches the interface. This minimum value, in some tests, approaches zero; i.e., the crack is almost arrested.
3. The crack then crosses the interface and starts to accelerate again, up to some maximum value of velocity. This second maximum value is attained close to the interface region.
4. When the crack passes the maximum value of velocity, it decelerates smoothly and tends to some constant velocity, which depends on the plate configuration and the applied strain rate.
5. Finally, when the crack approaches the opposite longitudinal boundary of the plate it decelerates or accelerates smoothly, depending on the sequence of phases hard-soft or soft-hard, respectively.

Then the slant interface of the plate acts as a *decelerator* of the propagating crack and, in some cases, as a *crack arrester*. However, its influence on the mode of propagation of the crack is not as intense as in the case of normal interfaces, where the crack is always arrested for a time between 5 and 50 μsec, depending on the properties of the bimaterial plate (Refs. 27, 30). The difference between normal and slant interfaces is only quantitative.

This difference is due to the fact that the influence of the normal interface on the behavior of the propagating crack is concentrated in a restricted zone, which extends from the tip of the crack to the normal interface. In contrast, the influence of the slant interface starts far from the point of the final meeting of the crack with the interface. Accordingly, it has been observed that a decrease of angle θ elongates and smooths the variation of the crack velocity.

Figure 23 shows a series of plots of the crack velocity, v_c, versus the instantaneous crack length, a, for a series of tests, indicating the variation of the crack velocity along the plate for different combinations of phases in the bimaterial plate. The trends of all these curves may be classified in two groups, one corresponding to a hard-soft succession of phases in the plate (Figure 23a) and the other to the inverse (Figure 23b). The crack tip velocities were calculated accurately from the position of each caustic, as explained previously and in Reference 39, and by using the interpolation method for a second-degree Lagrange polynomial.

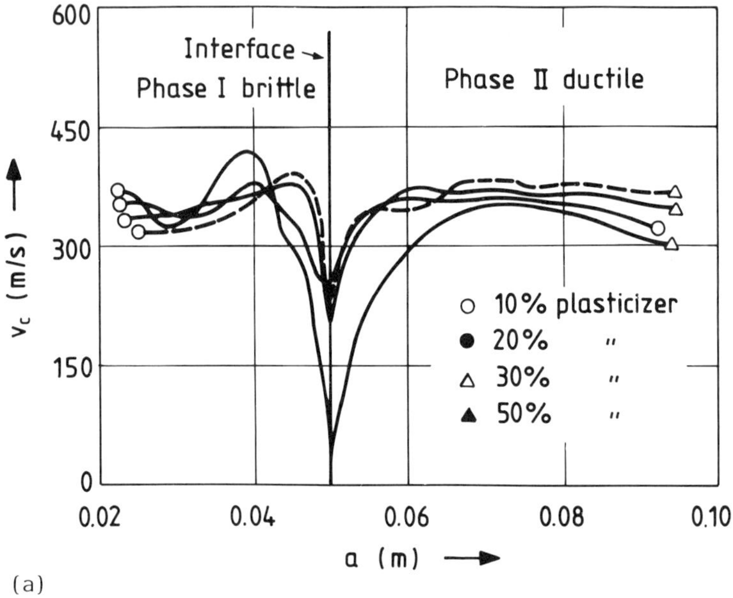

(a)

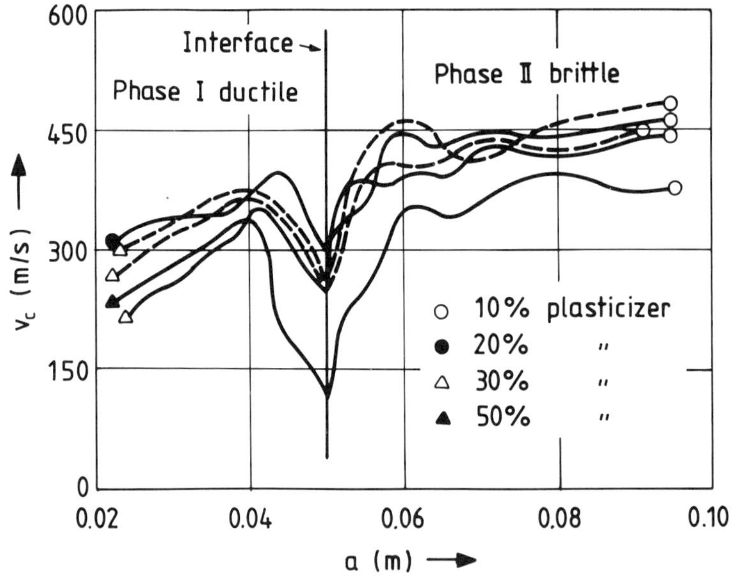

(b)

FIGURE 23 Variation of crack velocity v_c with crack length a: (a) hard-soft succession of phases, full lines for $\theta = 45°$, dashed lines for $\theta = 60°$; (b) soft-hard succession of phases, dashed lines for $\theta = 45°$, full lines for $\theta = 60°$.

Figure 24 presents a series of photographs which indicate the process of propagation of a crack in a bimaterial plate, with hard phase I (10% plasticizer) and soft phase II (50% plasticized), whose interface subtends an angle $\theta = 45°$ with the transverse axis of the plate. The plate was submitted to a stress pulse with a strain rate $\dot{\varepsilon} = 2\ \text{sec}^{-1}$. We observe from these photographs that, when the crack tip lies in the vicinity of the interface, the caustic progressively disappears, submerging in the interface zone, while its remaining part does not change shape. In photographs 3 to 6 phase II remains unchanged and seems not to feel the approaching singular stress field of the crack. Moreover, there is almost complete disappearance of the stress field in photographs 5 and 6 when the crack is arrested. Only after a time lapse of 30 μsec from the first disappearance of the caustic does a new caustic start to emerge in phase II. This caustic corresponds to a good approximation to a steady crack creating a singular stress field with a real stress singularity of order $r^{-0.635}$, evaluated according to Ref. 41.

Finally, the crack, after its initiation in phase II, starts to propagate again under the influence of the external tensile pulse, which is long enough to cover the whole procedure, up to the complete failure of the plate.

The disappearance of the caustic, which indicates the neutralization of the stress field around the crack tip, when the crack arrives at the discontinuity of the interface, may be explained by rapid absorption of the kinetic and strain energies stored at the propagating crack tip. Only after a time period, during which the external load is still applied, is strain energy again stored at the crack tip, which progressively increases up to some limiting value, sufficient for initiation and propagation of the crack in phase II.

Moreover, in the case of crack arrest at the interface, a detailed examination of the mesophase zone has shown that the surface deformations in the vicinity of the interface are considerable, creating an important obstacle to the propagation of the singular stress field around the crack tip. Experiments with biphase plates whose lateral phases were polished after casting and curing showed that the mesophase obstacle is considerably reduced, and there is a rather smooth transition of the crack from phase I to phase II.

These results allow the hypothesis that an important factor affecting the mode of transition of the propagating crack from phase I to phase II is the quality of the material discontinuity at the interface zone. The last depends to a large extent on the surface and structural anomalies of the mesophase layer surrounding the interface. The influence of the quality of this layer when the interface is parallel to the applied load was thoroughly discussed in the previous paragraphs.

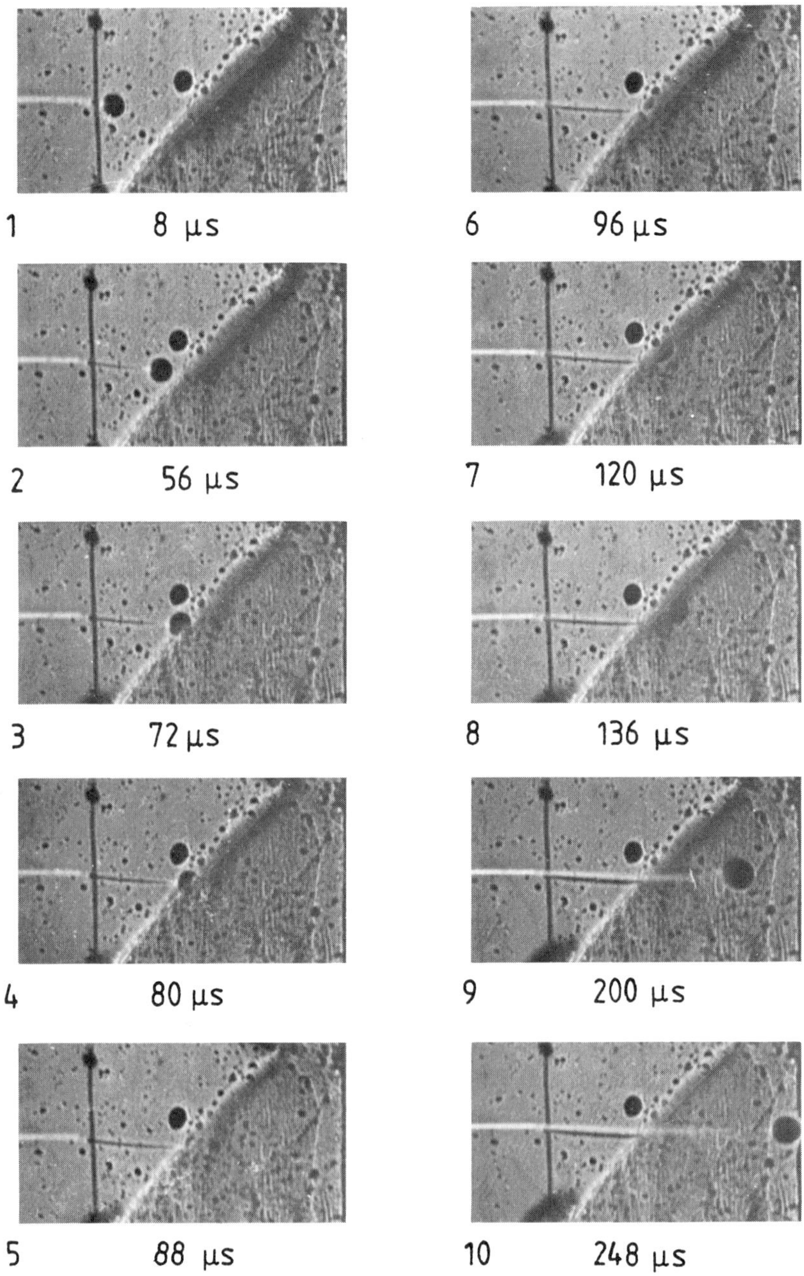

FIGURE 24 Crack arrest at an oblique interface (phase I, 10%; phase II, 50% plasticized).

We now consider in detail the mode of variation of the crack velocity in the plates. Another series of tests with a soft phase I (30% plasticized) and a hard phase II (10% plasticized), but with different angles of inclination of the interface (θ = 45 and 60°), was executed. The maximum value of the crack velocity for both tests was the same and equal to v_c = 357 m/sec, but it was attained at different places for different angles θ. Thus, for θ = 60° the distance between interface and position of maximum v_c was d_{max} = 0.008 m; for θ = 45° this distance increased to d_{max} = 0.010 m, indicating the influence of inclination of the interface on the behavior of the crack in phase I. After this maximum, the crack decelerated and its velocity attained a minimum, always lying on the interface. The values of decelerations measured in all similar tests were about the $b_a = 3 \times 10^6$ m/sec^2.

The crack transferred to phase II accelerated again rapidly, with average accelerations of the order of $b_a = 6 \times 10^6$ m/sec^2, until its velocity again reached a maximum. This second maximum was always about 10% larger than the respective maximum in phase I. Finally, the maximum velocity was followed by an almost constant-velocity zone, until the influence of the opposite end of the bimaterial plate interfered, creating an insignificant increase of velocity near the boundary.

Figure 25 shows the variation of the crack velocity v_c in terms of the instantaneous crack length for a plate with hard phase I (10% plasticized) and soft phase II (20% plasticized). In this plate the angle θ was 45° and the applied strain rate was $\dot{\varepsilon} = 2.4$ sec^{-1}.

For this sequence of phases the crack propagated in the hard phase with almost constant velocity and v_c increased near the interface, corresponding to a maximum velocity of the order of v_c = 425 m/sec. After this maximum the crack decelerated, reaching a minimum value at the interface. The decelerations measured were of the order of $b_a = 6 \times 10^6$ m/sec^2, and the minimum velocities were 20 to 30% smaller than the respective minimum velocities at the interface for the soft-hard sequence.

The average accelerations of the crack tip in the soft phase were $b_a = 2{-}3 \times 10^6$ m/sec^2, and the second maximum of the velocity was smaller than the respective maximum in phase I. Finally, the maximum velocity in phase II was followed by an almost constant-velocity zone, with decreasing tendency near the boundary.

The rather complicated behavior of the propagation velocity was due mainly to two factors. The first factor was the influence of the mesophase on the mode of propagation of the crack. This mesophase consisted of two strips on both sides of the common boundary of the constitutive phases of the specimen with different mechanical properties than the materials of the phases.

The second factor was that the crack in the biphase plates with slant interfaces always propagated in a nonhomogeneous dynamic

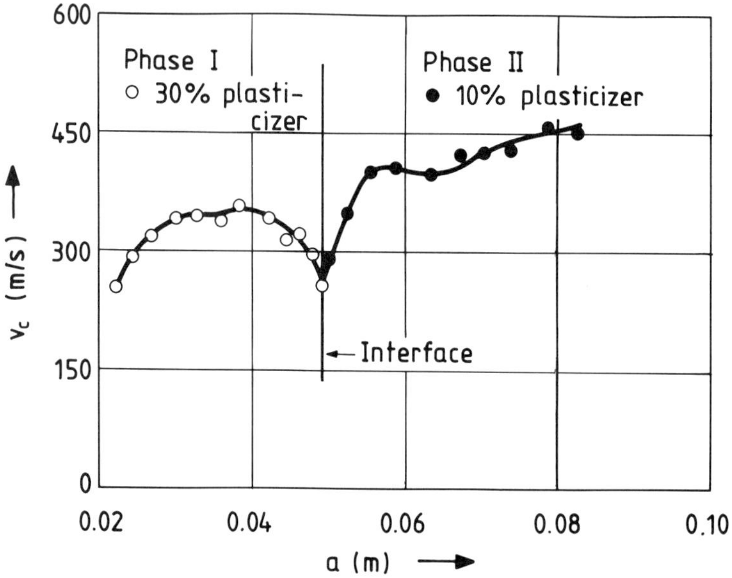

FIGURE 25 Variation of crack velocity v_c with crack length a for a specimen with $\theta = 45°$ and phase I 30%, phase II 10% plasticized.

stress field. Indeed, the crack, whose initial part always lay in phase I, propagated under the influence of a stress field created by the tensile pulse, applied at the bottom of the biphase plate, after it had traversed the interface. Along the mesophase the pulse was subjected to reflections in phase II and refractions in phase I, and therefore the stress pulse arriving at the crack tip was a compound pulse, consisting of a longitudinal and a transverse component. Because of the difference in the angles of refraction and reflection and the difference in velocities between longitudinal and transverse waves in the plates, we may assume that the stress field engendering the propagation of the crack was strongly inhomogeneous.

In order to have a simple image of the stress field in which the crack was propagating, we calculated the tensile $\sigma_y(x)$ stress in the biphase plate under the following simplifying assumptions:

1. The plate was submitted to a constant displacement v at its lower transverse boundary.
2. The plate was composed of independent bimaterial fibers parallel to the Oy axis with continuously varying lengths in phases I and II, depending on the position of the fiber in the plate.

With these assumptions it is valid that:

$$\sigma_y(x) = \frac{vE_I E_{II}}{h(E_I + E_{II}) + (E_I - E_{II})mx} \tag{22}$$

where h is the half-length of the plate, E_I and E_{II} are the moduli of elasticity of phases I and II, and $m = \tan\theta$.

Based on this simplified model, we may assume that the crack was propagating in a stress field which is variable, according to a law similar to that expressed by Eq. (22). Thus, the crack was propagating in an increasing tensile stress field when phase I was ductile ($E_I < E_{II}$), whereas it was propagating in a decreasing tensile field when phase I was brittle ($E_I > E_{II}$).

Finally, because of the small length of the trajectory of the crack relative to its initial length, the crack propagation velocities in phase I, as well as phase II, were always below the limiting velocities corresponding to the applied stress pulses and the respective geometries of the plates. These limiting values were found to be:

$$c_{T0} = 670 \text{ m/sec} \qquad \text{and} \qquad c_{T50} = 410 \text{ m/sec}$$

where c_{T0} and c_{T50} correspond to the shear wave velocities for an unplasticized and 50% plasticized epoxy polymer.

In conclusion, the variation of the crack velocity, which showed an increasing tendency in the case of a soft-hard sequence of phases, was the result of a superposition of two effects. First, the crack was propagating with much lower velocities than the limiting velocities of the materials, and second, it was propagating in a continuously increasing tensile stress field. For the inverse case of a hard-soft sequence of phases, these two effects counteract, resulting in an almost constant velocity.

Concerning the crack paths in the plates, it may be observed that in phase I the paths were curved, whereas in phase II the cracks were propagating in straight lines normal to the direction of loading. A difference in paths was also observed, depending on the sequence of phases. Whereas in the hard-soft sequence the crack path approached the interface tangentially, as if it was repelled by it, in the soft-hard sequence the crack path approached the interface normally, as if it was attracted by it.

In order to study the variation of the instantaneous values of the stress intensity factor, we assumed first that only the K_I component was operative, according to the fact that the crack propagates in both phases under approximate mode I conditions. Furthermore, assuming that the order of singularity all over the plate was $r^{-\frac{1}{2}}$, the values of the K_I factor were determined according to the method of caustics (Ref. 39).

To evaluate the K_I stress intensity factor in the mesophase zone use was made of the approximate evaluation of K_I as given previously for cracks approaching the interface normally. This is permissible since, for the angles subtended by the crack axis and the interface and the materials used in the bimaterial plate, the stress singularly at the crack tip is always real and very close to its value $r^{-\frac{1}{2}}$ for the isotropic material (Ref. 41). The variation of K_I was similar to the variation of the crack propagation velocity, and it was differentiated according to the sequence of the phases in the plate.

When the crack propagated from the soft to the hard phase, K_I diminished at the beginning of propagation and passed through a minimum value before any deceleration of the crack. This factor later increased progressively up to a maximum value in phase I at a region close to the interface. The maximum value was 20 to 30% higher than the average value of K_I. The passage of the crack through the interface was related to an abrupt reduction in the value of K_I, which was 5 to 10% lower than its initial value.

In the hard phase II, K_I oscillated at the beginning and then continuously increased as the crack velocity increased. The variation of K_I with the crack length is shown in Figure 26 for the case of soft phase I (30% plasticizer) and hard phase II (10% plasticizer) and angle $\theta = 60°$.

When the crack propagated from the hard to the soft phase, the K_I factor first showed unstable behavior with oscillations and later stabilized to a continuous increase up to the neighborhood of the interface. As the crack approached the mesophase and decelerated, the K_I factor increased abruptly to a maximum of about 40% above its initial value and then decreased rapidly to 10% less than the average values. As soon as the crack crossed the mesophase, K_I increased rapidly to a maximum of about 120% of its initial value in phase I and then started to decrease before the crack velocity attained its maximum.

Figure 27 shows the variation of K_I with crack length, a, for the case of a hard phase I (10% plasticizer) and a soft phase II (30% plasticizier) and angle $\theta = 45°$. We may conclude that the variation of K_I is related to the variation of the crack velocity by a nonlinear relation at the interface zone. Moreover, the K_I factor shows an unstable behavior which is more intense in the brittle phase of the plate than in the ductile phase.

Another interesting conclusion is that the interface behaves as a decelerator for the propagation of the crack. The influence of the interface on the propagating crack is related to the abrupt change in the mechanical properties of the plate there. The actual interface cannot be considered an ideal separating boundary between two different plates with different mechanical properties; rather, it is a

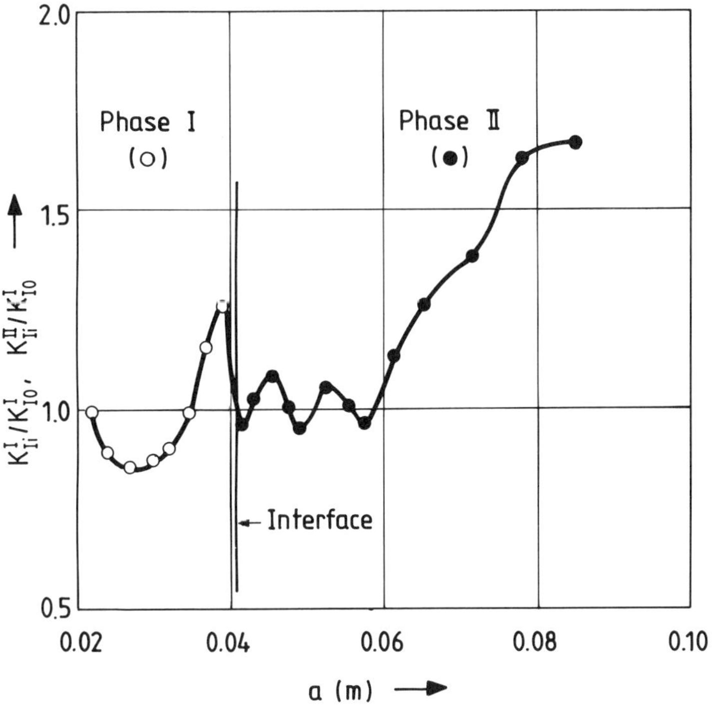

FIGURE 26 Variation of the stress intensity factor, normalized at each phase to its initial value at the initiation of the crack in phase I (K^I_{10}), as a function of crack length a, in the case of a soft-hard succession of phases (phase I, 30%; phase II, 10% plasticized).

mesophase region. Moreover, the singular stress field around the crack tip, as the crack approaches the interface, is a complicated field and cannot be analyzed by static conceptions. Concerning the variation of the influence of the mesophase on the propagating crack with angle θ, we find that this influence decreases with decreasing θ and simultaneously spreads out in a larger zone around the mesophase.

Finally, the propagation of the crack in both phases takes place under approximate mode I conditions. The crack follows a slightly curved path, with the crack axis adjusting continuously to the principal transverse axis of the stress field. The variation of the crack propagation velocity during fracture of a biphase plate shows a complicated behavior with oscillating velocities due to consecutive decelerations and accelerations.

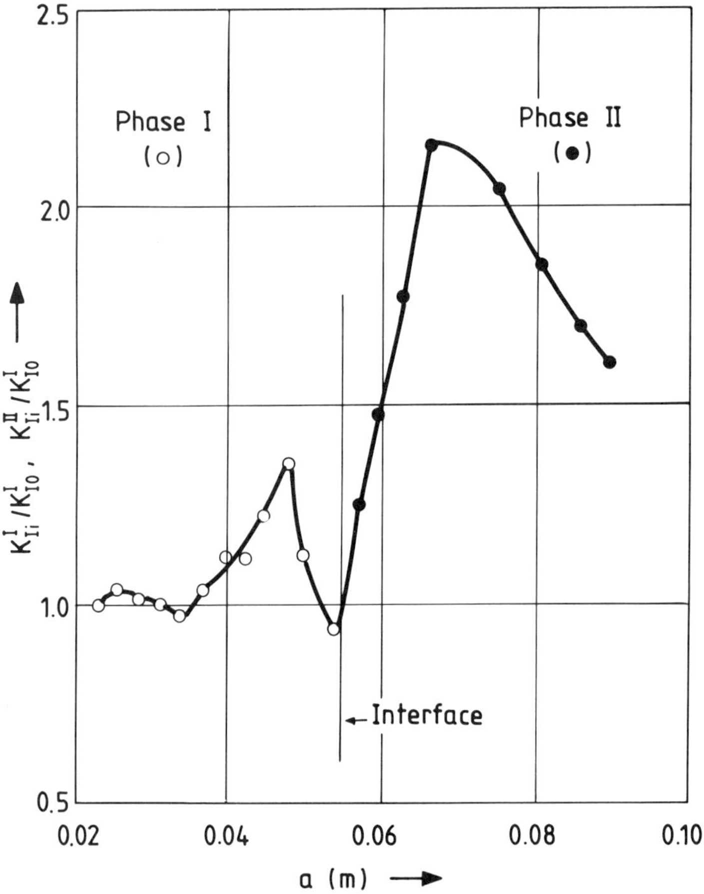

FIGURE 27 Variation of the stress intensity factor, normalized at each phase to its initial value at the initiation of the crack in phase I (K_{10}^{I}), as a function of the crack length a, in the case of a hard-soft succession of phases (phase I, 10%; phase II, 30% plasticized).

In general, we condlude that the accelerations and decelerations in the hard phase are higher than the respective values in the soft phase, and the influence of interface on the crack propagation velocity is also greater when the crack propagates from the hard to the soft phase than in the inverse sequence.

MACROSCOPIC MODE OF FRACTURE IN PARTICULATE COMPOSITES

There exist several modes of failure of composites deriving from broken fibers, from flaws and cracks in the matrix or the inclusions, from delamination at the interfaces, from weak bonds between phases, and from any combination of these modes. The initiation of a crack always appears in the zone of inherent defects of the phases of the composite or at their interfaces. It is not possible to incorporate all these failure mechanisms into a reliable analytic model, which should be a simple and effective one. However, fracture mechanics may be employed to determine and interrelate relevant geometric and material parameters of the composites on the macroscale for adequate prediction of failures due to unstable crack growth.

The problem of failure of composites by crack growth presents insurmountable difficulties because of the direct interaction of the crack with adjacent fibers. For particulate composites similar situations hold, because the inclusions in these materials force the crack to deviate from its least-energy path, which, under normal conditions, is the K_I path.

The method of caustics (Ref. 39) is a convenient one for studying the overall behavior of crack propagation in composites. As an example we have applied this method to determine crack propagation velocities and evaluate stress intensity factors in iron-epoxy particulate composites.

Specimens with three different low percentage volume fractions of filler were used, namely 3, 5, and 7%. It was found that the mean crack propagation velocity decreased linearly with increasing filler volume fraction. In addition, by using the method of caustics, the K_I stress intensity factor was calculated for each of the composite specimens separately (Ref. 57).

The matrix material was in all cases a cold-setting system, based on a diglycidyl ether or bisphenol A resin, cured with 8% triethylenetetramine, which is slightly less than stoichiometric. One particle size (150 μm) of iron powder was used for all mixtures. The properties of the composites used in the tests are given in Reference 57. All procedures for preparing the mixtures of the liquid-matrix phases and the metal powder, stirring these mixtures in a closed mold in order to obtain a uniform distribution of filler particles, for curing processes, etc., followed the respective standards. For the present investigation we used composite specimens with a length of 0.30 m, a width of 0.10 m, and a thickness of 2×10^{-3} m.

The process of crack propagation is characterized by the fracture zone developed at the crack tip. In composite materials the dimension of this fracture zone, when compared to the particle diameter, affects the mode of crack propagation.

Moreover, the changes in the average properties of the highly strained material in the failure zone must be considered as a second phenomenon relating the changes in strength with filler particles, which are small compared to the fracture zone at the crack tip (Ref. 58). The size of the fracture zone is greatly affected by the presence of the finely dispersed particles. When the size of the fracture zone increases, a microfracturing process takes place, and this may lead to an increase in the fracture energy.

In addition, the nature of the interfacial bonding between matrix and filler particles is of crucial importance, causing the creation of the mesophase zone. Perfect adhesion, corresponding to continuity of stresses and displacements at the interface, is a common assumption in analytical treatments. However, with real composites this condition is very seldom fulfilled, and cracks, voids, and flaws extensively distort the stress/strain fields developed in the composite. Forms of imperfect bonding, depending on the nature and/or size of the particles, on the manufacturing process, and on the conglomeration of particles when extremely complex situations appear in the composite, are hardly susceptible to analytical treatment.

An impedance mismatch—that is, $\rho_f v_f \gg \rho_m v_m$ with v_f and v_m being the velocities of propagation of the stress waves in the filler and the matrix, respectively—is a common occurrence in metal-filled plastics. This mismatch tends to create situations similar to those caused by imperfect bonding. Indeed, a stress wave does not enter the reinforcing particle, but simply circumvents it. The filler in this case only perturbs the stress fields by restraining the matrix, provided the strain is sufficiently high.

Figure 28 shows the crack propagation through a composite reinforced with 5% of 0.15-mm iron particles. The variation of the dimensions of the caustics created at the crack tip depends on both reflection and refraction of the stress waves at the rigid interfaces of filler particulates and on the variable size of the fracture zone. From the photographs in Figure 28 we see that, in all cases of crack propagation, either a local deviation of the direction of the crack or, in extreme cases, a bifurcation takes place.

It is well known that when a running crack in a macroscopically homogeneous and isotropic material reaches a critical velocity, the crack bifurcates into two or more branches. This is accompanied by a momentary deceleration of crack propagation, followed by a reacceleration. In homogeneous and isotropic materials, such as pure

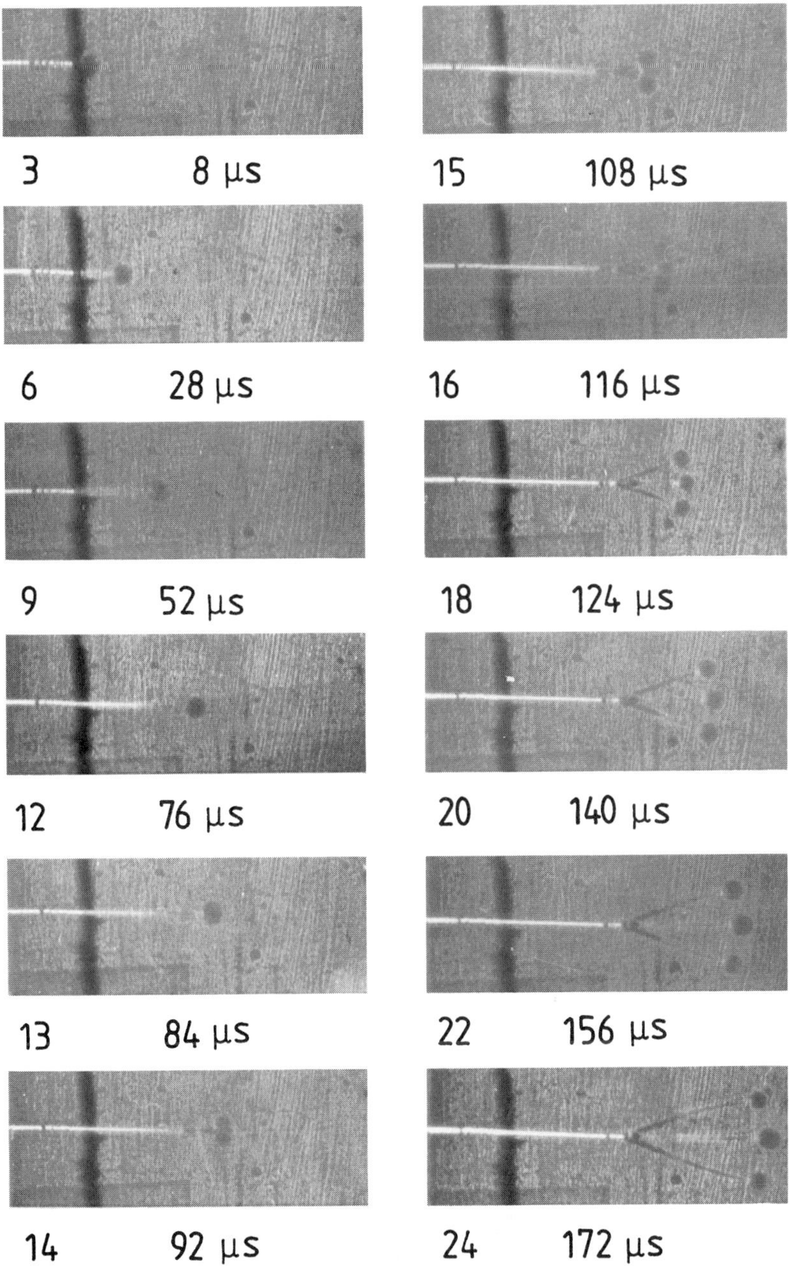

FIGURE 28 Series of photographs showing crack propagation in a 0.75-mm iron particle-epoxy composite with volume fraction 5%.

epoxies, where a global symmetry exists except for the critical velocity concept, the parameters on which this bifurcation depends are evantual localities of stress concentration, asymmetric loading of the specimens due to secondary bending at the extremities of the plates, and interference of stress waves with running cracks.

In the case of composites the most significant parameter affecting the bifurcation process is the microheterogeneity of the material network. In order to understand this, it is necessary to explain the fracture process in pure epoxies.

When a pure epoxy specimen is loaded in tension, a crazing process eventually takes place. Specifically, craze is accompanied by cavitation at a stress concentration in inhomogeneity which results in microvoid formation. Next, the microvoids enlarge and the polymer flows and orients in the direction of the applied stress, resulting in the formation of so-called fibrils, which are highly oriented macromolecules located within the craze. When the cavity formed attains a critical dimension, the crack propagates in a direction perpendicular to the applied stress direction through the craze. Moreover, in a macroscopically homogeneous and isotropic polymeric material there are regions characterized by a low cross-link density and others characterized by a high cross-link density. Because of this local microheterogeneity, there is a preferential deformation at regions of low cross-link density.

From the above mechanism of deformation of the polymer we can conclude that the microheterogeneity is of considerable importance in explaining the bifurcation phenomenon, much more than any other macroscopic contribution. Thus, in the case of particulate composites, where the presence of the filler particles develops more local microheterogeneities, the phenomenon of bifurcation must be more pronounced.

The bifurcation phenomenon is also closely related to the crack arrest phenomenon. It has been observed (Ref. 58) that when a crack reaches an inclusion surface, a crack propagation delay is always observed. This propagation delay is more pronounced as the distance of the crack axis from the inclusion center decreases. Thus, if we consider that, during the crack propagation process, the crack reaches the surface of an inclusion aggregate whose dimensions are large compared to the crack dimensions, this will lead to a long crack delay having as a result a crack bifurcation.

Figure 29 shows the dependence of the crack propagation velocity on the crack length for the particulate with 5% iron particles studied in Figure 28. From this graph we see that the crack propagation velocity fluctuates about a mean value. This mean value decreases linearly with increasing volume fraction of filler. This is due to the crack delay induced by the filler particles, which is more pronounced when the filler volume fraction increases.

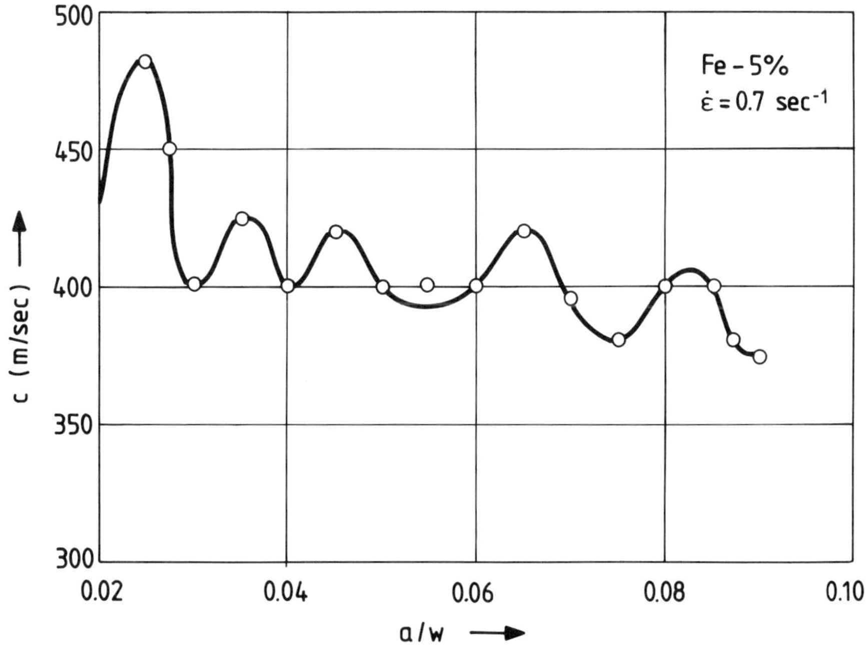

FIGURE 29 Crack propagation velocity versus crack length for the 5% iron particle-epoxy composite.

Figure 30 shows the variation of the dynamic stress intensity factor K_I^d for the three composite specimens, normalized to its initial value at the initiation of the crack, $K_{I,0}^d$, as a function of the crack length, a, normalized to its initial length, a_0. From this figure one can derive that the stress intensity factor increases by steps. This mode of increase is due to both the existence of the rigid interfaces of the filler particles, which play the role of barriers, and the different properties of the filler material.

It can be concluded from this experimental study that the main factors affecting the crack propagation modes in a particulate composite material are:

1. The ratio of the filler particle diameter to the dimension of the fracture core zone created at the tip of the moving crack.
2. The filler volume fraction
3. The properties of the filler material

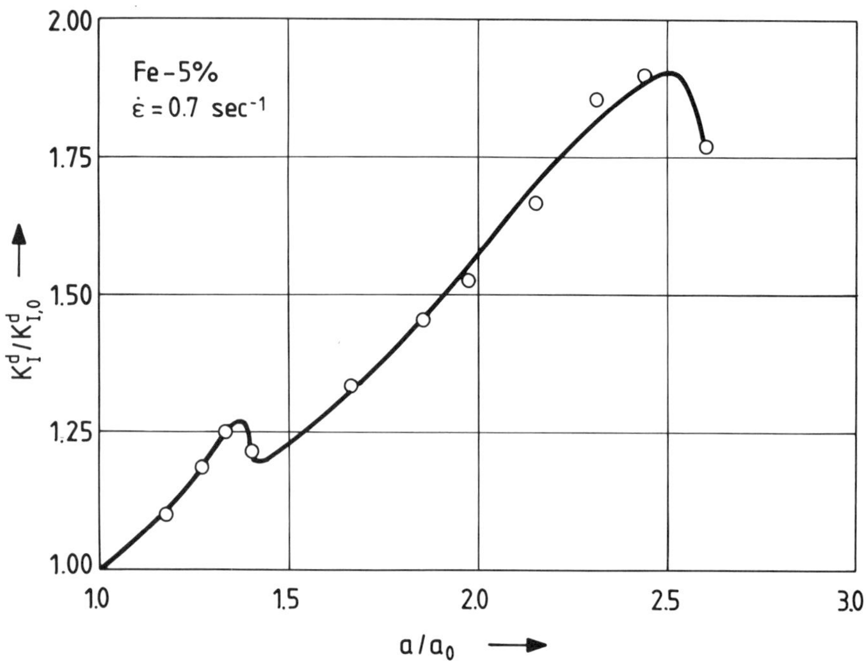

FIGURE 30 Normalized stress intensity factor versus normalized crack length a/a_0 for the iron particle-epoxy composite with volume fraction 5%.

4. The applied strain rate
5. The quality of adhesion between filler and matrix
6. The eventual existence of aggregates

These factors complicate the study of crack propagation in particulate composites and more sophisticate methods must be used to overcome these difficulties. The difficulties increase considerably for denser concentrations of particles. However, the method of caustics can be applied successfully in the case of particulate composite materials. Moreover, from the whole study it can be concluded that:

1. The mean crack propagation velocity decreases linearly with increasing filler volume fraction for small fractions of filler.
2. The dynamic stress intensity factor increases by steps with increasing crack length every time the crack tip meets a particulate inclusion and its mesophase.

3. The presence of the filler particles in the polymeric matrix results in a bifurcation or multifurcation effect. This effect is accompanied by a reduction in the critical crack propagation velocity with increasing filler volume fraction.

CONCLUSIONS

In this chapter the existence of the boundary layer constituting the mesophase, which is developed between the two main phases of a two-phase composite, has been taken into account in the study of their mode of fracture. Proof of the existence of the mesophase is ample, based on theoretical grounds and on experimental evidence. An additional proof of the influence of the mesophase on the mechanical behavior of composites is its contribution to the mode of their fracture.

Models consisting of biphase plates represented meridional sections of fibrous composites. The thin layers of the plates were assumed to be representative of the behavior of the respective sections. Cracks were initiated on either the soft or the hard phase of the plates and propagated under static or dynamic loadings along these phases and the consecutive phases, passing through the respective mesophases developed along the boundaries of neighboring phases.

The mechanisms of propagation of these cracks were studied by the experimental method of caustics, aided with a high-speed camera apparatus. The paths followed by the cracks in the phases and the mesophases, their instantaneous velocities, and their variations, as well as the variations of the respective stress intensity factors, were detected and evaluated, supplying an abundant amount of reliable information on the behavior of the composites in fracture.

It was shown qualitatively that the mesophases acted as decelerators and arresters of cracks, as reflectors of parts of their strain energy backward, and therefore as means of reducing the stored energy on the crack tip, with the effect, after successive interactions of these boundaries, of eventually arresting the crack and releasing the structure from a catastrophic failure.

Furthermore, it was shown that for mesophases between soft and hard phases an eventual branching of the propagating crack may occur; this has the advantage of separating into two or more parts the total energy of the main crack, which in such cases is unable to propagate further inside the material, so that the cracks are eventually arrested. In this way the soft-hard mesophases acted as bumpers in the further progress of fracture.

Finally, a criterion has been introduced, based on the T criterion related to the components of the distortional and dilatational energy density at the crack tip, which interrelated the mechanical properties of the main phases and the mesophase with the mode of fracture of the composites.

Although these models are not useful for a direct quantitative study of the fracture mode of composites, they yield an insight into the mechanisms of reaction of these materials in fracture, which may be helpful for the understanding and the appropriate design of convenient structures with these modern and efficient materials.

REFERENCES

1. Z. Hashin, J. Appl. Mech., *29*(2): 143 (1962).
2. Z. Hashin and B. W. Rosen, J. Appl. Mech., *31*(3): 223 (1964).
3. B. Budiansky, J. Mech. Phys. Sol., *13*(3): 223 (1965).
4. R. Hill, J. Mech. Phys. Sol., *13*(3): 213 (1965).
5. E. H. Kerner, Proc. Phys. Soc. London, *69B*: 808 (1956).
6. C. Van der Poel, Rheol. Acta, *1*: 198 (1958).
7. J. N. Goodier, J. Appl. Mech., Trans. ASMe, *APM-55-7*: 39 (1933).
8. H. Fröhlich and R. Sack, Proc. R. Soc. London Ser. A, *185*: 415 (1946).
9. F. R. Schwarzl and J. Eikhoff, Central Lab. TNO Rept. CL71/55 (1971).
10. F. Maurer, Ph.D. Thesis, Univ. of Duisburg, Federal Republic of Germany (1983).
11. G. Sagalaev and I. Simonov-Emiljanov, Plastinassy, *N2*: 48 (1973).
12. L. E. Nielsen, J. Compos. Mater., *1*(2): 100 (1967).
13. T. A. Kudykina and I. G. Pervak, Ukr. Fiz. Zh., *20*: 1664 (1975).
14. Y. Takano and A. Sakanishi, J. Appl. Phys., *44*: 4023 (1973); see also J. Appl. Phys., *45* 2811 (1974).
15. G. Spathis, E. Sideridis, and P. S. Theocaris, Int. J. Adhesion Abrasives, *1*(2): 195 (1981).
16. P. S. Theocaris, Proc. Natl. Acad. Sci. (Athens), *59*(I): 327 (1984).

17. M. Takayanagi, S. Uermura, and S. Minami, J. Polym. Sci., Part C, No. 5: 113 (1964).

18. F. Spathis, E. Kontou, and P. S. Theocaris, J. Rheol., *28*(2): 161 (1984).

19. E. P. Plueddemann and G. L. Stark, Proc. 35th Annu. Tech. Conf. Reinf. Plast. and Compos. Div. SPI Section 20B (1980).

20. A. D. Jonath and J. Crowley, in *Adhesion and Adsorption of Polymers* (L. H. Lee, ed.), Plenum, New York (1980), vol. 12A, p. 175.

21. P. S. Theocaris, J. Reinforced Plast. Compos., *3*(3): 204 (1984).

22. P. S. Theocaris, New Developments in the Characterization of Polymers in the Solid State, in *Advances in Polymer Science* (H. H. Kausch and H. G. Zachmann, eds.), Springer Verlag, New York (1985), vol. 66(6), p. 149.

23. P. S. Theocaris, Colloid Polym. Sci., *262*(12): 929 (1984).

24. P. S. Theocaris, J. Appl. Polym. Sci., *30*(2): 621 (1985).

25. P. S. Theocaris, Proc. 16th Int. Congress Appl. Mech., Copenhagen, Denmark (August 1984).

26. Yu. Lipatov, *Physical Chemistry of Filled Polymers* (R. J. Moseley, transl.), Intern. Polym. Sci. Tech. Monogr. 2, Khimiya, Moscow (1977).

27. P. S. Theocaris and J. Milios, Int. J. Fract., *16*(1): 31 (1980).

28. K. Ishikawa, A. K. Green, and N. E. Maiden, Exp. Mech., *12*(1): 32 (1972).

29. P. S. Theocaris and J. Milios, Eng. Fract. Mech., *13*(3): 599 (1979).

30. P. S. Theocaris and J. Milios, Int. J. Solids Struct., *17*(2): 217 (1981).

31. P. S. Theocaris and J. Milios, Int. J. Mech. Sci., *23*(7): 423 (1981).

32. F. Katsamanis, D. Raftopoulos, and P. S. Theocaris, Exp. Mech., *17*(4): 128 (1977).

33. J. Miklowitz, J. Appl. Mech., Trans. ASME, *3*(2): 122 (1953).

34. H. Kolsky, Wave Propagation Effects and Fracture, Brown Univ. Rept. 11 (1974).

35. K. Oi, Exp. Mech., *6*(2): 463 (1966).

36. V. M. Finkel, I. S. Guz', I. A. Kutkin, and A. Va. Volodarskii, Fiz. Tverd. Tela, *12*: 2300 (1970).

37. I. S. Guz' and A. P. Zotov, Probl. Prochn., *4*: 63 (1974).

38. F. Kerkhof, *Dynamic Crack Propagation* (G. Sih, ed.) (1972), p. 3.

39. P. S. Theocaris, in *Mechanics of Fracture*, vol. VII (G. Sih, ed.), Nijhoff (1981), chap. 3, pp. 189–252.

40. D. B. Bogy, J. Appl. Mech., Trans. ASME, *12*: 911 (1971).

41. P. S. Theocaris, J. Appl. Math. Phys., *26*(1): 77 (1975).

42. P. S. Theocaris and J. Milios, J. Reinf. Plast. Compos., *2*(1): 18 (1983).

43. J. Cook and J. E. Gordon, Proc. R. Soc. London Ser. A, *282*: 506 (1964).

44. E. P. Chen and G. C. Sih, J. Compos. Mater., *5*(1): 12 (1971).

45. G. A. Cooper and A. Kelly, J. Mech. Phys. Solids, *15*(3): 279 (1967).

46. A. S. Tetelman, Composite Materials: Testing and Design, ASTM STP460 (1969), p. 473.

47. A. Kelly, Proc. R. Soc. London Ser. A, *319*: 95 (1970).

48. G. A. Cooper and M. R. Piggott, Fract. Proc. Int. Confl. Fract. No. 4 (1977), vol. 1, p. 557.

49. A. S. Jayatilaka, *Fracture of Engineering Brittle Materials*, Applied Science, London (1979).

50. F. Erdogan and V. Biricikoglu, Int. J. Eng. Sci., *11*(4): 745 (1973).

51. M. I. Darby, Int. J. Fract., *15*(3): 311 (1979).

52. P. S. Theocaris, Eng. Fract. Mech., *19*(6): 975 (1984).

53. A. Nadai, *Theory of Flow and Fracture of Solids*, McGraw-Hill, New York (1950).

54. *P. S. Theocaris, Proc. U.S. Greece Symp. on Mixed-Mode Crack Propagation* (G. C. Sih and P. S. Theocaris, eds.), Sijthoff and Noordhoff (1981), p. 21.

55. P. S. Theocaris and N. P. Andrianopoulos, Eng. Fract. Mech., *16*(3): 425 (1982).

56. P. S. Theocaris and D. Pazis, Int. J. Solids Struct., *13*(7): 611 (1983).

57. P. S. Theocaris, C. G. Papanicolaou, and G. A. Papadopoulos, J. Compos. Mat., *15*(1): 41 (1981).

58. W. G. Knauss and H. K. Mueller, *Proc. First USA—USSR Symp. Fract. Compos Mater.* (G. Sih and V. P. Tamuzs, eds) (1978), p. 41.

6

Nonmetallic Fillers

JEAN-PIERRE REBOUL

Laboratoire d'Electricité
Université des Sciences et Techniques du Languedoc
Montpellier, France

INTRODUCTION

For the past 50 years, the use of polymers has been increasing and polymeric materials have competed with most other common materials. First, polymers were appreciated for their specific properties such as electrical insulating behavior, heat insulation, transparency, softness, flexibility, etc. Later, their use was extended to applications in which the opposite properties were required: electrical conductivity, heat conduction, light and radiation screening, hardness, rigidity, etc. The latter applications have been made possible by filling polymers in order to obtain composite materials with quite different properties from unfilled polymers.

Various kinds of fillers have been used, each of them fitting well with required applications or behavior.

Composites with metallic fillers are of interest when one or several of the following properties are required: hardness, high modulus of elasticity, heat conduction, antistatic behavior, electrical conductivity, radiation screening, etc.

However, for many applications, the required behavior may also be obtained with nonmetallic fillers, so that a wide range of applications may be provided by both kinds of fillers. In addition, nonmetallic fillers have their own range of uses. The following is a review of the principal kinds of nonmetallic fillers and their usual applications.

In the next section, we present the various carbon fillers, their use in rubber reinforcement, the analogy between mechanical and electrical properties, and the difference in electrical behavior of metal-filled and carbon-filled polymers.

The third section is devoted to glass fillers and to a new kind of conductive filler consisting of glass microspheres coated with silver. The results of some of our tests on the conductivity imparted by this filler are given.

The fourth section is concerned with silicon carbide powders as semiconductive fillers. Nonlinear electrical behavior and applications are mentioned.

The fifth section is an abstract of the more usual applications of mineral fillers, while the conclusions compare and contrast the properties imparted by metallic and nonmetallic fillers to polymer composites.

CARBON FILLERS

Various Kinds of Carbon Fillers

Carbon is always produced by the pyrolysis of its compounds. The partial combustion of most organic materials leads to formation of soot and carbon particles, called carbon blacks.

The size, structure, and properties of carbon blacks depend on the material which is burned and on the conditions of burning. Industrial processes are used to make carbon blacks from gas and oils. The channel process consists of burning natural gas jets against a moving metal channel. *Channel blacks* result from this process. Partial combustion of gas or volatilized oil in a furnace is another process which is now replacing the channel process and which leads to *furnace blacks*.

The various industrial carbon blacks are well classified and precisely designated, for example, as MPC (medium process channel) or HAF (high abrasion furnace). More details on the structure, manufacture, and classification of carbon blacks may be found in Medalia (Ref. 1) and Blanchard (Ref. 2).

Carbon blacks have been used for a long time as coloring pigments, but the major use is now in the rubber and polymer industries (tires and conductive plastics). The use of natural or artificial graphites as fillers is not usual, but a reference to graphite structure is of interest in the study and classification of carbon blacks. Indeed, carbon blacks are colloidal or polycrystalline. X-ray studies provide evidence of crystallites in the form of graphite layers more or less parallel to each other. Graphitization treatment may also be used to increase the size of crystallites. Some processes for making artificial graphite consist of a graphitization treatment of carbon blacks.

The layered structure of graphite is well known, but the degree of graphitelike order varies from one type of carbon black to another, influencing most properties. When the hexagonal pattern of carbon atoms is developed but the graphite layers are not absolutely parallel, the structure is called turbostratic.

The turbostratic structure is of interest for improvement of mechanical properties, which are close to those of graphite but with slipping along the graphite planes rendered much more difficult.

As a consequence, turbostratic structure is considered the optimum arrangement for carbon fibers. Carbon fibers are generally obtained by pyrolysis of organic polymer fibers, called the precursor. The process of pyrolyzing is long and must be carefully done. The precursor is not a raw material but an organic fiber, so that carbon fibers are always expensive to produce.

Carbon fibers are used in the reinforcement of various polyesters and resins. The major advantage is lightness. The major drawback remains the high price. Other details on carbon fibers may be found in Parker (Ref. 3), Gill (Ref. 4), and Delmonte (Ref. 5).

Uses and Properties of Carbon Black-Polymer Composites

The choice of carbon as a filler mainly results from three kinds of requirements: color, mechanical properties, and electrical properties.

All forms of carbon, diamond excepted, are black because they absorb light. This property may be of interest for coloring purposes or light screening.

Not all the mechanical properties of polymers are systematically improved by filling them with carbon, though carbon black may behave as an antioxidant and stabilizer. Stiffness is generally increased by mineral fillers, but fracture energy would be reduced, according to Kendall (Ref. 6), so that the composite becomes more brittle toward high loading levels. Besides, most mechanical behavior drastically depends on filler concentration and on careful mixing (Ref. 7). Fox (Ref. 8) found a melt viscosity related to the volume loading of Ketjenblack EC. He noted some striking similarities between mechanical and electrical properties, concerning their dependence on carbon loading. In particular, when the carbon black concentration is sufficient to allow the formation of a continuous conductive network throughout the polymer, an identical sharp change occurs in both shear stress and conductivity. The critical concentration thresholds for shear stress and conductivity are probably fundamentally related.

This is not surprising, as investigators have long observed similarities between the mechanical and electrical properties of unfilled polymers. From a comparison of electrical and mechanical relaxation of polyvinyl chloride and polyvinyl acetate, Caillon (Ref. 9) concluded that small lag time hysteretic phenomena, electrical as well as mechanical, were connected to the same molecular motions. In polystyrene, McCrum et al. (Ref. 10) mentioned the good correlation between mechanical relaxation and dielectric α relaxation, as shown by their identical locations on a frequency-temperature plot. In the same way, Mikhailov and Borisova (Ref. 11) found, in halogen derivative of polystyrene, a correlation between the β component of a dielectric peak and the β relaxation observed by Illers and Jenckel (Ref. 12). However, correlations between electrical and mechanical behavior are not yet fully understood, either in unfilled polymers or in filled ones.

As might be expected, not all the filled polymers are brittle. Their dynamic properties (Ref. 13) and elastic moduli (Ref. 14) have been studied theoretically, and it is well known that reinforcement of resins by metal fibers, glass fibers, or carbon fibers is a current procedure. The mechanisms of reinforcement with all kinds of fibers are similar. In contrast, reinforcement of rubber with carbon black seems to be specific to the rubber matrix-carbon black association. Mineral powders and other inert fillers may also have a reinforcing behavior but to a minor degree.

Theories on filler reinforcement (Refs. 15, 16) call attention to the increase of stresses around the filler particles. More recent investigations are concerned with finer parameters such as particle size distribution (Ref. 17), number of bonded particles (Ref. 18), surface slippage of rubber on carbon particles (Ref. 19), and hysteretic component of tear resistance (Refs. 20–22). Today, though the mechanisms of rubber reinforcement are not all explained, the technology of reinforcement by carbon black is perfectly controlled and tire manufacturers have considerably improved the mechanical and thermal behavior of tire rubber. Full details of the manufacturing processes are not revealed.

Concerning the electrical properties of polymers in greater detail, these may be completely reversed by filling them with carbon blacks or other conductive fillers. Indeed, polymers are usually good electrical insulators, and this behavior makes them very suitable for many uses such as insulating material in coils, wires, and cables or dielectric layers in capacitors.

In contrast, in mechanical applications, the insulating behavior may be unsuitable. For example, when static electrical charge formed by friction cannot be drained off, the consequences are attraction of dust, electrical discharge when handled, and incendiary sparks.

Filling the polymer with conductive powder is a method of canceling the insulating behavior. The conductive composites which are obtained are of interest in various applications such as antistatic materials, heating panels, shielding coats, anticorona layers in high-voltage cables, and even electronic devices or transducers. The list is not exhaustive and new applications are still being discovered.

However, canceling insulating properties of polymers requires large volume concentrations of filler, generally above 25%. Such high values are confirmed by experimental measurements and by theories. Indeed, probability calculations, for an insulating matrix containing randomly dispersed conductive elements, predict a transition from insulating to conducting behavior when the volume of filler reaches a critical value called the percolation threshold. The critical volume is dependent on the aspect ratio of the conductive particles. In a realistic model discussed by Kirkpatrick (Ref. 23) it would be 0.25.

In metal-filled polymers the percolation threshold is experimentally confirmed by a sudden increase of conductivity when the critical volume of metal is reached. Bakelite-silver mixtures studied by Gurland (Ref. 24) exhibit a typical percolation threshold. In polyvinyl chloride (PVC) filled with nickel or copper powders, studied by Kusy and Turner (Ref. 25), the critical volume is lowered because of a segregative distribution of the metal particles. This kind of composite is also of interest for thermal conduction (Ref. 26).

In positive temperature coefficient (PTC) materials (Ref. 27), concentrations of filler near the percolation threshold are required. Indeed, if the insulating matrix possesses a large thermal expansion coefficient, when the temperature is raised, the volume of the matrix increases while the volume of filler is practically unchanged so that the relative volume concentration of filler decreases. If it was just above the percolation threshold, it falls below it, and the composite becomes insulating. Carbon black-filled waxes made and studied by Bueche (Ref. 28) behave in this manner and are considered switching materials.

In carbon black-filled polymers, the transition from an insulating to a conducting state is more progressive than in metal-filled polymers. It is spread over a concentration range from 15 to 35%. In Norman's data (Ref. 29) on carbon black-filled rubber the transition is also progressive, but it depends on the type of carbon black. Chung et al. (Ref. 30) found metal-like electrical behavior in highly loaded PVC, depending also on the type of carbon black.

Hence, it seems that other electrical properties are present in carbon black-polymer composites in addition to those concerned with the percolation mechanism. Properties specific to the polymeric matrix-carbon black association, as in mechanical reinforcement, are, perhaps, responsible for the different behavior of metal-filled and carbon black-filled polymers. The model of vitreous structured domains around each carbon particle has already been proposed (Ref. 31). As we have discussed elsewhere (Refs. 32, 33), this model is of interest and it would need to take into account larger particles of filler comprised of a carbon black core coated with a vitreous layer. According to this model, a first percolation threshold concerned with semiconducting vitreous particles might be reached while the carbon black concentration is still under its own percolation threshold. So the succession of two percolation thresholds could explain the progressive transition from an insulating to a conducting state as the volume concentration of carbon black increases.

Because the transition is spread over a relatively wide range of carbon black concentration, many composites are neither completely insulating nor completely conducting and they may be classified as semiconducting. Besides, most of these semiconducting materials have nonlinear electrical properties. Figure 1 shows the logarithmic current-voltage characteristics of PVC samples filled with XC 72 carbon black. The volume concentration of filler is 20% in curve A and 30% in curve B. Figure 2 shows the logarithmic current-voltage characteristics of rubber samples filled with MPC carbon black. Volume concentration of filler is 20% in curve A and 30% in curve B. The curves show that Ohm's law is no longer verified as soon as the applied voltage becomes noticeable. The current-voltage characteristics may be divided into domains in which a dependence of current I on voltage U in the form $I = AU^n$, with $n > 1$ and A a constant, gives a good approximation, as also mentioned by Jachym (Ref. 34).

Various mechanisms have been proposed to explain this nonlinear electrical behavior, in particular tunneling between carbon particles (Ref. 35) and conduction by hopping (Ref. 36). It is certain that many conduction processes can take place in these mixtures, any one of which may be predominant in given circumstances. However, it seemed unlikely to us that such mechanisms might be the only bulk dc conduction phenomena responsible for nonlinear electrical properties.

Indeed, we have found, by using potential probes, that in some cases Ohm's law remains verified inside the body of the material while at the same time the overall sample exhibits nonlinear current-voltage characteristics (Ref. 32). Besides a study of the voltage variations along samples from cathode to anode when a constant dc voltage is applied, driving a constant dc current gave us evidence for an increase in electric field near the electrodes and consequently for a nonuniform electric field throughout the overall sample. This is not taken into account in many conduction models, which assume a uniform electric field. In contrast, the nonuniform electric field suggested a contribution from a mechanism concerned with space charge. In PVC filled with XC 72 carbon black, we have evidence, from a study of current responses to voltage pulses and to thermal pulses, for a contribution from ionic space charge (Ref. 37). A contribution of space charge was also found by Cashell et al. (Ref. 38) in carbon black-filled *cis*-polybutadiene. Bhattacharya et al. (Ref. 39) observed the restoration of ohmic behavior after annealing in copper PVC composites. Elimination of free ions by prolonged annealing was envisaged as a plausible mechanism by this school of investigators.

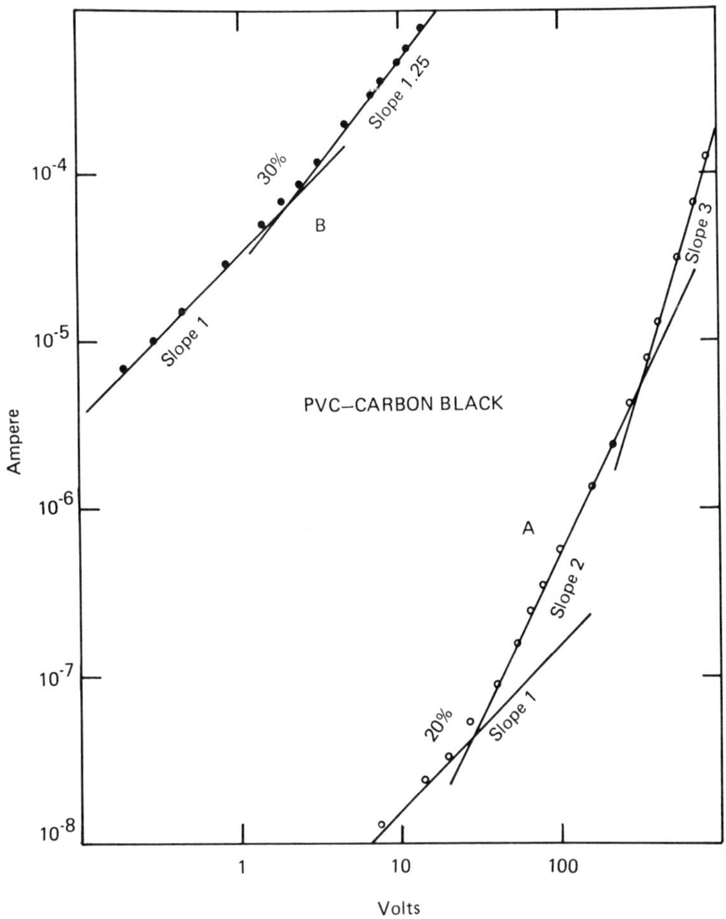

FIGURE 1 Current-voltage characteristics. PVC filled with XC 72 carbon black. Volume of carbon black: A, 20%; and B, 30%. Experimental conditions: sample thickness, 2.3 mm; sample section, 450 mm^2; temperature, 23°C.

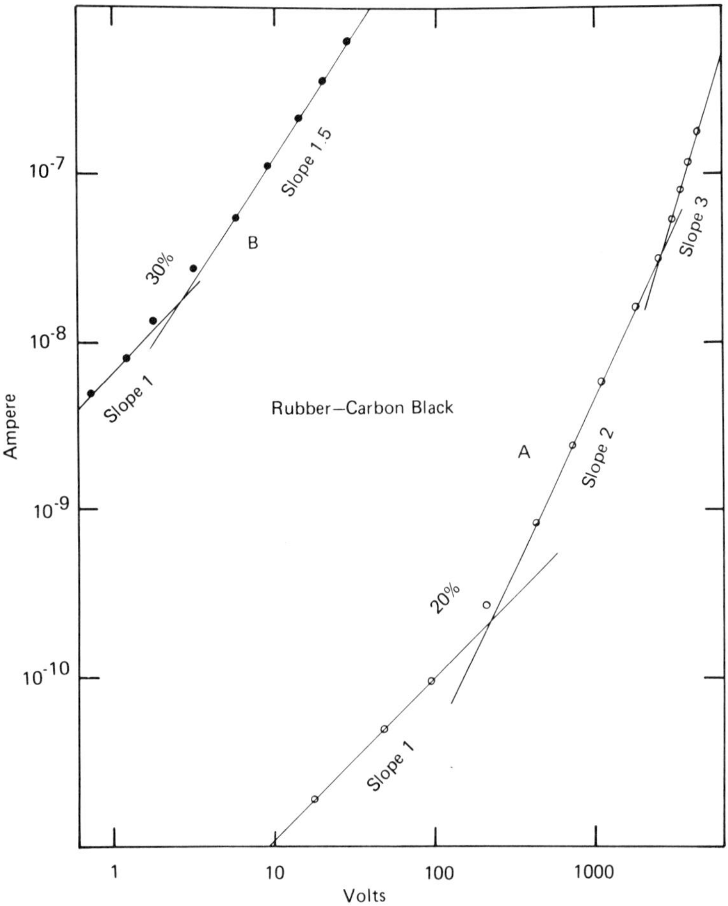

FIGURE 2 Current-voltage characteristics. Rubber filled with MPC carbon black. Volume of carbon black: A, 20%; and B, 30%. Experimental conditions: sample thickness, 2.2 mm; sample section, 450 mm^2; temperature, 23°C.

In brief, carbon black-polymer composites have, on the one hand, many properties quite comparable to those of metal-polymer composites, but, on the other hand, they exhibit a large range of typical properties, mechanical as well as electrical, which are probably related. Most of these properties involve mechanisms which are not yet completely understood but in which the linkage structure between carbon aggregates and polymer matrix probably plays a dominant role, so that carbon blacks remain privileged fillers, irreplaceable in many applications.

GLASS FILLERS

Various Kinds of Glass Fillers

Glass is used as filler in polymers in the form of powders, spheres, and fibers. When hardness and abrasive behavior are wanted, glass powder fillers are as suitable as metal powders. The transparency of the composition may be a determining factor in favor of a glass powder.

Glass spheres smaller than 1 mm in diameter are often used for optical purposes. The largest usage of composites containing glass spheres is in road signaling. Reflecting paints, reflectors, and materials which must be seen at night by drivers, are generally glass sphere-polymer composites. Indeed, a well-known property of the spheric lens is to reflect light in the direction of incidence.

Glass fibers are the most common form of glass fillers. Their use in reinforcement of plastic has developed over the past 40 years. Reinforcement of polymers by glass fibers improves their mechanical properties and yields materials capable of competing with metals and other common materials. Both thermosetting and thermoplastic polymers are convenient for making glass-reinforced composites, but the largest usage is in the form of glass-reinforced polyesters. Lightness, resistance to corrosion, transparency, and electrical and heat insulating behavior may be determining factors in the choice of glass fibers instead of metal fibers as reinforcing fillers.

Glass fibers are drawn from several kinds of glass: a high-alkali A-type glass, electrical grade E-type glass, and high-modulus and high-strength glasses.

More details on glass fiber manufacture and properties may be found in Parvin's work (Ref. 40).

Glass Spheres as Conductive Fillers

Glass is an electrical insulator and glass fibers were originally developed for electrical insulating purposes. However, a new kind of conductive filler was recently manufactured by Potters Industries in the form of glass microspheres coated by a silver layer. Sphere diameters are in the range 1–1000 μm. For spheres less than 44 μm in diameter, three thicknesses of silver are available, 260 Å, 2 × 260 Å, and 3 × 260 Å, corresponding to silver concentrations of 4, 8, and 12% by weight; these have reference numbers $S3000S_1$, $S3000S_2$, and $S3000S_3$, respectively. Obviously, the more silver, the more expensive is the filler. Other details have been published by Heinze and Ritter (Ref. 41) (Potters Industries) and are also mentioned by Bigg in Chapter 3.

Coating with silver or gold is a technique which is also employed with copper powder and other metal fillers to make them more conductive and to prevent high contact resistance connected with oxide layer formation.

Theoretically, the composite with silver-coated spheres must be as conductive as the composite with identical spheres of silver, since it is well known that, in conductive metal-filled polymers, the largest part of the resistivity is provided by the contact resistance between particles. If so, the quantity of silver necessary to render the polymer conductive may be considerably lowered by using thin silver coating.

In practice, the silver layer is brittle, chiefly when it is thin, and it may be stripped off during the mixing procedure. As a consequence, most contacts between spheres become insulating and the conductivity of the composite drops. Thus the quantity of coated spheres necessary to make a conductive network throughout the polymer is not quite independent of the thickness of the silver layer, and it is clear that very thin silver layers are unsuitable.

Moussalli (Ref. 42) studied the resistivity of epoxide resin samples as a function of filler concentration for the three kinds of spheres referred to above. Results are given in Figure 3. The percolation threshold is attained when the volume concentration of filler reaches the range 30–35%. The concentration required to render the composite conductive is not negligible, and, though less expensive than silver powder, the use of this kind of filler is limited by its cost.

In order to decrease the necessary quantities of silvered spheres Moussalli realized a segregative distribution of the filler in the insulating matrix. For this purpose, the insulating material consisted of insulating glass balls, much larger in diameter (about 200 μm) than silvered spheres, sunk into epoxy resin. In such a mixture, only the spaces between the glass balls, that is, the resin, must become conductive. From his results, presented in Figure 4, the critical concentration of filler is lowered by a factor of about 3 and its value is about 12%. Evidently other insulating particles or plastic balls may be used instead of glass balls in order to obtain segregative distributions of the filler, and in the case of loading levels

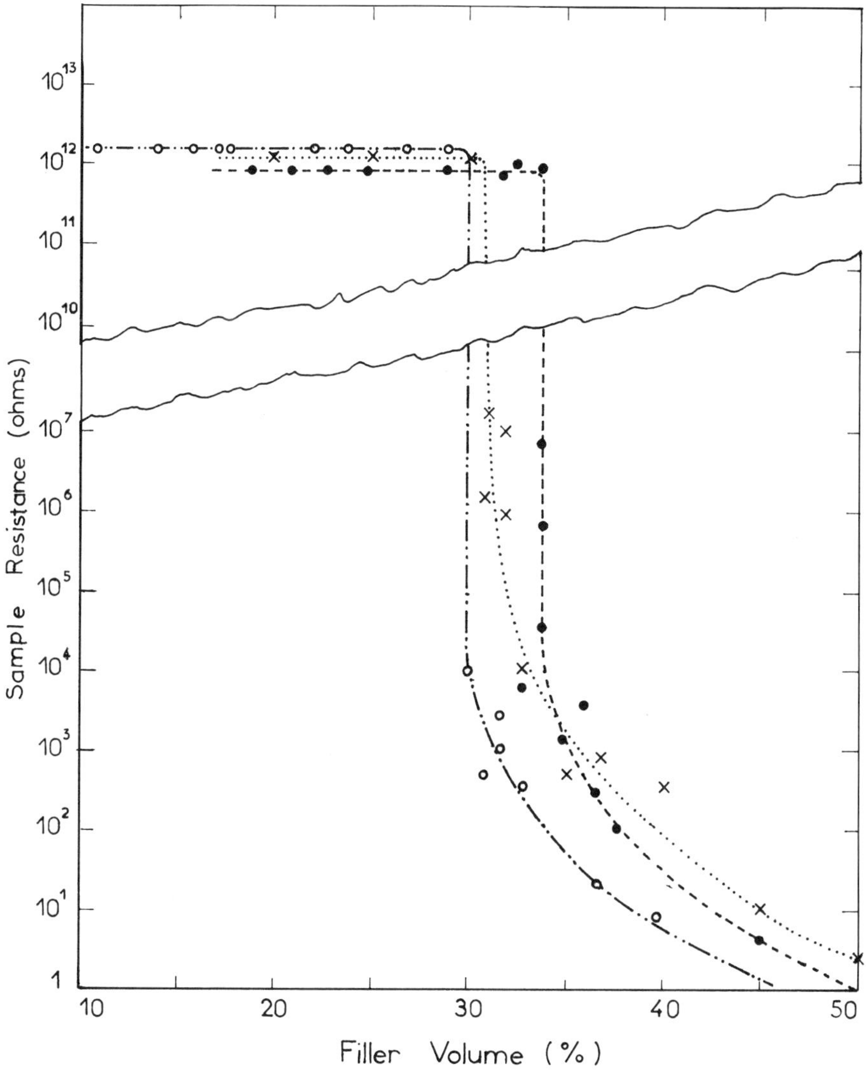

FIGURE 3 Sample resistance as a function of filler volume. Insulating matrix, epoxide resin; conductive filler, silvered microspheres, references $S3000S_1$ (×), $S3000S_2$ (•), and $S3000S_3$ (○). Experimental conditions: sample thickness, 8 mm; sample section, 240 mm^2; temperature, 23°C.

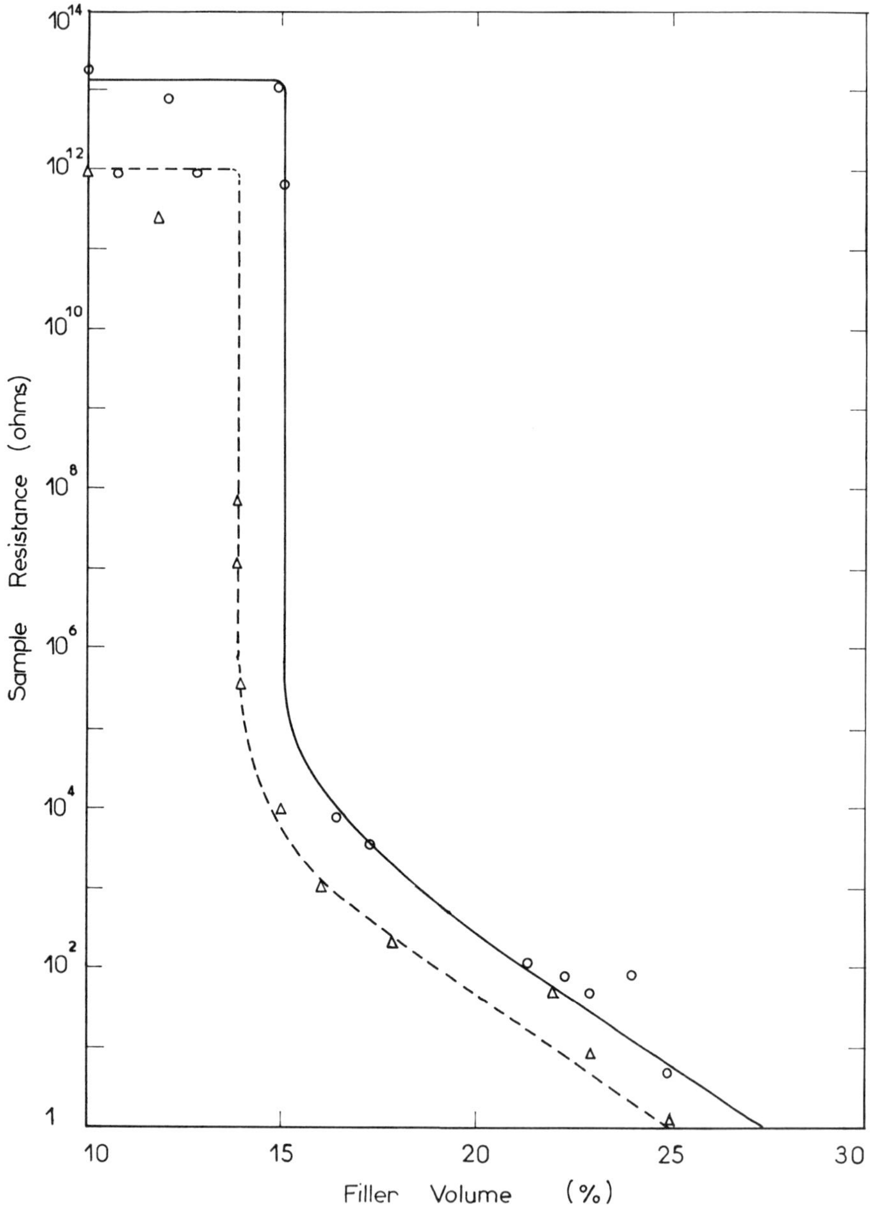

FIGURE 4 Sample resistance as a function of filler volume. Insulating matrix, epoxide resin 40%—glass balls 60%; conductive filler: silvered microspheres, references $S3000S_1$ (○) and $S3000S_3$ (△). Experimental conditions: sample thickness, 8 mm; sample section, 240 mm^2; temperature, 23°C.

lowered by segregative processes, this kind of filler becomes competitive in many more applications.

SILICON CARBIDE

Silicon carbide, also called carborundum and designated by the chemical formula SiC, is well known for its hardness, close to that of diamond.

The principal applications are concerned with this property, and carborundum may often advantageously replace emery. Various silicon carbide-polymer composites are used as abrasive materials, but most of them may be considered as carborundum grains joined by a polymeric cement rather than polymers filled with SiC. However, in addition to their mechanical behavior, the SiC-polymer composites have interesting electrical properties and have found applications as anticorona coatings in cables and leads.

Silicon carbide may be obtained by various processes. Heating a silicon-carbon mixture in an electric oven leads directly to SiC. Reduction of silica by carbon, above 1500°C, also leads to SiC. One of the industrial processes consists of heating a mixture of sand, coke, and sodium chloride in an electric oven. A core of SiC crystals surrounded by graphite is obtained in this manner.

Silicon carbide is available in the form of crystalized grains or powders. A particular electrical property of SiC grains is concerned with the resistivity. This is very dependent on impurities and other manufacturing parameters. It may vary from a few ohm-centimeters to a few megohm-centimeters. For electrical applications, high-resistivity or low-resistivity crystals may be chosen according to the required purpose.

Another particular electrical property of SiC powder is provided by the contacts between grains. Indeed, the contact is generally not conductive until the applied voltage reaches a threshold of about 0.5 volt between the two grains. As a consequence, SiC powders, and also polymers filled with SiC powders in adequate concentration, have nonlinear current-voltage characteristics. The larger the applied voltage, the more conductive is the sample. In SiC-polymer composites, this property is found only for large concentrations of filler, that is, when a continuous network is formed by the SiC grains throughout the polymer. Volume concentrations of filler above 30–35%, the percolation threshold, are generally adequate. For example a composition of 50% epoxy resin and 50% SiC powder (sample A) possesses typical nonlinear electrical properties, as we have verified.

Our results are represented by curve A in Figure 5. Sample A was in the form of a disk 30 mm in diameter and 7 mm thick. As

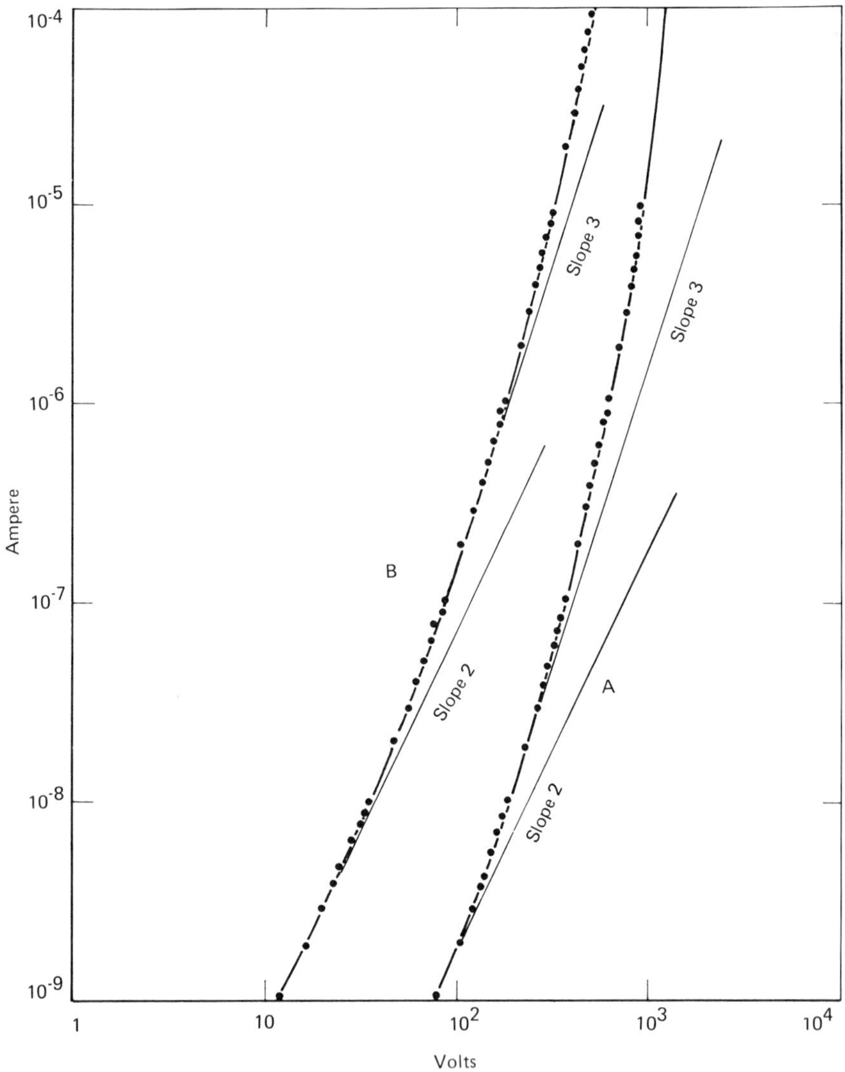

FIGURE 5 Current-voltage characteristics. Epoxide resin-SiC powder composites. Sample composition (volume): A; resin 50%, SiC 50%; B; resin 50%, SiC 25%, glass balls 25%.

shown by curve A, in the assumption of a law of the type $I = AU^n$ between current I and voltage U, n increases from 2 to 3 and more.

Curve B in Figure 5 corresponds to a composition of 50% epoxy resins, 25% SiC powder, 25% glass balls (sample B). Glass balls make passive insulating volumes which are responsible for a segragative distribution of SiC grains so that the effective concentration of SiC in the resin is raised to 33.3%. Samples A and B had identical shape and dimensions.

As shown by curve B, the dependence of current on voltage is of the same nature as in sample A, but for larger voltages. As a consequence, it may be thought that the nonlinearities are characteristic of the network of SiC grains, essentially the number of grain contacts in chains joining the electrodes.

In order to verify that the nonlinearities are not concerned with phenomena near electrodes, for example, space charge as in some carbon black-filled polymers, we measured the resistivities of cylindrical samples using a four-electrode method as shown by the scheme in Figure 6. Here F_1 and F_2 are the feeding electrodes and C_1 and C_2 are two intermediate electrodes in the shape of a collar. The voltage drop U between C_1 and C_2 is measured with an electrometer, while a voltage V is applied between F_1 and F_2, driving a current I through the sample. The curves U versus I and V versus I present identical discrepancies from Ohm's law.

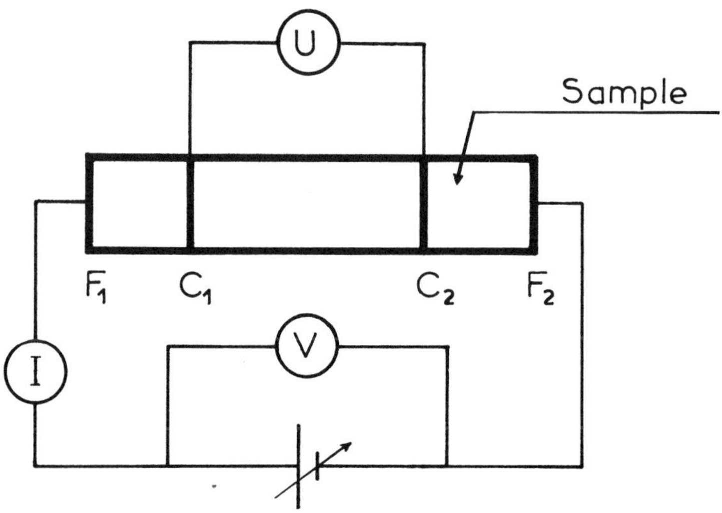

FIGURE 6 Diagram of the four-electrode method used to measure the bulk resistivity. F_1, F_2: feeding electrodes; C_1, C_1: potential electrodes.

The nonlinear electrical behavior of SiC-polymer composites is thus a bulk phenomenon, in contrast to contact space charge phenomena occurring in some carbon black-polymer composites. The assumptions concerned with conduction processes from grain to grain, such as tunneling (Ref. 43), which have been envisaged between carbon black aggregates, are quite plausible in the present case.

Interesting applications of the nonlinear behavior of SiC composites have been found in cable technology. For example, when an insulated cable has to pass through a metallic framework or shielding, the concentrated electric field between the edge of the hole and the cable conductor produces a strain on the cable insulator and leads to its breakdown.

Indeed, if U is the potential of the cable conductor, the shielding being at zero potential (Figure 7a), the surface of the insulator is practically at the potential U everywhere except in the vicinity of the hole, where the potential is zero because of contact with the metallic shielding. A large longitudinal electrical field and noticeable leakage currents are then concentrated on the insulator surface near

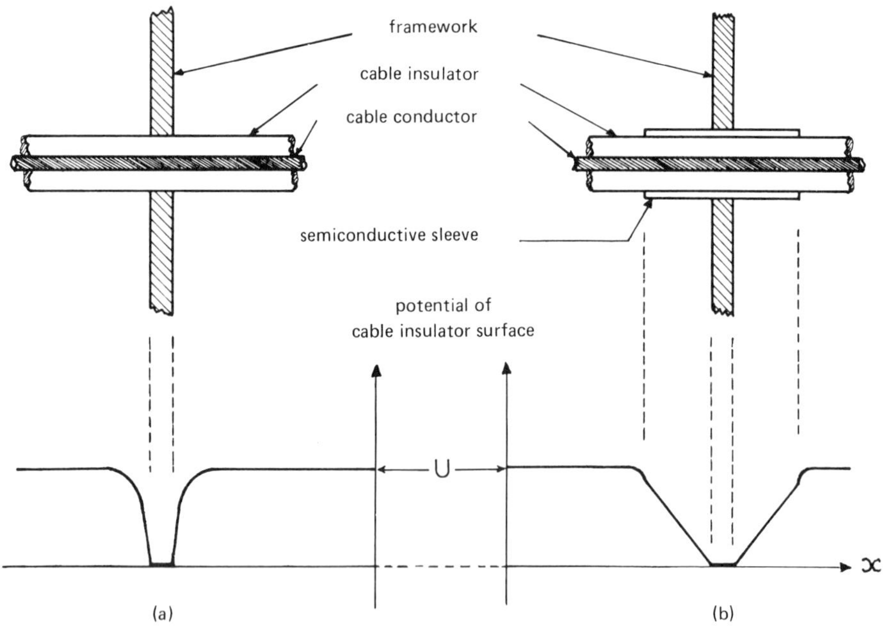

FIGURE 7 Potential of the cable insulator surface in the vicinity of passing through a grounded framework: (a) without protection; and (b) with a nonlinear semiconductive sleeve.

the metallic shielding. A coating or a sleeve made with a semiconducting nonlinear material distributes the electrical strain more uniformly over a larger region (Figure 7b). Near the shielding the leakage currents are large, but the resulting voltage drop in the nonlinear coating is weak. Far from the shielding the leakage currents are weak but the resulting voltage drop in the nonlinear coating is large. Thus a quasi-linear variation of the potential of the insulator surface may be obtained by using an adequate nonlinear sleeve.

Several patents concerning SiC composites and this kind of application have been filed (Ref. 44).

MINERAL FILLERS

A large variety of inorganic minerals is available in abundance in the ground, and even in their refined and purified forms, mineral fillers remain generally cheap. Thus a first and interesting effect of filling polymers with mineral powders is to make the composite less expensive than the unfilled material. Mineral fillers, like other fillers, may also enhance or modify properties of the unfilled polymer, such as rigidity, hardness, heat resistance, thermal expansion, electrical behavior, density, color, and opacity. It is relevant to close a chapter on nonmetallic fillers with some words on the more current applications of mineral fillers.

Calcium carbonate fillers find the largest use in the PVC industry. They facilitate extrusion and calendering. In the form of ultrafine powder, they increase viscosity even at relatively low concentrations, so that the consistency and flow properties of the mixture may be adjusted to facilitate molding or calendering. They also reduce the shrinkage in curing procedures.

Alumina is used as filler when high electrical resistivity and thermal conductivity are required. Emery and other alumina grains in the form of coarser corundum are also used to impart hardness and abrasion resistance.

Clays and aluminosilicates find applications in increasing viscosity, absorbing exothermic heat during curing, and also as cheap bulking fillers.

Silica in the form of sand or flour is appreciated for its low coefficient of thermal expansion, high thermal conductivity, electrical insulating behavior, transparency, and hardness. Besides, it does not excessively increase viscosity, so that it is very convenient for epoxide castings.

Many other mineral fillers find applications in the polymer industry but this subject cannot be developed further here. The interested reader will find more details on mineral powders and asbestos in Whisson (Ref. 45) and Wicker (Ref. 46), respectively.

CONCLUSIONS

In this chapter we compare and contrast the properties imparted to polymer composites by metallic and nonmetallic fillers. Linkage structure and the nature of the filling material itself are probably responsible for the differences. Among similarities one finds: (a) mechanical reinforcement by both kinds of fillers, (b) increased impact resistance and abrasion resistance by both kinds of fillers, (c) machineability, (d) increased viscosity, and (e) electrical conductivity. The imparted properties which are more specific for nonmetallic fillers are: (a) heat insulation, (b) electrical insulation, (c) specific reinforcement of rubber by carbon black, and (d) nonlinear electrical properties.

Both enumerations are not exhaustive. Many other properties leading to less usual applications may be added. Besides, the parameters which today influence the choice of this or that filler for such applications may be drastically modified in the future, owing to a change of price or an improvement in processing. It would then be difficult to set up a rigid frontier line between the applications reserved for the different kinds of fillers.

REFERENCES

1. A. I. Medalia, in *Carbon Black-Polymer Composites* (E. K. Sichel, ed.), Dekker, New York (1982).
2. A. F. Blanchard, in *Fillers for Plastics* (W. C. Wake, ed.), Iliffe, London (1971).
3. D. B. V. Parker, in *Fillers for Plastics* (W. C. Wake, ed.), Iliffe, London (1971).
4. R. M. Gill, *Carbon Fibres in Composite Materials*, Iliffe, London (1972).
5. J. Delmonte, *Technology of Carbon and Graphite Fiber Composites*, Van Nostrand Reinhold, New York (1981).
6. K. Kendall, Br. Polym. J., *10*:36 (1978).
7. D. M. Bigg, Polym. Eng. Sci., *15*:684 (1975).
8. L. P. Fox, in *Carbon Black-Polymer Composites* (E. K. Sichel, ed.), Dekker, New York (1982)
9. P. Caillon, Onde Electr., *35*:463 (1955).
10. N. G. McCrum, B. E. Read, and G. Williams, *Anelastic and dielectric effects in polymeric solids*. John Wiley and Sons, London (1967).

11. G. P. Mikhailov and T. J. Borisova, Vysokomol. Soedin., *4*: 1732 (1962).

12. K. H. Illers and E. Jenckel, Rheol. Acta, *1*:322 (1958).

13. S. A. Paipetis, G. Papanicolaou, and P. S. Theocaris, Fibre Sci. Technol., *8*:221 (1975).

14. T. S. Chow, J. Appl. Phys., *48*:4072 (1977).

15. J. Rehner, Jr., J. Appl. Phys., *14*:638 (1943).

16. E. Guth, J. Appl. Phys., *16*:20 (1944).

17. A. E. Medalia, Rubber Chem. Technol., *34*:1134 (1961).

18. A. F. Blanchard, *Applied Science of Rubber* (W. J. S. Naunton, ed.), Arnold, London (1961).

19. B. B. Boonstra, in *Reinforcement of Elastomers* (G. Krauss, ed.), Interscience, New York (1965).

20. L. Mullins, Trans. Inst. Rubber Ind., *35*:213 (1959).

21. H. W. Greensmith, J. Polym. Sci., *21*:175 (1956).

22. J. A. C. Harwood and A. R. Payne, J. Appl. Polym. Sci., *12*:889 (1968).

23. S. Kirkpatrick, Rev. Mod. Phys., *45*:574 (1973).

24. J. Gurland, in *Plansee Proceedings* (F. Benesovsky, ed.), Springer Verlag, Vienna (1962), p. 507.

25. R. P. Kusy and D. T. Turner, Nature Phys. Sci., *229*:58 (1971).

26. R. P. Kusy and R. D. Corneliussen, Polym. Eng. Sci., *15*: 684 (1975).

27. J. Meyer, Polym. Eng. Sci., *13*:462 (1973).

28. F. Bueche, J. Appl. Phys., *44*:532 (1973).

29. R. H. Norman, *Conductive Rubbers and Plastics*, Elsevier, Amsterdam (1970).

30. K. T. Chung, A. Sabo, and A. P. Pica, J. Appl. Phys., *53*: 6867 (1982).

31. M. L. Deviney, Adv. Colloid Interface Sci., *2*:237 (1969).

32. J. P. Reboul, in *Carbon Black-Polymer Composites* (E. K. Sichel, ed.), Dekker, New York (1982).

33. J. P. Reboul and G. Moussalli, Int. J. Polym. Mater., 5:133 (1976).

34. B. J. Jachym, in *Carbon Black-Polymer Composites* (E. K. Sichel, ed.), Dekker, New York (1982).

35. E. K. Sichel, S. I. Gittleman, and Ping Sheng, in *Carbon Black-Polymer Composites* (E. K. Sichel, ed.), Dekker, New York (1982).

36. E. O. Forster, IEE Trans. Power Appar. Syst., *PAS 90*:913 (1971).

37. J. P. Reboul, J. Appl. Phys., *46*:2961 (1975).

38. E. M. Cashell, J. M. D. Coey, G. E. Wardell, V. J. McBrierty, and D. C. Douglas, J. Appl. Phys., *52*:1542 (1981).

39. S. K. Bhattacharya, S. Basu, S. K. De, A. K. Pal, and S. Chowdhury, J. Appl. Phys., *49*:3001 (1978).

40. K. Parvin, in *Fillers for Plastics* (W. C. Wake, ed.), Iliffe, London (1971).

41. R. E. Heinze and J. Ritter, Verre Text. Plast. Renf., *14*:40 (1976).

42. G. Moussalli, Thèse de Docteur Ingénieur, Montpellier, France (1977).

43. P. Sheng, Phys. Rev. B, *21*:2180 (1980).

44. For example, L. G. Virsberg et al., U.S. Patent 3,066, 180 (1962).

45. R. R. Whisson, in *Fillers for Plastics* (W. C. Wake, ed.), Iliffe, London (1971).

46. G. L. Wicker, in *Fillers for Plastics* (W. C. Wake, ed.), Iliffe, London (1971).

Index